PENGUIN BOOKS

TINDERBOX

Craig Timberg is the former Joha ... _gton Post._ From his position, he visited twe ... dozens of major stories about AIDS. He is ... :chnology writer.

Daniel Halperin, Ph.D., is an epidemiologist and medical anthropologist at the University of North Carolina, Chapel Hill and has taught at Harvard and the University of California, Berkeley. He was a top technical adviser in the U.S. government's PEPFAR program to combat AIDS.

Praise for _Tinderbox_

"Remarkable . . . reads like a detective novel." —_The New Yorker_

"Gripping." —_The Nation_

"A strong warning to those who would disregard the cultural specificities of those one is trying to serve." —_The New York Times_ (Editors' Choice)

"_Tinderbox_ will help readers understand . . . why the period ahead is so critical in fighting the epidemic." —_The Washington Post_

"A fascinating account." —United Nations News Service

"In addition to a useful history of the disease, Timberg and Halperin examine how to confront it and develop more effective ways to fight it. . . . [Timberg and Halperin] present a forceful case with which future students of HIV and AIDS will have to reckon." —_Kirkus Reviews_ (starred review)

"This absorbing interdisciplinary study of HIV/AIDS explores how the West inadvertently unleashed the AIDS epidemic and then failed to combat it effectively. . . . This timely exposé advocates practical solutions to a seemingly intractable problem." —_Publishers Weekly_

"Extensively researched, eminently readable and accessible, Timberg and Halperin's work is a notable and invaluable addition to the AIDS canon." —_Booklist_ (starred review)

"Essential for understanding a relentlessly urgent issue." —_Library Journal_

"A remarkable volume. With rare clarity, _Tinderbox_ lays bare the origins of the AIDS virus, and then reveals the often hapless and delinquent responses of the international community. It's a fascinating read: relentlessly honest, sometimes scathing, always principled." —Stephen Lewis, director of AIDS-Free World, former UN Special Envoy on AIDS in Africa

"A fascinating chronicle of how human foibles led to missed opportunities to control the pandemic and how unintended outcomes probably raised the death and disease toll substantially." —Health Affairs

"[Tinderbox] is laced with science—virology, epidemiology, the mechanics of circumcision—but the writing remains crisp and clear. The story is full of real, live people that help the reader understand African cultures, colonialism, and the devastating effects of disease." —The Concord Monitor

"Highly readable and informative, highlighting the importance of prioritizing HIV prevention in sub-Saharan Africa. Their analyses trace the social and cultural origins of the HIV epidemic and provide convincing arguments on population-wide measures needed to slow the spread." —James Curran, dean, Rollins School of Public Health, Emory University

"Randy Shilts's And the Band Played On was the first—and for decades the best—book on AIDS. Craig Timberg and Daniel Halperin's Tinderbox is every bit as good, revealing the same denial, the same story of politics trumping science, and the same tragedy. This time, it is about the whole world, not just San Francisco. Read it!" —Malcolm Potts, author of The AIDS Reader and Ever Since Adam and Eve

"The sometimes glorious, often tragic constellation of science, politics, and personalities in the fight against AIDS comes to life in the masterful storytelling of an energetic journalist and a passionate scientist." —Arthur Allen, author of Vaccine

"An excellent read. Tinderbox brilliantly outlines the successes, failures, and missed opportunities in the battle of HIV prevention over the last thirty years." —Elly Katabira, M.D., president, International AIDS Society

"Tinderbox is an indictment of Western ineptitude and meddling and lost opportunities to prevent millions of infections and deaths. But it also contains valuable prescriptions for making change—and it's an important read for anyone who cares about Africa." —Stephanie Nolen, author of 28: Stories of AIDS in Africa

"Timberg and Halperin have been challenging conventional wisdom (and behavior change skeptics like me) for years. Their book is entertaining, thought-provoking, human, and in the end, hopeful for a continent that craves some answers after two decades of HIV prevention failures." —Francois Venter, M.D., president, Southern African HIV Clinicians Society

HOW THE

WEST SPARKED THE

AIDS EPIDEMIC

TINDERBOX

AND HOW THE

WORLD CAN FINALLY

OVERCOME IT

CRAIG TIMBERG *and*

DANIEL HALPERIN, PH.D.

Penguin Books

PENGUIN BOOKS
Published by the Penguin Group
Penguin Group (USA) Inc., 375 Hudson Street,
New York, New York 10014, U.S.A.

USA / Canada / UK / Ireland / Australia / New Zealand / India / South Africa / China
Penguin Books Ltd, Registered Offices: 80 Stand, London WC2R 0RL, England
For more information about the Penguin Group visit penguin.com

Penguin Books Ltd, Registered Offices: 80 Strand, London WC2R 0RL, England

First published in the United States of America by The Penguin Press,
a member of Penguin Group (USA) Inc., 2012
Published in Penguin Books 2013

THE LIBRARY OF CONGRESS HAS CATALOGED THE HARDCOVER EDITION AS FOLLOWS:
Timberg, Craig.
Tinderbox : how the West sparked the AIDS epidemic and how the world can finally
overcome it / Craig Timberg and Daniel Halperin.
p. cm.
Includes bibliographical references and index.
ISBN 978-1-59420-327-5 (hc.)
ISBN 978-0-14-312300-2 (pbk.)
I. Halperin, Daniel. II. Title.
1. Acquired Immunodeficiency Syndrome—epidemiology—Africa.
2. Acquired Immunodeficiency Syndrome—etiology—Africa. 3. Colonialism—
Africa. 4. HIV Infections—epidemiology—Africa. 5. HIV
Infections—etiology—Africa. 6. Western World—Africa. WC 503.41
614.5993920096—dc23
2011040206

Printed in the United States of America
1 3 5 7 9 10 8 6 4 2

DESIGNED BY MICHELLE MCMILLIAN
MAPS BY JEFFREY L. WARD

To all who have suffered from HIV,
and to all who might yet be spared.

AUTHORS' NOTE

This book is the product of two authors: Craig Timberg reported and wrote most of the text; Daniel Halperin developed the ideas that form its scientific backbone and also worked extensively on developing the manuscript through its many iterations. Both of us occasionally appear as observers as well as characters in the narrative. For ease of readability, Timberg speaks in the first person, as "I." Halperin is referred to in the third person, as "Halperin." We are a bit of an odd couple. One of us is a journalist, the other a scientist. And while we've faced our share of challenges in crafting a book we believe speaks to both a general audience and a scientific one, we hope you will agree that in tackling this urgent and frequently misunderstood topic, we are more than the sum of our parts. What has consistently bonded us together is a fascination with human experience, a refusal to accept easy answers, and a determination to contribute, however modestly, in the quest to defeat the AIDS epidemic.

tin·der·box:
noun/ˈtindər͵bäks/
"A thing that is readily ignited"

CONTENTS

Prologue   *1*

BOOK I: SILENT SPREAD

1. Francistown   13
2. Searching for the Beginning   23
3. One Tiny Speck of Truth   30
4. A Tale of Two Viruses   37
5. The Lion and Dr. Livingstone   42
6. *Femmes Vivant Théoriquement Seules*   53
7. The Gift   61
8. The Big Bang   67

BOOK II: AN EPIDEMIC OF POLITICS

9. Americanizing AIDS   77
10. It Can't Be Here Already!   85
11. *Attention na SIDA*   92
12. You Won't Believe   100
13. Fear Worked   109
14. Born in Africa   119
15. The Condom Code   126

16. The Beat-up   136

17. Things Just Fell Apart   152

BOOK III: THE HUMBLING

18. X Factor   169

19. The Interests of the ANC   181

20. Poverty Trap   192

21. A, B, and C   198

22. On the Jericho Road   206

23. Gordon and Thandi   213

24. A Marshall Plan for Botswana   224

25. What Shall We Do?   234

26. Raymond the Great   243

27. *Makhwapheni Uyabulala*   256

28. The Flood   263

29. Mother and Son   272

30. What Shall We Do? Part II   280

Epilogue   297

Appendix: How the AIDS Epidemic Can Be Overcome   *305*

Acknowledgments   *318*

Notes   *321*

References to the Appendix   *388*

Additional Suggested Readings   *403*

Index   *406*

PROLOGUE

There once was a place deep in the forest where few people dared go. Trees a dozen stories high loomed overhead, blocking out all but the faintest dapples of sunlight. Below a riot of wildlife—parrots, antelopes, chimpanzees—ranged freely. The few humans to travel routinely through this remote corner of creation were Baka "Pygmies," who were so small and unaccustomed to the trappings of civilizations that other people regarded them as little more than animals themselves.[1] Man and beast all lived together, not in harmony, but in a rough kind of stability, never having reason to take notice of the strange sickness that claimed the occasional chimpanzee. It was the course of nature, in which births and deaths were capricious, unremarkable events. And yet, in the case of some of the dying chimpanzees, there was a difference.[2] Silently, invisibly, a virus with no name lurked in the blood, lymph nodes, and semen of the chimps, waiting for the chance to fulfill its fate as one of the planet's most terrible killers. For centuries, perhaps millennia, it never did.

This uneasy peace among human, chimp, and virus broke with the pounding of exhausted feet against the ground. Those who came crashing through the jungle did not do so of their own accord. They were porters, human pack animals pressed into service by a strange new race of pale-skinned men carrying metal sticks that exploded with fire. Any Africans who resisted lost their hands, or their lives, or their wives and children to prison camps. Those who complied fared little better, as they were force-marched through unfamiliar land fifteen miles or more a day, carrying the white man's gear, the white man's gray gobs of rubber, the white man's creamy tusks of ivory. Exhausted and in a place none of them wanted to be, the porters scavenged for something beyond the gruel ladled out by their masters. Among the few sources of meat were jungle animals, including, when they could be trapped, those same chimpanzees in whose bodies the virus hid.

We are unlikely ever to know all of the details of the birth of the AIDS epidemic. But a series of recent genetic discoveries point strongly toward a moment such as this, when a connection was made from chimp to human that changed the course of history. We now know where the epidemic began: a small patch of remote southeastern Cameroon. We know when: within a couple decades on either side of 1900. We have a good idea of how: somebody caught an infected chimpanzee for food, allowing the virus to pass from the chimp's blood into the hunter's body, probably through a cut during the butchering.[3] As to the why, here is where the story gets even more fascinating, and terrible. We typically think of diseases in terms of how they threaten us personally. But they have their own stories. Diseases are born. They grow. They falter, and sometimes they die. In every case these changes happen for reasons. For decades nobody knew the reasons behind the birth of the AIDS epidemic, though theories tended to focus on growing transport links, shifting social values, and rising sexual experimentation in a world where traditions were giving way. But the new genetic discoveries have lent new precision to what were once only educated guesses.

It is now clear that the AIDS epidemic's birth and crucial early growth happened amid the massive intrusion of new people and technology into a land where ancient ways still prevailed. European powers engaged in a feverish race for wealth and glory blazed routes up muddy rivers and into dense forests traveled only sporadically by humans before. The most disruptive of these intruders were the thousands of African porters forced to cut paths through the same areas that researchers have now identified as the birthplace of the AIDS epidemic. It was here, in a single moment of transmission from chimp to human, that a strain of virus called HIV-1 group M first appeared. In the century since it has been responsible for 99 percent of all of the world's deaths from AIDS—not just in Africa but in Moscow, Bangkok, Rio de Janeiro, San Francisco, New York.[4] All that began when the West forced its will on an unfamiliar land, causing the essential ingredients of the AIDS epidemic to combine. It was here, by accident but with motives by no means pure, that we built a tinderbox and tossed in a spark.

The epidemic spent the next half century following the disastrous contours of modern African history, as Westerners remained crucial, if unwitting, agents in spreading HIV around the continent and eventually the world. The virus exploited the rise of colonial cities, the growth of prostitution, and the creation of fast transport routes that linked far-flung communities together into robust new pathways for sexual transmission of disease. And when Africans finally regained their independence in the 1960s, the chaos of the years that immediately followed, with their sweltering new slums and rampant instability, only helped HIV toward its grim destiny.

But on a deeper level, the story of AIDS is not just a story of a disease and its many victims. It is a story of how greed and arrogance set the world in motion in ways that those responsible couldn't have foreseen and those swept up in the consequences had little power to reverse. With Bibles and Gatling guns and great rivers of concrete, the West tried to remake Africa and ended up igniting a disastrous new epidemic. This is a story that has never been fully told, aside from a few references in the academic press. We all know about the abandoned orphans and suffering widows, as well as the doctors, politicians, and celebrities who are determined to help them. But who considers the outbreaks of sexually transmitted disease and infertility that followed each step of "progress" into Africa? Who recalls the disarray that followed the abrupt retreat of European power? And who would have guessed that as the West tried to look away from what it had wrought in Africa, an epidemic incubated during the colonial era would somehow escape, finding its way into our common future?

Yet the backstory of AIDS is essential, because some of the legacies of colonialism continue to undermine today's efforts to fight the epidemic. It's not fair to compare those fighting the disease today, motivated overwhelmingly by the desire to help, with the Europeans (and in some cases Americans) who brutally exploited Africans a century or more ago. But it's unfortunately true that something of the relationship between the West and Africa has endured from that earlier era. Where once there were conquerors and the conquered, now the relationship too often is understood narrowly, in terms of saviors and the saved. Power, meanwhile, remains concentrated where it has always been. The result has been a war against AIDS run mainly by people who see disease through Western eyes and spend the largest sums of money on the Western biomedical tools in which they have the most faith. Similar approaches have often been successful against other global health calamities, such as polio and smallpox, but reversing the spread of HIV has proven more elusive, more entwined with cultural subtleties that outsiders have struggled to understand. Africans provide the images for the endless stacks of reports churned out by what has become a $16 billion a year industry, but the strategies emanate from the same places where the money does—from Geneva, London, Washington.

Such problems are common to even the most well-meaning foreign aid initiatives, as chronicled in William Easterly's *The White Man's Burden* and Dambisa Moyo's *Dead Aid*, which show the distortions created when billions of dollars are pumped into places that lack the capacity to spend the money effectively. In the countries where HIV is most severe Westerners have overlooked some homegrown initiatives in favor of ones that better fit their own training and experiences, and often their own ideologies. Meanwhile, a handful of scientists who strive to ground their ideas in deeper understandings of the societies ravaged by AIDS often have been ignored and sometimes mocked. For years some of the very ideas now regarded as the best available tools for slowing HIV in Africa, home to at least two thirds of the world's

MAIL BOX COMBINATION

ID_____

NAME_____

BOX #____168_____

To open, turn dial to the left at least

3 times, stopping at #   14

Turn dial to the right, passing the first #

Once, stopping at #   3

Turn dial to the left, stop at #   31

Twist dial to the right until it stops, then pull door open

infected people, were mired in fractious debates driven by Western politics and sensitivities.[5] This was a new kind of colonialism: rule by rich-world donor dollars.

The result has been a series of missed chances to slow the spread of HIV. Men and women in some parts of Africa are still becoming infected nearly as rapidly as they ever did. South Africans turning fifteen today have a roughly fifty-fifty chance of contracting HIV in their lifetimes.[6] There have, of course, been dramatic strides in treating people once they are infected, extending the lives of millions. In this effort, each day is its own miracle, another chance for a mother to hold her children or a father to bring home grocery money from a factory. And recent studies have raised hopes that further expanding access to HIV drugs could also slow the spread of the virus. But while AIDS treatment programs have prolonged many lives, most prevention campaigns have fallen short of expectations. The fight against AIDS has become like a battle against lung cancer in which resources were devoted mainly to chemotherapy and surgery while little useful was done to curb smoking. The result is ever more infections, and ever larger demands for expensive treatment dependent on the continued largesse of foreign donors. And it is here that our failure to understand the forces that shaped the epidemic has carried such a high price: Many millions of people have been left more vulnerable to infection with an incurable virus.

I first began to sense the cracks in the Western world's tidy portrayals of the AIDS epidemic after I moved to South Africa to become *The Washington Post*'s Johannesburg bureau chief in 2004. If HIV supposedly preyed on the poor and war ravaged, I couldn't figure out why the poorest and most war-ravaged places I visited, such as Darfur and Sierra Leone, had high rates of many other diseases but relatively few people with AIDS.[7] I also had the uneasy feeling that all the talk about condoms and "safe sex" obscured something deeper about how the epidemic spread. I remembered the terror of AIDS dating back to my high school years in the 1980s, and the resulting fixation on sexual caution. But in the most affected parts of Africa I struggled to find either the terror or the caution among people in far more peril than my friends and I had been in back home. When my warm, handsome guide to the South African township of Soweto introduced me to his steady girl-friend, then to two other women he was regularly sleeping with, I knew I was in unfamiliar territory.

My growing sense of unease led me to Daniel Halperin, an epidemiologist and medical anthropologist who was then working for the U.S. Agency for International Development in Swaziland. His job was to help that tiny nation and others in southern Africa battle their horrific HIV epidemics.[8] Halperin is tall, dark haired, and lean, and his smile is disarming. This is good, because he has made a career of telling people that most of what they think they know about HIV is wrong. When I met him while working on a story in 2005, he immediately departed from the script of most experts I had met. While their conversations had focused mainly on condoms, HIV testing, or the distant hope of a breakthrough vaccine, Halperin veered quickly into the realm of the impolite. He insisted that the two most important factors in understand-ing the spread of AIDS through African societies were sexual behavior and male circumcision. In lands where polygamy was a common practice before Westerners began campaigning against it, substantial numbers of both men and women, whether married or single, still often had more than one sex partner at a time. This created sprawling networks that spread disease with rare efficiency. This was especially true, he said, in societies where the ancient ritual of circumcision, which removed the foreskin tissue most vulnerable to HIV, was not widely practiced. And where men were more likely to contract the virus, they were more likely to pass it on to their wives or girlfriends—and these women were more likely to pass it on to their husbands or other boyfriends.[9] Most African societies had been circumcising boys as part of coming-of-age rituals for thousands of years, but in some areas the practice had either not taken hold or had gradually fallen out of fashion, often at the behest of Christian missionaries, who regarded the initiation ceremonies as heathen. It was in those places, mainly in a swath of the continent running from East Africa's Great Lakes region down into the southern cone, where HIV now raged out of control.

As we talked, I began to sense that Halperin was an original thinker and, more important, a truth teller. In this he reminded me of people I had known on other beats I had worked. They were city planning officers, or government lawyers, or legislative staffers. They talked to me on the understanding that I wouldn't routinely quote them directly but would draw on their unvarnished accounts of things to get stories right. As I began to incorporate Halperin's insights into my work I had occasional pangs of doubt. I was not alone. Some of his conclusions seemed so startling, so at odds with the prevailing wisdom on AIDS, that many of those in power dismissed his ideas as eccentric, a threat to the intellectual edifice they had spent years building. How could condom promotion fail to reverse the spread of AIDS in Africa? Or HIV testing? Or antiretroviral drugs? Could rich people really be more vulnerable to HIV than poor ones? Should African women with HIV really breast-feed their babies? And what was all this nonsense about foreskins?

But I gradually realized that Halperin's insights were grounded in a way of thinking about epidemics that while new to me was as old as the study of public health itself. It framed disease in terms not of individual causes but of societal ones. The potential remedies were collective as well, aimed not only at people already sick but at those who might soon be if the outbreak were not brought under control. This approach proceeded from an elementary premise: From the standpoint of a whole community it is as important to stop new cases it is to treat old ones.[10] This can be as straightforward as finding a contaminated water source spreading cholera or as complex as understanding the chronic poverty that causes malnutrition. In the case of HIV, this inquiry is complicated by our nearly universal discomfort in discussing the main means of transmission: sex. Few people discuss their sex lives openly with strangers, making it difficult for scientists to study the related diseases. It is even harder to understand the consequences of billions of individual sexual decisions made by millions of people across entire societies. The other major mode of HIV transmission, injecting drug use, though relatively rare in Africa, makes the overall picture still more complex. Yet this complexity does nothing to diminish the urgency of the problem. This is, after all, not a moral investigation but a scientific one. And its goal is nothing less than overcoming one of the world's greatest killers.

A tinderbox is a metal container once commonly used to start fires. Steel, flint, and cloth worked together to create the initial spark and sustain the flame until it was big enough to spread. Once it got going the fire relied on a new mixture—fuel, air, and heat—to sustain itself. Take away any element, and it burned out. In the same way, the AIDS epidemic started when human actions inadvertently combined specific elements into a dangerous new mixture. And it kept going because the key ingredients became steadily more available. Had this not happened, in the time and place where it did, AIDS might have remained forever unknown.

This book relies on a range of contemporary scientific discoveries to tell the story of how AIDS grew from a small, localized outbreak into a devastating epidemic. And it shows how certain factors, especially patterns of sexual behavior and the extent of male circumcision, were decisive in whether societies had minor outbreaks, serious but confined ones, or consuming disasters that can rightly be compared to the worst plagues in history.[11] The virus was virtually the same everywhere, but the impact was not, meaning that our understanding of the path of HIV across the planet must be rooted in the ways that host communities vary.[12] The fast-lane gay lifestyles that arose in many American, European, and Latin American cities in the 1970s and 1980s featured sexual practices that readily transmitted disease—initially familiar ones like gonorrhea and syphilis, and eventually the virus that causes AIDS. African societies were not as freewheeling as these urban gay enclaves, but the legacies of polygamy and colonialism helped produce sexual cultures that were particularly vulnerable to the spread of HIV—especially where routine male circumcision was uncommon.[13] The result, across parts of the continent, has been networks for sexual transmission of disease far more robust than in most other societies. The reluctance of the global health community to address these admittedly awkward factors has undermined the increasingly massive and well-funded efforts to turn the tide against HIV. This is a key reason why, even as AIDS treatment has become increasingly available across the world, so many HIV-prevention campaigns have not succeeded.

But this unfortunate past need not determine the future. *Tinderbox* also tells the story of what might happen now if those who care most about the epidemic turned serious attention to the core factors underlying its spread. It's the story of an epidemic that could be smaller—perhaps much smaller— if these crucial lessons could be learned.[14]

They are not simple, and we don't presume to possess more intelligence or wisdom than the tens of thousands of men and women who have been valiantly fighting AIDS for many years. Compared to most people in key positions of the war against the epidemic, we are relative latecomers and, still to a substantial degree, outsiders. Halperin got immersed in HIV research in the mid-1990s, after an idiosyncratic career path that included stints as an aspiring jazz saxophonist and as one of San Francisco's most overeducated cab drivers. And I knew little about AIDS before joining the *Post*'s foreign staff. Yet perhaps because we came to the battle later than some, and had the benefit of hindsight, we see in the epidemic a story of lost opportunity, of unforeseen consequences, and of (mostly) good intentions gone awry. And yet, we also share a battered but persistent hope: The epidemic has depended on human action for its birth and spread, and so too could human action finally overcome it.

BOOK I

SILENT SPREAD

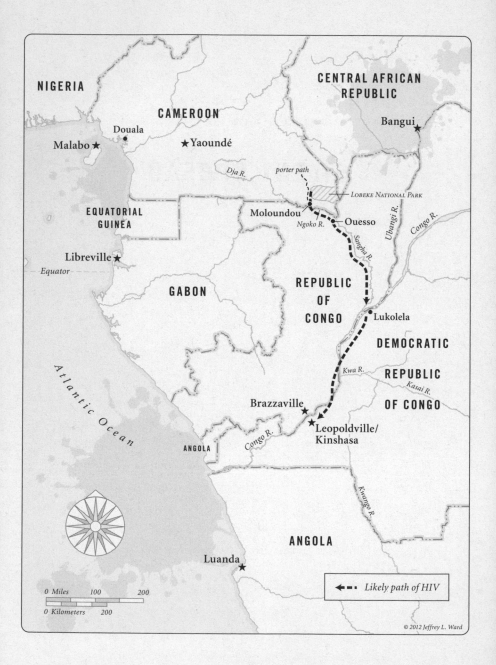

[1]

FRANCISTOWN

I was among those who wanted the AIDS epidemic to be straightforward, in line with our instincts about how the world worked. Clear plot lines often make for good journalism. If poverty and strife were at the root of the spread of HIV, and if greedy drug companies were the main barrier to stopping it, those were stories I knew how to write. Suffering lay in almost any direction I traveled from my home in Johannesburg. There was dictatorship in Zimbabwe, unrest in Congo, outrageous inequality in Angola. Even South Africa was struggling awkwardly—and often unsuccessfully—to emerge from the long shadow of apartheid. But Halperin had helped open my eyes to the contrary story that the science was actually telling. HIV was bad in many places, but nowhere was it worse than in peaceful, prosperous Botswana—bastion of good governance, haven of stability, favorite of foreign donors. And within Botswana, nowhere was it worse than in the industrious northern mining region whose biggest city, Francistown, was home to 85,000 people. If I wanted to understand the limits of money and good intentions in fighting AIDS, here was the place to start. [1]

After a short flight to the nation's capital, Gaborone, I rented a car and drove up a nearly featureless desert highway. By Botswana's standards, Francistown had a reputation as an unruly city, and a 2004 government survey put the adult HIV rate at about 40 percent.[2] More shocking still was the rate among women in their early thirties; two out of every three had the virus. The factors included, according to various experts, a bustling sex industry and an influx of desperate migrants from nearby Zimbabwe swelling the city's ranks of petty criminals and prostitutes.[3] As I approached the city limits, I spotted the first sign of trouble. On a sandy rise to the left, a red billboard had giant white block letters spelling out WARNING. This was a couple of years into my posting as a foreign correspondent in Africa, and I had visited some notorious hellholes already. So I slowed down my car in hopes of taking the measure of the city before me, a place treacherous enough that somebody had posted a giant alert to unwary travelers. But as I got closer, the smaller letters below the WARNING came into view: "Francistown now protected by Security Systems." This was not a marker of the medical disaster ahead; it was an advertising come-on straight out of Madison Avenue.

My disorientation grew as I entered Francistown and the austere desert highway transformed into a four-lane shopping corridor that could have been on the outskirts of any American city I knew back home. I passed gas station minimarts, big-box retailers, a Wimpy's fast-food joint. The city's redbrick downtown, meanwhile, bustled with urban energy, as people shopped and ate and lined up for cash at ATMs posted in front of stolid bank buildings. The Francistown Chamber of Commerce and Industry even had a fountain out front that, unusually for such extravagances in most of Africa, apparently had been maintained flawlessly and kept a burbling column of water spraying steadily into the air. This was no hellhole.

But, I wondered, what was happening here? Francistown wasn't poor. It wasn't war-torn. There were no signs of mass rapes or child soldiers or great throngs of refugees sweeping back and forth across some embattled patch of scrubland.[4] There was no evidence of the squalid shantytowns that had sprung up near most African cities. Francistown, like the rest of Botswana, even had bounteous amounts of the world's preferred weapons against the AIDS epidemic, courtesy both of the nation's treasury and an unusually heavy dose of donations from the U.S. government, the Bill and Melinda Gates Foundation, and the drug company Merck. HIV testing and counseling centers were widely available, and doctors had begun ordering tests of their patients on routine medical visits. Condoms were available in virtually every clinic, shop, and bar, and national reported usage rates were the highest in Africa, and probably of anywhere in the world.[5] Even the expensive drugs that could usually control the worst ravages of AIDS were free and easy to get at health centers across the city. If the Western world's playbook for fighting HIV in Africa was working, Francistown should be an unquestioned success. Yet the data made clear it wasn't.

Looking for answers, I turned into the nicest of the several shopping centers built to either side of Blue Jacket Street, the main drag through Francistown. After parking my car in a lot paved to flawless perfection, I headed for a bar with a familiar name. The Hard Rock Cafe was not part of the international chain, but its sensibility was decidedly First World, with posters of British soccer stars and rows of European beers chilling in glass-fronted refrigeration towers. The bartender was a tall, gregarious guy, Brian Khumalo, who was twenty-five and had a ready, gap-toothed smile and plenty of tales about what he observed each evening as he opened beers for the Hard Rock Cafe's many patrons. As I gently raised the subject of the epidemic raging through his city, Khumalo nodded knowingly and gestured toward a young, spiky-haired woman in a corner booth. "She's new around here, so every guy is going to talk to her," he said. "She will be with me today. Tomorrow she will be with my best friend. And I will be with somebody else."

Khumalo said he had moved here the year before from Gaborone and was startled by the accommodating sexual culture he discovered in Francistown. His first night, Khumalo said, he slept with a woman he had just met. He did the same the second night, and the third. Casual sex was hardly unknown in Gaborone, where Khumalo had been a student at the University of Botswana, but in Francistown it was at the heart of the rambunctious social scene.

He tried his best to use condoms for these casual hookups. But when an old friend visited from out-of-town, he was disturbed to see Khumalo indulging so eagerly in risky behavior. Instead of merely giving him a lecture the friend drove Khumalo to Nyangabgwe Hospital, on the southern side of downtown, and pulled up near a low-slung white building. It was where people lined up to collect supplies of the antiretroviral medicine that kept their HIV infections at bay. The number of patients startled Khumalo, but so did the lack of visible signs of illness. Most appeared healthy. Many were young. He was chilled in particular at what appeared to be so many "beautiful women going to get pills."

The experience was enough to convince Khumalo to swear off casual sex for a few weeks. But temptations continued to arrive almost nightly. So Khumalo settled on a half measure he hoped would make a difference. He began having sex only with a confined group of women he knew fairly well. The relationships were not monogamous, but they were steadier. Even after seeing the lines of healthy-looking young women getting medicine, Khumalo still imagined he could spot the one who would give him HIV before it happened. By the end of that year, Khumalo estimated, he had slept with one hundred women in Francistown.

Discussions of sexual behavior and AIDS have been entwined since before the disease even had a name. The first sentence of the original alert by U.S. federal disease monitors, put out in June 1981, referred to the original cluster of victims as "active homosexuals." And early reports of AIDS in Africa frequently described "promiscuity" as a key factor in how it spread. The focus on sexual behavior has waxed and waned in the discussion about HIV ever since, and in many quarters the subject had become controversial. But Halperin had convinced me that to understand AIDS I needed to understand why some places developed epidemics that were so much worse than others. And that conversation inevitably returned to differences in sexual cultures. Even if Brian Khumalo was an extreme case, or a braggart, or imbued with the heedlessness of youth, something was happening that I couldn't ignore, no matter how impolitic it was to suggest that people here were not merely victims of an epidemic but also helping to cause its spread. I also began wondering: Was it a coincidence that the AIDS epidemic was especially severe in a place whose atmospherics otherwise felt so familiar, so Anyplace, USA?[6]

Over the next couple of years Francistown became a touchstone. I interviewed businessmen, students, teachers, shopkeepers, and office workers. And however controversial it was to discuss sexual behavior and HIV in polite company in the Western world, the connections were so stunningly obvious as to be unavoidable in Francistown. One young doctor who had seen more than her share of AIDS patients casually acknowledged that she kept up ongoing relationships with three different men. A worker for a major company complained about how a tendency among Botswana's employers to routinely transfer people to offices hundreds of miles away made long-term monogamy seem impossible. Several of the women I interviewed at the Hard Rock Cafe, most in their twenties or thirties, said they had all but given up on finding men with whom they could share exclusive relationships. In a country where only about one in six of adults were married, many had come to regard the idea of happily-ever-after to be quaint, something from storybooks or movies but impractical in the real world.

There were exceptions. A teacher I met at Hard Rock Cafe, Faruk Maunge, had traveled extensively overseas and was astonished to discover that while he was away the onslaught of AIDS had failed to change a social scene he regarded as profoundly dangerous. "They are just a lost bunch," Maunge said, in a deep, smoke-cured voice filled not with condemnation but concern. "They are very, very reckless."

Maunge, who was in his late thirties, had long dreadlocks, a goatee, squared-off glasses, and a yellow WWJD bracelet—for What would Jesus do?—around his wrist. But Maunge was no angel. He had run with a fast crowd as a younger man, and now could count at least twenty dead friends. To drive home the toll he invited me to his concrete-block home in the outskirts of town. When I arrived the next day Maunge ushered me into a room that had a television, rows of shelves, and two couches. He urged me to sit down on one while he pulled out a green plastic first-aid box filled with

Faruk Maunge, a teacher in Francistown, points to pictures of friends of his who have HIV, or have died of it. The city has one of the highest infection rates in the world.

photos. As he flipped through the pictures Maunge offered a tour of life and death in Francistown in the Age of AIDS.

"This one is gone," Maunge said, pointing to a faded picture of a woman in a red top who was nibbling her fingernails. Moving deeper into the pile, he continued: "This one is gone, Mooketsi. And this one is gone, Themba. This one is gone, too, this one on the far left. This one is [HIV] positive."

When he got to the picture of one man, frustration crept into Maunge's voice. "He's sleeping around again." He also grew irritated at a picture showing a friend with AIDS who seemed to father a child—he was awaiting his fourth—with every girlfriend.

"Praise God, I've been lucky," he said. "It's like you have ten bullets going through you and none hits you."

The story that struck me hardest in Francistown was from a grandmother whom I will call Angela.[7] She was smart and gentle, with long, lovely braids that fell down past her slender shoulders. I had already filled several notebook pages with her observations on how HIV spread here. Not only did she have the virus; so did her husband, and he had come within days of death before getting the antiretroviral drugs that revived him. Most of Angela's friends, who like her were in their forties, had HIV too. And Angela had a good grasp of one of the main reasons why.

"We've got multiple partners here in Botswana," Angela said simply. "A girl who's not working, there's going to be a boyfriend who's going to pay the rent, [another] boyfriend who's going to pay the water, buy the groceries."

Such relationships were not strictly economic. Angela had a young boyfriend on the side in what she called "a small house," a slang term common in some parts of southern Africa. She clearly enjoyed the romance, but it also terrified her to the point that she started having nightmares. The man, who was more than a decade younger, did not have HIV—yet. And though they tried to use condoms for every encounter, Angela knew this required discipline that few couples managed consistently over the long term. As a consequence, Angela's nights were filled with troubling dreams in which she would give the virus to her boyfriend. Then he would transmit HIV to another woman, and further into Francistown's sexual networks. And then, eventually, HIV would find its way to Angela's own son or daughter. In this way, her unwillingness to forego the passion of youth would deliver early death to her children, or at least permanently alter the trajectories of their lives.

The popular image of African women with AIDS is of some combination of poverty, victimhood, and ignorance. Angela challenged the stereotypes on all three. She dressed well, in a button-down blouse and skirt. She may have gotten HIV from her husband but, given the facts of her own sex life, the transmission easily could have gone in the other direction.[8] And she understood how HIV spread, saw its terrible consequences, and yet persisted in behavior that she knew kept the virus moving.

To add to the irony, Angela and her boyfriend had met through their activism with an AIDS group. They were helping deliver food packages to sick people and gradually grew closer. Angela had tried to break off the affair but kept drifting back into it over the course of several years. Angela had no expectation that her boyfriend was confining his affections to her.

"It's just a natural thing," she said, in a soft, wistful voice. "I cannot control it. It's such a difficult thing. I take it that it's just human nature."

In one sense, of course, it is. Humans are wired to want sex. And cultural and religious expectations that we limit our intimacies to mutually monogamous relationships do not short-circuit the desire to be with others. But it is equally obvious that sexual cultures vary across time and place. Even within a single community, rules get rewritten as the generations pass, as shifting faiths, technologies, and other forces alter sensibilities about what is acceptable and what is not. These changes have consequences, in births, in deaths, in the lives lived in between.

Medical science has advanced to the point where those with access to its wonders tend to view illnesses as discrete misfortunes. But pulling back the camera, from the individual level to the societal one, reveals a picture considerably more complex. Epidemics result from an accumulation of millions of collective actions, rather than individual moments of poor judgment or bad luck. Africans on average die much younger than people in richer parts of the world and often from afflictions easily treated by modern medicine, or prevented entirely through better diets or access to clean water. These sturdy facts have long infused the discussions about AIDS, fueling conclusions that this disease, like so many others in Africa, is a consequence of poverty, of ignorance, of poor health systems. And that conclusion has dominated thinking about how to fight HIV, framing the battle as fundamentally one about resources, especially biomedical resources, such as drugs, condoms, or vaccines.

But what of Angela? With access to resources, and information far beyond that of most Africans—and of most people in the world overall—why does she still indulge in behavior that she knows could spread it to others, even the ones she cares for most?

There is another axis upon which questions about HIV turn: Time. Francistown in the early twenty-first century is an AIDS disaster, but the disease was all but unknown there a quarter century earlier. And two centuries earlier, a major outbreak of HIV would not merely have been unthinkable. It would have been all but impossible.

Botswana's sexual culture was not monogamous in the era before Christian missionaries began traveling widely through the region in the late 1800s. It was polygamous. The most powerful and accomplished men were supposed to have more than one wife, and to father as many children as possible in a harsh desert environment that killed so many of them before they reached adulthood. In this traditional culture—remnants of which survive today—there also was a tacit acceptance of sexual relationships outside of marriage, so long as they were conducted discreetly.[9] Ludo Margaret Mosojane, a stern-faced judge in Francistown's customary court that adjudicates violations of cultural norms, told me that men traveled frequently to busy trading centers, to the remote cattle posts where they maintained herds of livestock, and then back home to small village homesteads. It was not uncommon to have a sexual partner in each place. Nor was it uncommon for women to have boyfriends while their husbands were away. Mosojane explained men were not supposed to return home directly from travels but instead to spend the first night with a brother or cousin. Whatever the origin of the tradition, it had the practical benefit, she said, of giving women time to send away boyfriends and to sweep away incriminating footsteps that might have accumulated in the sandy paths to their homes. Setswana, the main language here, even lacked a word for "fidelity" in the sexual sense, Mosojane said. Conveying the concept would take eight words that translated roughly as "to be with one person and no other."

But several other elements essential to spreading HIV were missing. There were few highways or cities. Trade moved at the speed of cattle or donkeys ambling on desert paths. And despite Mosojane's account of flexibility in marital arrangements, sex had visible consequences in this era before easy access to birth control. Having sex with dozens of women, as Khumalo claimed to have done in contemporary Francistown, was a path to ruin unless a man had vast personal resources. Parenthood carried universally accepted responsibilities in precolonial Botswana, as it did throughout most of Africa.[10]

This meant men and women had to be careful to avoid pregnancy outside of traditional family structures.

And so young people undergoing traditional Setswana coming-of-age ceremonies were discreetly taught practices such as "thigh sex" and early withdrawal.[11] Such village-based rules, which could also have helped slow the spread of HIV, began to change when Western missionaries and adventurers began ranging widely through Africa in the nineteenth century, and it accelerated with the development boom that followed the discovery of Botswana's rich diamond deposits in the 1960s. The transformation of these places, from societies where AIDS was improbable to ones where it was rampant, is part of a much larger tale about the forces that shape the lives of individuals in ways they barely perceive. But telling this story properly requires something that long has been elusive in discussions of the AIDS epidemic: It requires that we find the beginning.

SEARCHING FOR THE BEGINNING

The longing to understand disease is likely as old as we are as a species. And throughout that history, even when maladies were thought to result from angry gods or jealous ancestors, the potential remedies were understood to flow directly from the causes. Appease the god, placate the ancestor, and a return to health was possible. Modern medicine has attempted to replace those ancient stories with new ones about microscopic germs, errant DNA, and invisible toxins, but the underlying concept has not changed: Only if we solve the riddle of an illness's cause can we hope to defeat it.

Few diseases in modern times have rivaled AIDS in their ability to disrupt that equation. After its discovery in 1981, researchers urgently sorted through clues searching for the reasons behind its spread. Did AIDS result from the inhaled nightclub drugs called "poppers"? Or was it a well-known pathogen that had morphed into something new and terrible? Or a cluster of sexually transmitted diseases working together to mount unprecedented onslaughts on the immune system? During those first hectic years, even the idea of a vengeful god gained currency among those craving the moral verities of an earlier age.

The identification a few years later of HIV, the virus that causes AIDS, provided an essential clue. Blood and sexual fluids—and, crucially, not mosquitoes, saliva, or some man-made substance— carried HIV from victim to victim. This put scientists on track to eventually develop a powerful new class of drugs to combat the virus in the bodies of those already infected with it. But the insights offered less to those attempting to stop it from entering the human body in the first place. On a planet where the exchange of sexual fluids was an essential part of life, finding the pathogen did not answer the biggest questions about stopping AIDS. It was now clear that HIV was the central character in an awful drama. But what was the setting? What obstacles had it overcome? What obstacles still lay in its path? And to what extent could human ingenuity bend the narrative toward a better ending?

There had long been evidence that HIV, which on a genetic level closely resembles chimpanzee strains of the virus SIV (for simian immunodeficiency virus), emanated from Africa. And this understanding, such as it was, spawned ridiculous speculation in some quarters about Africans having sex with chimps.[1] It also fueled suspicions among many Africans that the CIA or some other shadowy Western entity created HIV to kill black people.[2] Given the well-documented role of the United States and European powers in assassinations, coups, and other scourges of postcolonial Africa, it's easy to see how biological warfare seemed not entirely far-fetched. The West's allies in the battle against communism included the apartheid-era South African government, whose scientists did in fact experiment with ways to limit the nation's black population, through poisons distributed in common food sources such as beer, cigarettes, and peppermint candies.[3]

In this stew of suspicion and speculation landed a provocative thesis, articulated in a 1992 article in *Rolling Stone*, that HIV may have grown from polio vaccination campaigns that Belgian colonial authorities carried out in Congo in the 1950s. One of the scientists involved in the vaccination effort sued *Rolling Stone* for defamation, and the magazine ultimately retreated from its most ambitious claims in a carefully worded clarification. But the idea that Western hands had started an epidemic that was killing millions of Africans got a profound boost. The article also drew the attention of a British journalist, Edward Hooper, who went on to spend years trying to prove the thesis. The result was a massive book, topping one thousand pages, called *The River: A Journey to the Source of HIV and AIDS*, published in 1999. It fell short of proving the vaccine hypothesis but produced enough circumstantial evidence to unnerve mainstream scientists who long had doubted the idea. The key was the suggestion—denied by the main researchers involved in the program—that the kidneys of chimpanzees were used in making the oral polio vaccine distributed in the Belgian Congo in the late 1950s, and that some of these kidneys could have been infected with SIV.[4] Hooper also detailed how colonial health officials distributed these vaccines widely in places where HIV later became a major health problem. The book had enough well-reported detail that it made the vaccine argument hard to dismiss any longer. Some scientists were sufficiently intrigued to demand the kind of rigorous research necessary to finally test the idea properly: by investigating whether there were any possible connections among chimpanzees, vaccine production, and HIV.

Perhaps the most prominent of these was Oxford University scholar William D. Hamilton, a legend in the rarified world of evolutionary biology. The field was devoted to the study of how different life forms evolved and now were connected across webs of creation that spanned eons. Hamilton had made his name with research into the evolution of sex and the biological rationale for altruism and vengeful behavior in animals. And the idea that mass vaccination campaigns triggered the AIDS epidemic clearly called to Hamilton, as did the suggestion that nefarious forces may have conspired to keep the true story from being told. In a powerful foreword to *The River*, Hamilton gushed, "Everyone should read this book."

One who did was Michael Worobey, a young Rhodes scholar from Canada. Worobey had a fascination with Africa, and at one point spent two months bicycling alone through its western bulge, from the Sahara desert to the Atlantic Ocean, a thousand miles away. He also was an evolutionary biologist with an interest in the origin of HIV, but from a newer generation that relied less on grand inspiration than on exacting analyses of the microscopic world of genomes. All living things carry DNA that dictate how they grow and change and that also, when analyzed with the right tools, tell the story of their evolution. It was Worobey's aptitude in this kind of work that formed the basis of the unlikely partnership between the two scientists. Hamilton, the graying eminence, wanted to test his latest big idea. But to pass scientific muster, he recruited a partner in Worobey who was nearly four decades younger and—as would gradually become evident—far more cautious about the idea that vaccines were related to HIV's birth.

Before Hamilton and Worobey departed for Africa in January 2000, they spent months talking through the dilemmas of the research. Hooper's work pointed to a Belgian lab in Kisangani, on the fabled Congo River, where chimpanzees had been kept in captivity for medical purposes. But that had been decades earlier. How would Hamilton and Worobey now find the right chimpanzees for testing? How could a useful sample of SIV be extracted without hurting them? Surely wild chimps would not sit calmly for the simian equivalent of an HIV blood test. So the scientists decided to search for SIV in the urine and feces excreted by the animals, adopting a technique pioneered by others. Collecting the waste was the only problem, though Hamilton had an intriguing idea for how the process might work. After the long journey from London to Kisangani, as the two scientists were still settling in to their new research environment, they went to a Congolese market for supplies and bought a large, multicolored umbrella. With this, he imagined, they could sidle up to unsuspecting chimpanzees, open it upside down like a giant specimen cup, then slide it beneath the beasts' backsides during their moments of private repose.

Hamilton and Worobey figured that the chimps who had lived at the Kisangani lab likely were captured nearby, so the scientists hired several local hunters to help them track the animals in the densely forested areas where groups of wild chimps still lived.[5] The hunters were happy for the work, but when Hamilton pulled out the umbrella during an early trip out into the forest, they looked puzzled. Then they burst into laughter. Chimps are easily startled, the hunters explained, and would not sit still as a person opened an umbrella. Once their laughter subsided the hunters demanded that it be stashed away, and the team continued its journey into the dense equatorial forest.

Progress at first was hard to discern. For several days and nights, there was no sign of chimpanzees, and the scientists began suspecting that they had been duped by the hunters. But eventually, they began hearing occasional chimps' cries in the distance, coming and going in no clear pattern. The group moved the camp deeper into the forest, closer to what they believed was the origin of the noises. As the sounds got stronger, the hunters began making overnight forays even more deeply into the forest to track the animals. Night after night this continued. Until early one morning, the hunters could see a chimp they had been following for several hours. After it emptied its bowels and bladder, the hunters scooped the precious trove of genetic material into plastic containers and, beaming with pride, returned to camp.

That first successful collection bolstered the group's confidence. The chimp feces were easy enough to get into a specimen cup, and the broad leaves on the jungle floor acted much as the scientists had hoped the umbrella would, pooling the urine for easy collection. As the hunters gradually became more skilled in the art of collecting chimp waste, Hamilton and Worobey began at last feeling hopeful about finally being able to test the vaccine theory of HIV's birth. But then, their luck turned. Worobey had impaled his left hand on a thorn on their first day in the forest and now, a week later, it was dangerously infected. Half of his hand had turned black, and he was developing fevers from what appeared to be blood poisoning. Worobey had to make an emergency hike out of the forest to seek medical attention, carrying with him the first batch of chimp waste. A few days later, as he was recovering at a clinic run by Doctors Without Borders, Hamilton unexpectedly appeared with another batch of samples in tow. But all was not well. Soon after they returned to Kinsangani, Hamilton spiked a fever of his own, along with a worrisome case of chills. A disease even more prolific than AIDS had gotten him— malaria.

Hamilton and Worobey soon began an emergency retreat to Britain, retracing their steps back to London as their scientific adventure abruptly turned into a medical nightmare. Hamilton's illness became huge news in the world of evolutionary biology, and Worobey spent weeks shuttling back and forth to the hospital, sitting by the bedside of his friend. But in the end, Hamilton succumbed to a gastrointestinal hemorrhage, perhaps caused by medicine he had taken to combat the malaria. He died on March 7, 2000, at age sixty-three. It was just two months after he had departed on his final, thrilling intellectual journey. The questions that drove him to Central Africa remained, for all Hamilton's brilliance, beyond his mortal reach.

The loss devastated Worobey. Hospital visits, then the funeral and its aftermath, sidetracked the research project for months. But he eventually pushed onward with the lab work. When it was completed, Worobey discovered that none of the chimp feces that Hamilton and the hunters helped collect showed signs of SIV. The same was true for the vials of chimp urine—save for two. These samples, containing just a few drops each, reacted weakly to an antibody test for HIV-1. It was the first tantalizing evidence of SIV in the jungle near Kisangani.

Viruses have reproductive lives that faintly resemble those of humans. Each generation goes through changes that make it distinct from the one before. In humans that change is caused by births. In viruses it is caused by mutations that happen when viruses make copies of themselves, or when they merge together in what even scientists call, with a smirk, "virus sex." The resulting mutations occur at a fairly predictable pace. So if you can determine the relationship between two copies of virus—whether they are first cousins or fourteenth cousins—you eventually can assemble a family tree that offers clues about their shared history. The goal is to find a common ancestor, a viral Adam or Eve, from which both strains descended. If they have a common ancestor, and you can get a handle on how many generations ago it lived, you can estimate a birth date for the virus.

The SIV-infected chimp urine that Worobey found was not enough to definitively test Hooper's theory. But Worobey returned in February 2003 and assembled a much larger operation for collecting chimp waste so that he could submit it for sophisticated genetic tests. After months of work, his team of chimp stalkers assembled nearly 100 samples from the area, including enough versions of SIV to test whether HIV may have been a direct descendant from a virus harbored by chimps at the Belgian lab in Kisangani.

When Worobey completed the lab work, and had built a genetic family tree showing the relationships among the virus samples he had collected, it offered answers that might have disappointed his fallen colleague. The strains of SIV common in the region of the Kisangani lab were only distantly related to the strain of HIV that caused most AIDS deaths. The time frame for the polio vaccine theory also was implausible. Worobey concluded that there was no way SIV could have given birth to HIV where Hooper proposed it had, nor could the jump from chimp to human have happened as recently as the 1950s, when the polio vaccine campaigns were taking place. Hamilton had lost his life on a wild goose chase. Worobey published his findings in a 543-word "brief communication" in an April 2004 issue of *Nature* that, he wrote, "should finally lay the [vaccine] AIDS theory to rest."

But laying to rest Hooper's theory hardly resolved the mystery surrounding HIV's birth. Finding more convincing answers about the murky history of the AIDS epidemic seemed more urgent than ever.

ONE TINY SPECK OF TRUTH

A virus that thrives in chimps can adapt to also thrive in people. Wild chimps and humans don't interact much, but humans in Central Africa have hunted chimps for many centuries. Anyone stopping by roadside markets in parts of Congo will see dead primates—generally billed as "bush meat"—for sale. There likely is no definitive way to prove it, but scientists now widely accept that HIV was born when a hunter killed a chimp infected with SIV and cut himself during the bloody business of butchering the animal. That would have allowed the virus to enter the hunter's body and begin the original HIV infection. But the theory needed more than a means of transmission. To provide a reasonably full picture of HIV's birth, it also needed a time and a place.

As the work of Worobey and Hamilton effectively ruled out central Congo as the birthplace of HIV, another group of scientists were closing in on a more plausible alternative. SIV was common not only among chimps but among many types of simians. Gorillas got it. So did mandrills and mangabeys.[1] But as the twentieth century drew to a close, researchers had zeroed in on a single, highly lethal strain of virus called HIV-1 group M. It was responsible for nearly all the AIDS deaths in the entirety of the epidemic, in every part of the world. Genetic analysis made it clear that this virus came from a strain of SIV harbored by chimpanzees, and not just any chimpanzees. The source was the so-called Central Chimpanzees that lived in a broad swath of the Congo River basin. And the community of Central Chimps that harbored the version of SIV that most resembles HIV-1 group M was concentrated in southern sections of Cameroon, a sprawling country with bustling Atlantic Ocean ports, populous highlands, and a lightly developed southern region where relatively few people live even today.

The final bit of narrowing took a remarkable degree of scientific ingenuity. An international research team led by Beatrice Hahn of the University of Alabama at Birmingham and Paul Sharp of the University of Edinburgh developed an elaborate project that involved collecting chimp feces across southern Cameroon. To find a strain of SIV that was, on a genetic level, essentially indistinguishable from the most lethal form of HIV, the research team set up ten stations across the region. Two of the stations were in the particularly remote southeastern corner of the nation, as far as possible from major population centers. It was in these two stations alone that Hahn and Sharp's team discovered the samples of SIV that most closely resembled the dominant strain of HIV in humans. When the scientists held sheets of the genetic markers side by side, they were almost impossible to tell apart. This discovery, published in *Science* in 2006, also came with a ready-made explanation for how HIV could have found its way out of the Cameroonian forest after the initial infection.[2] Not far from where these chimpanzees lived was a major river, the Sangha, flowing toward the heart of Central Africa.

This section of the Sangha was not ideal for navigation because of its ribbons of sandbars and the dense vegetation along its banks. Variations in rainfall even caused the lower reaches of the Sangha appear to sometimes reverse directions, from a south-flowing river to a north-flowing one, depending on the time of year. In the especially treacherous middle section, near where Hahn and Sharp's team found the viral ancestor of HIV, few major human settlements ever developed. But there were numerous communities on the Sangha's more accessible stretches. And due south, past riverside trading towns, was the mighty Congo itself. That meant that once SIV made the jump from chimp to human, a single infected person could have carried HIV down the Sangha, onto the Congo River, and into Kinshasa, the biggest city in Central Africa and one with the size, density, and dynamism to brew up a substantial epidemic.

A final, powerful bit of evidence supported this theory.

Scientists studying the dominant strain of HIV gradually found many related varieties, each with a slightly different genetic structure and path through the world. But as they plotted out the histories of these varieties and built an extensive family tree for HIV, they all appeared to have spread themselves from a single explosion, a Big Bang of the AIDS epidemic. Ground zero was Kinshasa.

Dating the birth of the AIDS epidemic was, if anything, more difficult than locating it. Evidence pointing to a particular era gained and lost scientific support, but little of it proved definitive. There was the English sailor who came down with something very much like AIDS in the 1950s. And a fifteen-year-old-boy in St. Louis developed, in 1969, a terrible run of infections that many later believed resulted from AIDS. Journalist Randy Shilts, in *And the Band Played On*, famously pointed to an international convergence of tall ships in New York Harbor in 1976 as a key event, along with the later travels of a sexually voracious air steward whom federal disease trackers dubbed Patient Zero.

Amid the mass of contradictory evidence, one piece stood out for its irrefutability. Researchers looking for possible genetic underpinnings to malaria infection collected more than one thousand blood samples in several locations in Central Africa in 1959 and took the cargo back to Seattle for analysis and, eventually, deep-freeze storage. A quarter century later, when another generation of scientists were investigating HIV's history, they unfroze the samples and found a single copy of HIV from samples drawn in Kinshasa.[3] It soon became the North Star in a confusing constellation of data about the history of AIDS.

That was 1986, and for the two decades that followed no one managed to refute the authenticity of this discovery, nor could anyone discover another piece of HIV that was nearly as old. Two points make a line, two notes a rudimentary melody. But one of anything is an orphan. The virus from 1959 definitely was part of the story. Yet anyone seeking to determine whether it was at the heart of the drama or a bit player, whether it was one of the first examples of HIV or a product of decades or centuries of mutation, were doomed to nothing more than guesswork.

As Worobey moved beyond his disastrous experience in Kisangani, he began pondering whether he might be able to find a second piece of historic virus from Kinshasa. The existence of one sample from the era made it clear that HIV was in the city at the time, raising the likelihood that some other evidence of it remained behind as well, waiting to be discovered. And though a chaotic, troubled megalopolis, Kinshasa had a colonial legacy that for all of its evils also had bequeathed a legacy of modern medical facilities, including research universities with archives dating back many decades. Worobey knew that doctors routinely ordered biopsies when patients arrived at their offices with strange symptoms. The curious combination of fevers, rashes, and wasting caused by AIDS certainly would have puzzled doctors in the 1950s and 1960s. And if a blood sample had survived from that era, Worobey figured that biopsy samples might have as well. He e-mailed several scientists both in Belgium and at the University of Kinshasa, the oldest and biggest medical research institution in Congo, and they confirmed the existence of archival biopsy samples. The problem was that there were thousands of them at the university, each encased in chunks of paraffin wax about the size of a human toenail. His newfound collaborators packed up several hundred of the wax-encased samples and sent them off for the long journey from Kinshasa to Worobey's lab in Arizona, where he had taken an assistant professorship after completing his work at Oxford. But he knew it would take years to examine them all properly. Scientists can't just draw blood from a forty-year-old piece of desiccated flesh and search for telltale antibodies. Nor was it clear if useful genetic material could survive so long in dusty storage rooms.

Worobey pondered these problems, then decided to do some triage. The most useful biopsy samples would be the oldest. So he winnowed the collection down to the ones between 1958 and 1960. Then he asked himself what kind of tissue most likely would contain HIV. A piece of skin or muscle might be of no use, because the virus doesn't tend to concentrate in these places. But Worobey knew that lymph nodes were a favorite hiding place for the virus. Worobey also included some biopsies of livers and placentas, which can harbor reservoirs of HIV as well. These samples—twenty-seven bits of human tissue in all—became the focus of his research.

Searching for remnants of the virus itself was no easier. The process took months of what scientists call "amplification," in which just a few molecules are multiplied enough times so that they can be analyzed. But finally Worobey had a digital rendering of several genetic sequences of the virus. After years of work—with all the sweat and loss it involved—the moment of discovery was banal. The genetic coding for HIV simply popped up on Worobey's computer screen. Remnants of HIV, one tiny speck of truth, had survived in lymph node tissue cut from a twenty-eight-year-old woman in 1960. That was thirteen years before Worobey was born.

Such elaborate, delicate work carried a high risk of contamination. Stray molecules can get into the sample, causing false readings. So Worobey ran all the tests again. The results were the same. Then he sent out another piece of the same wax-encased lymph node to one of the few other labs in the world that can do such sophisticated work, at Northwestern University in Chicago. When the results came back months later, they too were the same. Finally Worobey accepted the discovery. Against steep odds, he had found a second piece of historic HIV in Kinshasa.

Worobey called his discovery DRC60, for the name of its originating country, the Democratic Republic of Congo, and the year of the biopsy. Once Worobey compared the genetic structure of DRC60 to the one found years earlier, a surprising new story emerged. The two copies of HIV, though originally from roughly the same time and place, were genetically very different. This bolstered the theory that Kinshasa was at ground zero of the Big Bang of the AIDS epidemic. It also made clear that HIV was older than most scientists thought. To produce varieties of virus as different as these two required many decades of mutation. For years the leading theory had been that HIV had started spreading in the 1930s. But by mapping out the number of mutations necessary to produce the two samples, Worobey dated the virus's creation to the beginning of the twentieth century, and perhaps beyond. Sometime between 1884 and 1924, Worobey calculated, the dominant strain of HIV was born.

Many scientists would have stopped there. Worobey had found a time window that was more precise than anyone had produced before. Yet here was where Worobey's previous experience in Africa paid off. These places and dates meant something to him, and he suspected that a larger narrative was at play—an impression only heightened by his travels in Congo with Hamilton. Worobey decided that, before reporting his discovery, he should attempt to understand why HIV started where and when it did, and why its spread around the globe had been so halting and uneven.

"There were certain areas on the globe that were like tinderboxes," Worobey later explained in an interview. "There were other places that were like wet moss."

He read *Through the Dark Continent*, explorer Henry Stanley's two-volume history of his coast-to-coast journey across Central Africa in the 1870s. He read *Heart of Darkness*, the classic novel by riverboat-captain-turned-author Joseph Conrad. And he read *King Leopold's Ghost*, historian Adam Hochschild's masterful account of Belgium's disastrous colonization of the Congo River Basin. This was enough to convince Worobey that the birth of the AIDS epidemic and colonialism had more than an incidental connection.

The chronology of HIV's early movements, he learned, fit the region's historic narrative to an eerie degree. Kinshasa had its roots in a colonial outpost that was founded in 1881 and quickly grew into the first inland city in Central Africa. Worobey's calculations suggested that HIV was let loose there not long after. This same period also brought the first steamships up the Sangha River to southeastern Cameroon, opening a new era of fast, long-distance trade routes on which a fragile young virus could have hitched a ride. Without this transformative technology and the new cities built by the region's European masters, HIV might never have found its way out of southeastern Cameroon. Worobey, though not a historian, did not shy away from this remarkable conclusion when he reported his findings in the British journal *Nature* in October 2008. In describing HIV's early period of slow growth, he wrote:

> This pattern, and the short duration between the first presence of urban agglomerations in this area and the timing of the most recent common ancestor of HIV-1 group M, suggests that the rise of cities may have facilitated the initial establishment and the early spread of HIV-1. Hence, the founding and growth of colonial administrative and trading centres such as Kinshasa may have enabled the region to become the epicentre of the HIV/AIDS pandemic.[4]

In the same issue, Hahn and Sharp coauthored an article declaring the search for HIV's birth story complete. While few scientists would put it so bluntly, the accumulation of discoveries was revealing something troubling: Colonialism had had the effect of transforming the region into a tinderbox capable of creating the AIDS epidemic. Then it fanned the flames.

A TALE OF TWO VIRUSES

Scary images of HIV are staples of AIDS prevention campaigns. They often show balls bristling with spikes, like the deadly end of the medieval weapon called a "morning star." Those craving realism draw the spikes more like plungers. Either way, the message is the same: If HIV finds your flesh, it will stick. Soon you will die. Given the huge toll, with tens of millions of people dead and tens of millions more infected, this makes sense. Yet it's also wrong. We all know that we can't get the virus from a mosquito bite. Nor are we in danger from sharing a glass of water or a kiss with an infected person. But what's most surprising is that even most forms of sex are inefficient ways to spread HIV. Scientists estimate the rate of transmission from unprotected vaginal intercourse, depending on various factors, at roughly one infection per one thousand acts. The odds are much smaller for oral sex.[1] In millions of marriages, years of routine sexual contact have failed to spread the virus between spouses. When the suave, sweet-faced basketball superstar Magic Johnson announced he had HIV in 1991, many assumed that his wife—who was pregnant at the time—did too. But she didn't. The Johnsons had sex often enough for her to get pregnant but not HIV.

Even when an infection appears to spread between two regular sex partners, the reality is not necessarily what it seems. Studies of discordant couples—meaning those in which one partner has the virus and the other does not—have found that most of the uninfected ones remain free of HIV for years, even when condoms are not routinely used. Several of the more recent studies, conducted in Africa, used genetic markers to track the source of new infections when they did occur. In about 30 percent of these cases HIV came not from the spouse but from a sex partner outside the marriage.[2]

Few pathogens so consistently upset what we think we know about the world, and so relentlessly expose the gulf between the lives we display to others and the messier realities. A national survey in Zimbabwe reported in 2006 that about half of the HIV-positive unmarried women ages fifteen to twenty-four said they were virgins when interviewed (typically at home, often with relatives or neighbors nearby).[3] There is no reason to believe that Zimbabwean women are less truthful than other people. Obfuscation appears to be nearly a universal instinct when it comes to matters of sexual behavior, and even more so when it might be spreading a terrible disease. That problem, as much as any other, has made it hard to grasp the basic facts of the AIDS epidemic. Somewhere between idealized versions of monogamy and hyperactive Hollywood depictions of human sexuality there exists a wide, muddy, middle ground. Few aspects of behavior are more difficult to study. But the virus itself has left a trail of uncomfortable truths.

The human immunodeficiency virus is a tiny genetic package only one hundred nanometers across. It would take fifty copies of the virus lined up in a row to reach across a strand of spider's silk. At that scale, HIV does indeed bristle with what look like spikes or plungers for attaching to target cells. And once inside a human body, the virus lives up to its reputation. It is aggressive and destructive. It seeks out cells that are crucial to the functioning of our immune systems and kills them, but not before firing its own genetic material into these cells and forcing them to churn out new copies of itself. The damage to a human body is not immediately obvious, though many people newly infected with HIV will experience what seems like a case of the flu. Then typically a long period will pass with no symptoms at all. But over about a decade the millions of copies of HIV gradually overwhelm a victim's immune system, causing AIDS, for acquired immune deficiency syndrome.[4] Then fevers, diarrhea, skin rashes begin. Pneumonia and tuberculosis often take hold. Once this process starts only drug treatment can stave off death.

But while HIV cannot be entirely vanquished once inside a human body, the virus remains fragile outside of it. Flu viruses can spread with a sneeze, and anthrax spores can survive in soil for decades. But air kills HIV almost instantly. The virus needs a ride in blood, semen, or women's genital secretions to get from victim to victim. Condoms usually block this by keeping infected fluids from moving between sexual partners.[5] Yet latex is not the only effective barrier to HIV. An intact vaginal lining, much more often than not, will keep HIV from infecting a woman.[6] So will the saliva in a human mouth. So will unbroken skin on a man's penis. These don't always work. Some sexually transmitted infections create sores that may make it easier for HIV to get through. Sex without adequate lubrication also creates tiny cuts that ease the way for the virus.[7] That is especially true with anal sex; the lining there rips easily. Yet generally HIV faces long odds. It is forever attempting to infect new people. The vast majority of the time, it fails.

Strangely, the slowness of HIV's spread is one key to its power. If it moved easily, if it killed quickly, the first group of victims might have been the last. The Ebola and Marburg viruses can kill dozens of people in a few weeks. Then outbreaks typically burn themselves out. Those infections spread so quickly, and death follows so spectacularly, that people recognize the danger and seek safety—often simply by running away. This is crucial, because an outbreak grows only if each infected person, on average, infects at least one other person. Otherwise the epidemic itself dies out.[8]

Imagine if 10 people had a fatal, incurable new virus. But because it is hard to spread, only half of these people infect new victims. Those five then infect three others. Those three then infect two others. Those two infect one other. That person infects nobody. Those 21 people will die, but the virus has killed itself off too. If all this happened in a remote African forest, it likely would pass from the earth without the outside world taking notice. Now imagine the same 10 people instead had a different fatal, incurable new virus—only it spreads much more easily. Those 10 people end up infecting

20 people, who infect 40 others, who infect 80 others. The death toll already is 150, and there's no end in sight. Soon the whole world will know about this new virus.

So which one is HIV? Amazingly it is both. Most of the time it spreads too slowly to cause many deaths. That is what Michael Worobey meant when he said that the spark of the virus often landed in wet moss. In those settings, the conditions weren't right for a major HIV outbreak. And such inhospitable conditions prevailed in most places, at most times, throughout human history. In fact, Worobey and other scientists estimate that HIV moved from chimpanzees into humans dozens, and perhaps hundreds, of times during the many centuries that people have hunted chimpanzees for food. But according to the genetic record, only one of those occasions sparked the AIDS epidemic we know today. Other times, the virus could not reach new victims more quickly than it killed off old ones. And one time—a single cursed moment—HIV landed in a tinderbox.[9]

One of the essential lessons of epidemiology is that pathogens work in concert with human societies. For HIV to become one of history's most feared killers—to go into the overdrive mode that allowed it to find new victims faster than it killed old ones—it required a combination of factors that once were rare in the world but have become increasingly common over the past century. In other words, the AIDS epidemic needed social change to catch fire. The same is true for many great scourges. The bacteria that caused the Black Death in fourteenth-century Europe rode in the fleas that rode on the rats that rode in the ships that carried trade all the way from Asia. The smallpox virus that killed untold millions of American Indians arrived in the bodies of European settlers determined to claim a vast, verdant land for their own. The Spanish flu virus that produced history's biggest epidemic spread with the help of troops returning home after the first global military clash, World War I.

The history of AIDS has long been more elusive. But now that Worobey, Hahn, and Sharp have given us a time and a place for its birth, it's possible to begin understanding the story of HIV and to study that story for clues to its possible defeat. We now know the epidemic started small and stayed that way for many years. It grew larger only when conditions changed. And the disaster we know today developed only after many decades of slow, quiet spread. Those are, of course, only the broad outlines of a plot that already has sprawled across a century, and certainly has decades more to go. To get at the details—to understand exactly what caused a fragile, hard-to-spread virus to race through certain societies at certain times—requires understanding a narrative older even than the AIDS epidemic itself.

Europeans had plied Africa's ports for centuries, trading goods and resupplying ships for journeys to India and beyond. They also, of course, ran massive and tremendously destructive slave-trading operations along the Atlantic coast of Africa, as did Arabs along the Indian Ocean coast. But for sheer disruption even these did not compare with the influx of explorers, missionaries, and miners who began pushing deeply into Africa in the mid-1800s. Before, outsiders were among many players in a continental history driven mainly by tribal rivalries, burgeoning trade, disease, migration, and other indigenous forces. But as the twentieth century approached, Westerners used their superior technology and firepower to claim nearly all of Africa and divvy up its spoils for themselves.

The appeal was clear. There long had been rumors of vast treasures of gold and jewels, and port traders offered clear evidence of valuable stocks of ivory. The mythical headwaters of the Nile, one of geography's most coveted prizes, lay hidden somewhere within the continent. And in an era when Europeans and Americans had established hegemony across much of the world, Africa offered a challenge that many found hard to resist. It was the ultimate mystery, a "heathen darkness" to which Westerners felt certain they could bring light—and from which they could extract profit. And most did not stop to consider that such intrusions might have anything but benevolent consequences.

THE LION AND DR. LIVINGSTONE

n 1844, a Scottish missionary with dark hair and a bushy mustache trained his rifle on a massive lion that had been harassing a village near the present-day border between South Africa and Botswana. The lion had been devouring the village's precious cattle, and had become so brazen that it attacked even during daylight. It already had evaded the efforts of one hunting party, but on this day the missionary was determined to end the threat to the village. He fired two bullets into the lion's flesh. But the beast did not die. Instead, the wounds enraged it. As the missionary attempted to reload his weapon the lion leaped at him, knocking him to the ground, and crushed his left shoulder with massive jaws. The missionary knew death was near, and a strange calm overtook him as he prepared to meet his creator. But before the lion could finish the job, another man from the village intervened to distract the terrible monster. The lion let go of the missionary, and instead attacked first the villager and then another African man.

The odds looked poor for all three men. The lion was a sleek, golden miracle of muscle, teeth, and bone, and it had the taste of human blood in its mouth. Yet it also was carrying two bullets in its body, and before it could finish off the men, the lion weakened and fell to the ground. The shots from the missionary belatedly had done their job, killing the lion. All three of its victims—the missionary and both of the villagers—had serious injuries but survived.

The missionary in this story was Dr. David Livingstone, and this account added a memorable dose of adventure to the first chapter of his first book, *Travels and Researches in South Africa*.[1] In that same memoir Livingstone became the first white man to visit a wondrous African waterfall.[2] He opened a remote missionary station. He learned the local language. He impressed the villagers with his physical stamina. And he baptized a local chief who converted to Christianity, even though it meant renouncing several of his wives out of respect for Livingstone's admonitions against polygamy. Amazingly, Livingstone managed to render these tales, and many others as well, with humility. When an admirer later asked Livingstone what he was thinking about when the lion closed its jaws on his shoulder, he replied without the hint of a boast: "I was thinking what part of me he would eat first."[3]

And so pious Dr. Livingstone became a legend in Victorian England. Part missionary, part healer, part explorer, he exemplified the values of his age as few did. Readers in Britain and across much of the world devoured his books. Fans thronged his rare public appearances back home. When Livingstone later seemed to disappear into the African bush, a New York newspaper publisher dispatched one of his most tenacious correspondents, Henry Morton Stanley, to find him. Stanley succeeded, meeting Livingstone near the shores of Lake Tanganyika, in East Africa's Great Lakes Region. This rendezvous in 1871 produced the enduring opening line—"Dr. Livingstone, I presume?"—and burnished his legend to even greater brilliance.

After Livingstone's death two years later, thousands of Europeans and Americans attempted to follow in the great man's footsteps, to answer his clarion call to bring "Christianity, Commerce and Civilization" to Africa. In a rush lasting just a few decades, European powers that long had expressed only vague interest in the vast continent to the south suddenly tried to swallow it whole in what historians call the "Scramble for Africa."[4] Soon nearly every African man, woman, and child would be claimed as the subject of some European power. Many forces came together to create the scramble, but it was a young Dr. Livingstone—rifle trained on a ferocious, man-eating lion—whose tales inflamed the Western imagination about a mysterious continent.

Most of those inspired by Livingstone displayed little of the humility or respect that were among the missionary's signature qualities. In his account of his encounter with the lion, Livingstone already spoke an African language, and was acting to protect African lives and well-being. His writings consistently expressed admiration for Africans, including his friend Mebalwe, "a most excellent man," who saved Livingstone's life by distracting the lion. And though Livingstone was determined to change things he disliked about the cultures he encountered—his main job, after all, was introducing a foreign religion—he showed respect for existing power structures and customs. Livingstone didn't blast his way into villages, despite access to superior firepower. He didn't trick chiefs into signing over their lands in treaties they couldn't read. He didn't enslave Africans to collect ivory, rubber, or anything else. Those virtues would set Livingstone apart from most of those who followed him.

Among the first and most destructive was Stanley himself, who moved from journalism into full-time exploration in the 1870s, developing along the way an infamous reputation for terrorizing people living within the regions he traveled. His books, like Livingstone's, were wildly popular in Europe and America, even though Stanley's boastful, purple prose displayed little of Livingstone's restraint. In Stanley's stories, he was always the hero, and the guns often were trained on Africans. Using these more aggressive tactics, Stanley soon became one of the first white men to traverse the vast Congo River Basin. When King Leopold II of Belgium moved to claim this area as his personal fiefdom, Stanley became the king's agent on the ground; he founded the region's first major inland city and named it Leopoldville after the king. (Later, it would be renamed Kinshasa.) Here and elsewhere in a colony seventy-five times larger than Belgium itself Stanley and Leopold pioneered a type of colonialism so disastrously cruel that reports of it eventually would shock the conscience of Europe. But for several decades, as European traders steamed their way up the Congo River and its tributaries, there were few checks on their brutality as they forced Africans to collect rubber sap and ivory. A private army loyal only to Leopold shot those who refused, or cut off their hands, or lashed them with the *chicotte*, a fearsome whip made of hippopotamus hide.[5]

Fueling the scramble was rising consumer demand in Europe and the Americas for raw materials available in Africa. Ivory may seem a touch quaint today, but in its heyday it was seen as beautiful, versatile, and essential to many everyday products. It was used to make billiard balls, jewelry, and cutlery. Furniture makers incorporated it into their cabinets, artists into their statues. Bagpipe makers used ivory for mounts, ferrules, buttons, and mouthpieces. When supplies of ivory gradually grew short, as colonial agents killed the once plentiful elephants by the thousands, rubber took its place as the economic lifeblood of colonialism in the Congo Basin. The first inflatable rubber tires for bicycles became popular in the 1890s. Mass production of cars soon spiked demand for rubber tires again. The only obstacle to European companies reaping huge profits in Congo was the massive labor required to collect both ivory and rubber. Getting ivory from an elephant required stalking it, killing it, and cutting off its tusks.[6] Getting rubber from vines required slashing them, collecting the oozing white sap, and drying it—sometimes on the collector's own skin. The solution to the manpower demands soon became obvious: Leopold, Stanley, and other colonialists created what was essentially slavery: cheap muscle at the point of a gun.

This approach was not confined only to collecting ivory and rubber. These industries created tremendous new needs for infrastructure to get goods to oceangoing ships along the Atlantic coast. That meant African porters had to carry materials and supplies anyplace the steamboats couldn't reach. Workmen were needed to build railroads, trading stations, dormitories. And somebody needed to operate the steamboats, load the railroad cars, carry the tusks or gobs of rubber in from the jungle. When workers became unruly the colonial companies deployed heavily armed soldiers to keep the cogs of these vast enterprises moving. All these roles were filled by Africans, many imported from hundreds or even thousands of miles away. Those who resisted felt the *chicotte*, or worse. African life here was beyond cheap. It was disposable. Contemporary accounts by journalists and missionaries told of how colonial officials across the Congo Basin ordered mass slaughters and the torching of rebellious villages while creating forced settlements that resembled nothing so much as concentration camps. The new mobility in the region also touched off epidemics. Most of the millions of deaths in the Congo River Basin under European control resulted not from guns or whips but from diseases spread by colonialism.[7]

Many people look at maps of Africa and see a puzzling mosaic of countries. Gambia appears as a horizontal finger at no point wider than thirty miles. The Guineas have names that forever confuse the rest of the world: Guinea-Bissau, Guinea-Conakry, and Equatorial Guinea. "The Congo," meanwhile, can refer either to the Republic of Congo or the Democratic Republic of Congo. As if that's not difficult enough, the Democratic Republic of Congo used to be Zaire, which used to be Congo-Leopoldville, which used to be the Belgian Congo, which used to be the Congo Free State. The Republic of Congo, usually called Congo-Brazzaville today, used to be French Congo. All these crazy-quilt borders and overlapping names are the fossilized remnants of European colonial designs. They seem absurd today, but just a century ago they were at the core of competing imperial ambitions.

The most important, from the point of view of understanding the birth and first movements of the AIDS epidemic, was that of the Germans. In the 1880s, Western powers sliced up Africa into most of the countries we know today, and the Germans ended up with control of Cameroon, which included the northwestern corner of the Congo Basin. At first they exerted colonial authority mainly along the Atlantic coast, but they gradually extended their control into Cameroon's densely wooded, lightly inhabited southeastern corner, all the way to the Sangha and Ngoko rivers, deep in the interior of Central Africa.

The Germans were fierce commercial competitors with other European powers and displayed the brutality toward Africans that was typical for Westerners on the continent. A German judge investigating reports of abuses in 1891 found that colonial forces—typically consisting of a few European officers leading dozens of African mercenaries—obliterated villages that resisted their rule.[8] In one called Toko troops killed indiscriminately as they burned down every hut and destroyed every useful plant, even cutting down the village's palm trees. When the slaughter was complete the soldiers retired to a nearby beach, and according to the German judge's report, each brought along the severed head of a victim "as a souvenir."[9] In other cases the bodies of rebellious villagers were cut up and their hands kept to memorialize victorious battles. Women, meanwhile, were often given as tributes to conquering soldiers. Such barbarity was not restricted to junior officials and their unruly charges. In 1893 the acting governor of the German colonial government responded to an attempted mutiny by African troops with an unforgettable demonstration of viciousness: After defeating the rebellious men he ordered twenty of their wives publicly stripped and whipped with the *chicotte*. At least some were raped as well. These tactics were not enough by themselves to pacify Cameroon. In many places villagers fought back—sometimes successfully, taking control of areas for months at a time. But over time the German quest for glory and profit, bolstered by firepower that the Africans couldn't hope to match, won out.

In December 1895 colonial authorities got reports that Cameroon's southeastern corner contained fabulously rich ivory and rubber stocks, awaiting exploitation. The only problem was that while the Germans nominally controlled this area, French, Dutch, and Belgian traders already had moved into the territory. And they circumvented German authority along the coast by sending steamships full of the lucrative bounty down the Ngoko and Sangha rivers, onward to the Congo, and out to the world.[10] The Germans soon after gave authority to a colonial company to take control of the vast region by force. Over the next four years they extended their power all the way through southeastern Cameroon and established a trading station on the Ngoko River, about seventy-five miles upstream from where its waters merged with the Sangha. In the wedge of land defined by these two rivers, HIV had either just been born, or soon would be.

The trading station was called Moloundou, and a busy border town remains there today. But at the time it was almost unimaginably remote. Few human settlements had developed among these forbidding forests. And there were only two practical ways out: by steamship to the Congo River; or over-land by foot to the Atlantic. The river route was the easier of the two, and steamships transported the bulk of the ivory and rubber collected in south-eastern Cameroon. But overland routes were necessary to connect Moloun-dou with other trading stations and inland areas rich with rubber and ivory. For these journeys the bounty was borne by Africans pressed into service as porters, who carried loads averaging fifty-five pounds each.[11] At the peak of the foot traffic that would develop between inland areas and the coast, the busy way station recorded more than a thousand porters passing by on a typical day.

There is a profound shortage of direct accounts of what life was like for Africans during Cameroon's early colonial period because few had access to a written language that could record their suffering. But historians have gath-ered bits of oral history that illuminate the era. One, recounted by historian Peter Geschiere, tells how the Germans appeared suddenly in an interior vil-lage, looking so pale that they were at first mistaken for ancestral spirits.[12] But the villagers knew that spirits would not come carrying sticks and guns, nor would they bring along a barking dog. The Germans had learned a lan-guage used in a neighboring village and presented themselves as friends. "I have come to make you rich, and I have brought presents with me," the leader said, according to the legend. He then offered each family living there five lumps of salt—a rare luxury that the women there soon stirred into pots of banana puree, producing delicious results. "Where do they come from?" one man asked about the strangers. "This is the very best salt, the very best salt!"

The price soon became clear. When the Germans demanded food for themselves and the other hungry soldiers, the villagers complied by slaughter-ing their livestock and cooking more bananas. After the feast the Germans rounded up the village men and ordered them to carry bags and other gear onward to the next village. Only one man in the story, named Miague, resisted, saying, "This is no business of mine; I can't carry the basket of some-one else; what did he give me?" The soldiers responded by laughing. Then they grabbed Miague, threw him to the ground, and beat him with their rifles. Miague soon found a heavy bag tied to his back. When the Germans left the village he walked among them as a porter.

The Germans' own accounts from the time record the amazing scale and brutality of the porter traffic through Cameroon. Africans often were chained together. Reluctant men had their wives and children seized as hostages. Those who escaped that fate were forced into becoming collectors of ivory or rubber sap. Village chiefs soon began complaining about famished caravans of porters raping and pillaging their way across the countryside. This appears to have been built into the business models of the German companies; porter teams frequently left bases without nearly enough food to complete their journeys.

Geschiere compared the rubber trade to a "hurricane" that blew through swaths of Cameroon, causing massive upheaval before moving on as supplies of sap ran low. To meet German demands desperate collectors often resorted to cutting down vines rather than tapping them with the careful slices that allowed sap to flow—albeit more slowly—while letting the vine survive. German lieutenant Hans Dominik, who was celebrated by colonialists for his leadership of campaigns to suppress unrest in Cameroon, wrote:

> I know from my own experience what the state is of an area where the struggle for gold of the jungle [rubber] has been waged. The lust for quick money, the fear of competitors vanquishes all scruples and the indigenous people are intoxicated as well: they slaughter their beasts, do not sow or harvest anymore but only make rubber, rubber. Happily for the country the wave rolls on when the rubber is exhausted. But then follow the caravans marching through, then food shortages arise and then follows violence.

It was 1908, two years after Dominik wrote those words, that Worobey would later single out as the likeliest moment for the historic transmission of HIV from chimp to human. Though it could have happened as early as

1884 or as late as 1924, his analysis put the odds on a bell curve, with the most likely year 1908. This time frame coincides almost exactly with the most rapacious period of German colonial exploitation in the region—a time when the quest for "gold of the jungle," as Dominik called it, was rolling furiously through southeastern Cameroon.

Ominously, something else followed the rubber trade through Cameroon: disease. Sleeping sickness, smallpox, and skin infections were the most obvious. Colonial authorities attempted mass inoculation campaigns for smallpox and set up quarantine zones that restricted where the porters were allowed to travel. But the historical record is equally clear about the massive spread of syphilis, which seems to have arrived with the Europeans. In just a few years it reached epidemic proportions along porter routes and riverside trading posts in Cameroon and throughout the Congo River Basin. It's impossible now to determine how much of this spread resulted from rapes as opposed to other kinds of encounters, but it's clear that colonial commerce created massive new networks of sexual interactions—and massive new transmissions of infections.

To fully grasp how the birth of AIDS fits into this picture it's worth remembering where Hahn, Sharp, and the other scientists later found the chimpanzee virus that so closely resembled the dominant strain of HIV. One of the collection stations was in what is now Lobeke National Park, in Cameroon's southeastern corner; the other was in a nearby village. To the east of them both is the Sangha River. To the south is the Ngoko River. And just to the west, running almost along the border of Lobeke, is a rutted dirt road that the Cameroonian government has ambitiously dubbed National Highway P4. A century ago it was just a porter path trod by uncountable thousands of African men and women through the thickest, darkest of equatorial forests as they walked to and from the German trading post at Moloundou.

So HIV's first journey looked something like this: A hunter killed an infected chimp in the southeastern Cameroonian forest, and SIV entered his body through a cut during the butchering, becoming HIV. Something like this likely had happened many times before, during the centuries when the region had little contact with the outside world. But now thousands of porters—both men and women—were crossing through the area regularly, creating more opportunities for the virus to travel onward to a riverside trading station such as Moloundou. One of these first victims—whether a hunter, a porter, or an ivory collector—gave HIV to a sexual partner. There may have been a small outbreak around the trading station before the virus found its way aboard a steamship headed downriver.

For this fateful journey south, HIV could have ridden in the body of these first victims, or it could have been somebody infected later: a soldier or a laborer who tossed wood into the steamship's hungry engine. Or it could have been a woman: a concubine or a trader. It's also possible that the virus moved down the river in a series of steps, maybe from Moloundou to Ouesso, then onward to Bolobo on the Congo River itself. People climbed on and off the craft every time steamships docked, creating the possibility of new transmissions to other sexual partners. There might even have been a series of infections at trading towns along the entire route downriver, as the waterway flowed through areas of German, French, and Belgian control. Yet even within these riverside trading posts HIV would have struggled to create anything more than a short-lived, localized outbreak. Most of this colonial world didn't have enough potential victims for such a fragile, hard-to-spread virus to start a major epidemic. To fulfill its grim destiny the virus needed a kind of place never before seen in Central Africa but that now was rising in the heart of the region—a big, thriving, hectic place jammed with people and energy, where old rules were cast aside amid the tumult of new commerce. It needed Leopoldville, where the Belgian king's vicious style of colonialism had turned Livingtone's benevolent vision on its head.[13] It was here, hundreds of miles downriver from Cameroon, where HIV began to grow beyond a mere outbreak. It was here that AIDS grew into an epidemic.

Laying the scientific story alongside the historical one offers one final revelation. The time frame illuminated by Worobey's research opened with the beginning of massive new intrusions by Europeans into the Congo River Basin. But perhaps as surprisingly, the time frame closed—at least for southeastern Cameroon—about the time that the period of commercial frenzy

there slowed. The onset of World War I in 1914 included skirmishes between the Germans and French on the Sangha River, and the Germans lost
all of their African colonies when they surrendered to Allied forces four years
later. The French took control of Cameroon, but colonial commerce in the
far southeastern reaches never returned to prewar levels.[14] In the 1920s, the
Sangha River's strategic importance continued dwindling as global rubber
prices collapsed. Plantations and synthetic products soon began serving the
world's needs more efficiently than Central Africa's wild rubber vines ever
could. Regular steamboat traffic on the Sangha slowed and eventually ended.

So the improbable journey of the killer strain of HIV became feasible for
only a few hectic decades, from the 1880s to the 1920s. Without the scramble, it's hard to see how HIV ever could have made it out of southeastern
Cameroon to eventually kill tens of millions of people. Even a delay—caused
perhaps by a hungry lion eating a charismatic missionary before his daring
tales could get history rolling in a certain direction—might have caused the
killer strain of HIV to die a lonely death deep in the forest.

FEMMES VIVANT
THÉORIQUEMENT SEULES

E dward Glave was the kind of young man every empire needs: brave, quick-witted, a good shot, and utterly convinced of his nation's inherent superiority. In his memoir, *In Savage Africa*, an ink portrait shows Glave looking just a few years removed from a British prep school. He wears a jacket and tie, his eyes are full of purpose, and his short brown hair is parted neatly on the left. But the image of schoolboy innocence wouldn't last. Forty-five days after departing Liverpool, his steamship, the *Volta*, suddenly began struggling through the turbulent waters flowing from the mouth of the Congo. The image, laden with metaphorical power as Glave began his push into the Dark Continent, left an impression. "[F]ar out into the blue Atlantic we could see the turbid, muddy stream thrusting its way and refusing to mingle with the waters of the ocean," he wrote in his memoir. "And soon the Volta was plowing her way through a mass of tropical vegetation littered over the surface of the sea in every direction, the waters growing tawnier and darker as we steamed slowly in toward shore."[1] A few days later, as the river's notorious rapids forced Glave and his fellow passengers to continue the journey on foot, he for the first time "donned the traditional dress of the explorer," a helmet, leggings, and a belt for his revolver.[2]

Glave's writings provide a glimpse of colonial Congo on the verge of its fateful encounter with HIV. He arrived in 1883, a decade after Livingstone's death and just a year before the convening of the Berlin Conference, during which European powers would gather in the German capital to slice up what King Leopold II called that "magnificent African cake" into new nations. It was also near the beginning of Worobey's estimated window for HIV's birth. What Glave found in his travels was a world already in the throes of transformation as the Belgians—to a degree beyond even what the Germans managed in southeastern Cameroon—were beginning to harness the jungle into an engine of Western profit. When Glave and a group of fellow Europeans approached villages on the Congo River they were greeted not with showers of poisoned arrows but by throngs of Africans eager to trade for goods such as cloth, metal trinkets, and liquor. And a British vice, snuff, had taken hold, courtesy of newly robust trade links with the outside world. Again from Glave's memoir:

> The one feature common to the people we met on the march was their snuffy condition. They were all inveterate snuff-takers; they bake the tobacco leaf perfectly dry and mix about an equal quantity of wood ashes with it; having ground this to a fine powder, they carry it in cloth pouches, and when a pinch is required, a thimbleful or so is emptied in the palm of their left hand and stirred with the blade of a long knife to insure its being of the requisite fineness. The needful amount is then conveyed on the blade of the knife to the nose, but so clumsily that mouth, chin, cheek, and nose are all smeared with the brown powder, which they do not attempt to brush away; in fact it seems "good form" in that land to possess such facial adornments. This snuff must be rather powerful, judging from the prodigious sneezes it causes, and the watery, blood-shot eyes of those addicted to its use.

When one Congolese man repeatedly pestered Glave for snuff, he responded as a master would to an ill-behaved dog, with a memorable punishment. He surreptitiously gave the man a concoction made from white pepper instead of tobacco. Glave later reported,

> [T]he next morning, as our caravan filed out of the village, he who had tested my "mixture," watched our departure with blood-shot eyes; and his general appearance of bewildered exhaustion showed plainer than words could tell how deeply he must have regretted his persistency.[3]

Glave's journey up the Congo River eventually brought him to Leo-poldville, which Stanley had founded two years earlier, hundreds of arduous miles in from the coast. Upon arriving, Glave immediately spotted Stanley himself pacing back and forth outside his one-story clay house, under a ve-randa made of thatch. The two men shook hands, and Glave quickly fell under the charismatic spell of this journalist-turned-explorer who now had started a third career, as a colonial potentate. Over the next few days Glave did clerical tasks and oversaw gangs of African men doing menial labor. There was plenty of work, for Stanley had been shrewd in locating Leopold-ville. It was perched just upriver from a massive waterfall that flowed into the treacherous lower stretches of the Congo. But the city itself—at this point really no more than a burgeoning frontier town—sat on the southern shore of a massive bulge in the river called the Stanley Pool. Here was the key to a navigable stretch of the upper Congo that continued inland for nearly one thousand miles.

Stanley's house sat on a hillside near ornamental gardens and the shade provided by the broad leaves of banana trees. White officers had well-built quarters, not far from a sturdy storehouse. At the bottom of the hill were rows of thatch huts for African workers. Farther away, by the riverbank, were workshops "in which the ringing of the blacksmith's anvil and blowing of the wheezy bellows mingled with the mournful but melodious singing of the gangs of Zanzibaris."[4] The Zanzibaris, of course, were men from Zanzibar, an island off the East African coast, who had been dragged seventeen hun-dred miles across the belly of the continent to work for Stanley. And beyond the workshops, at the edge of the Stanley Pool, were three steam-powered riverboats. The next few years would bring many more, both to Leopoldville and its twin city just across the river, Brazzaville, the capital of the French Congo territories.

Reading Glave with a modern eye the strict social and racial hierarchies jump out, as does a swashbuckling sense of adventure for those who took it upon themselves to tame—and exploit the riches of—this wild land. Yet there was something curiously missing in his descriptions of Leopoldville: women. Glave made no mention of quarters for them, or of the places they might have appeared as evening fell and weary men looked for female companionship. Maybe this new kind of commerce hadn't yet appeared, or speaking of it would have violated Glave's Victorian sense of propriety. But if it wasn't there during his visit (he soon departed for an even more remote outpost hundreds of miles farther up the Congo River), it arrived soon after. For as Leopoldville rose on the banks of the Congo, so did a burgeoning industry—prostitution.

Before colonialism arrived the Congo River Basin was one of the least urbanized regions of a sparsely populated continent.[5] Neighboring communities typically had different cultures, even different languages, likely limiting the establishment of sexual relations. Traditional polygamy also tended to circumscribe the options for women, and to some extent for men as well. Fertility was highly prized, and producing children was seen as an essential function of adult life. A man with several wives lived with the social expectation that he regularly sire new children, and he may have had less incentive to stray. Women, meanwhile, generally owed their reproductive potential to their husbands. That is not to say that these rules were never bent.[6] Clearly they were. But they were rules nevertheless, and there could be consequences for getting caught breaking them.

But as HIV made its historic journey down the Sangha and Congo rivers, colonialism was dramatically reshaping African life. Leopoldville was among a new generation of boomtowns created by Europeans to serve European visions of the economic destiny of Africa. They were places like South Africa's gold-rush town of Johannesburg, Kenya's thriving railhead of Nairobi, and Cameroon's ivory trading hub of Yaoundé.[7] Most were on upland plateaus—high enough to avoid the malaria-carrying mosquitoes that historically had taken such a toll on the European colonialists and their families—and all were strategically situated for the collection of lucrative raw materials for distribution to markets around the globe.

These new urban dynamos worked remarkably well at fattening the bank accounts of Western companies. The value of Cameroon's rubber exports doubled between 1891 and 1912, then doubled again, and doubled again.[8] For those Africans living in and around the new cities, life changed with terrible speed. Authority long vested in tribal chiefs became the province of heavily armed foreign powers. Traditions and rituals honed over centuries concerning family, marriage, sex, death, kinship, child rearing, work, property, money, and trade—all were undermined. The lure of better lives—and often outright force—pushed men to leave land that had nourished their families for generations, and into new homes in squalid dormitories where women and children weren't welcome.

Yet the men of colonial Leopoldville—and there were plenty of them, outnumbering the women by a factor of more than two to one—did not lack for sexual opportunities, according to research by American historian Nancy Rose Hunt. European men frequently kept Congolese mistresses called *ménagères*, quaintly translated as "housekeepers." Even in a Catholic colony the *ménagères* were all but officially sanctioned, especially in the era when Congo was regarded by most Belgians as too primitive for the delicate constitutions of white women. In 1906, a colonial doctor in the Congo River town of Basoko lamented the outbreak of sexually transmitted diseases sparked whenever steamboats of colonial dignitaries passed through. If the Europeans aboard "could travel with their *ménagères*, perhaps we could avoid the transmission of these noxious diseases in the posts served," the doctor wrote.[9]

Sexual license was one of the few freedoms in Congo not limited to Europeans. The most Westernized Congolese—the so-called *évolués*, a term translated as "the evolved men"—came to practice an updated form of polygamy that, unlike the traditional version from the countryside, was fundamentally secretive. These informal multiple wives, called *supplémentaires*, allowed *évolué* men to publicly conform to Belgian admonitions about monogamy while fulfilling traditional cultural roles in which the most successful men sought to maximize their production of children by keeping several regular sex partners at a time. The vast laborer class of African men—which the Belgians called *basenji*, for "savages"—was separated from wives or potential spouses for most of the year but had access to the city's plentiful prostitutes.[10]

Official efforts to control the sex industry began in 1909. In the areas where Europeans lived, police kept registries of known prostitutes, who were required to undergo regular gynecological exams and carry medical cards. Those found to harbor illnesses were confined to hospital beds until cured.[11] In this era before antibiotics, syphilis was a particular problem in Leopoldville. "It is by the prostitution of black women, especially, that these diseases are propagated. It is therefore against prostitution that the efforts of authorities must be directed," said a government circular in 1912. Colonial officials were nearly as unhappy about traditional polygamous marriages. Belgian officials worked to impose Christian-style monogamy as inherently superior but eventually discovered that the obvious solution—"liberating" women from polygamous marriages so that they could enter monogamous ones—did not have the intended effect. One government circular in 1914 cautioned against this aggressive form of social intervention:

> Civil servants must use prudence and discernment in matters of liberation. They will not forget that the native woman is not yet prepared, in most cases, for an independent life. It will serve nothing to liberate women for whom one has not assured the destiny in advance by facilitating a marriage with a monogamist. The woman free of all conjugal or pseudoconjugal ties becomes simple food for prostitution. Our agents will not proceed then

with liberation, if they do not have the assurance that the liberated women will be reclassified easily.[12]

The government eventually decided on a subtler approach to correcting what it saw as the misguided ways of Congolese women. It taxed them. Any wife after the first one (and up to thirty) was the subject of a polygamy tax, payable of course by her husband. Single adult women also had to pay a tax, based on the presumption that most were earning money through sexual relationships. For this category of women authorities settled on a term that was worthy of the Orwellian nature of the colonial Congo state: *femmes vivant théoriquement seules*, "women theoretically living alone."[13] All unmarried adult Congolese women residing in Leopoldville, even widows, had to pay it. The only way to avoid what amounted to a sin tax was to marry—and stay married to—an avowedly monogamous man.

As church officials fretted openly about the souls of women lured into Leopoldville's sex industry, secular authorities also grew concerned. The colonial economy required cheap, plentiful labor, and the Belgians often complained that Congolese women were not producing enough sons to keep the engines of commerce humming. Today historians blame the era's low birth rates—across a vast swath of Central Africa that came to be called the "Infertility Belt"—in part on colonial policies, including forced labor and taxation policies that contributed to chronic hunger. Malnourished women were less fertile, and underfed men often stayed away from home, searching for food. Men with jobs, meanwhile, spent most of their time living in single-sex company dormitories at factories or plantations, with few chances to see their wives living back in rural villages, much less make them pregnant.

But the most important cause of low birth rates was the era's rampant sexually transmitted diseases—not just syphilis but also gonorrhea and chlamydia. As in Cameroon, doctors throughout colonial Congo struggled to control the outbreaks that flowed along porter paths and other transport routes, wherever riverboats, trucks, or trains stopped for the night. One railroad town hundreds of miles from Leopoldville reported in the 1930s that

three in four Congolese men had gonorrhea. Both that disease and chlamydia often caused sterility when spread to women. Syphilis, meanwhile,
caused many miscarriages. Birth rates began recovering only in the 1950s,
after antibiotics arrived in the region on a mass scale. History offers scant
record of how the Congolese themselves experienced this disastrous spread
of disease and sterility. But a fertility cult formed in Congo in 1935 adopted
a telling slogan: "Let us give birth like before the whites" arrived.[14]

THE GIFT

Leopoldville grew into Central Africa's largest city and, in the tidy areas where Europeans lived and worked, one of its most Westernized. A Belgian priest promoted soccer in the city and later oversaw the construction of tennis courts, stables, and a pool. One of Leopoldville's manicured city parks even had a replica of the famous Belgian Manneken Pis Fountain, depicting a naked boy urinating into a basin below. Old black-and-white photos of Leopoldville show buildings that were boxy but with a hint of art deco style. The streets were wide, tree lined, and so clean it looked like workers swept them twice a day. One image from 1955 shows café tables on a spacious sidewalk. The waiters were dark skinned and the customers all white, many in button-down shirts and ties as they smoked pipes, sipped coffee, read newspapers. The store on the opposite corner, Au Petit Louvre, presumably offered a touch of continental atmospherics in a patch of Africa far removed from Joseph Conrad's primordial landscape, where "vegetation rioted on the earth and the big trees were kings." A photo taken from an airplane in 1958 shows the elegant Boulevard Albert I straight as a rifle shot past the city's best business and residential addresses. Also visible were high-rise buildings, a brick-and-stained-glass cathedral, cranes by the waterside, and factories in the distance. Blink, and it could almost be South Florida.

A street scene from Leopoldville, 1955.

These pictures, of course, didn't show the sprawling *cités indigènes* where Congolese men and women lived in conditions considerably bleaker. Their densely packed homes and businesses sat back from the swank riverfront areas, beyond the park, the zoo, and the golf club that separated European and African life in Leopoldville. And what's also missing from these images, and from the historic and medical literature from the era, are clues that the AIDS epidemic was growing in this city. Worobey's calculations suggest that after HIV's arrival from Cameroon the epidemic grew in a way that roughly tracked the growth of Leopoldville. As the population of the metropolitan area went from thousands to tens of thousands to hundreds of thousands in the first half of the twentieth century, the number of HIV cases went from dozens to hundreds to maybe a few thousand.[1] That suggests substantially less than 1 percent of Leopoldville's adult population harbored the virus in the years before independence arrived in 1960. Given that the symptoms of AIDS, including wasting, fevers, and diarrhea, can be caused by many other tropical diseases, it's not surprising that nobody appeared to have noticed a new epidemic emerging in their midst.

The Congo River forms a massive arc that people in the region have used as their major transport artery for centuries. Branching from the river are thousands more miles of waterways, but passage in many spots traditionally was limited by rapids, sand dunes, and fluctuating water flows. And where river travel wasn't possible, dense forest also could make travel tough. That's one reason why the section of the Sangha River basin that gave birth to HIV had so few inhabitants before the Germans arrived. There was another lightly populated area south of the main bend in the Congo River, and to the east, beyond the falls at Stanleyville, human density dwindled as well. In the precolonial era, a traveler setting off from the spot that would become Leopoldville eventually would have hit natural obstacles in almost every direction—jungles, waterfalls, mountains, and the Atlantic Ocean itself—that served as firewalls against the spread of diseases.

Steamboats began changing that, but the geography of the epidemic remained concentrated for decades after the historic strain, HIV-1 group M, traveled to Leopoldville from Cameroon. That didn't mean the virus wasn't growing and changing within the cauldron of this bustling new city. As it moved to other victims HIV gradually mutated into several identifiable new subtypes—the children and grandchildren of the group M family. But to make its next leap, the epidemic again needed human intervention. This came in the form of a historic influx of steel, concrete, and fossil fuels. The Belgians often chortled about how, before King Leopold II took over, many Congolese didn't appear to have adopted the prehistoric innovation of the wheel. In the eight decades that followed the Belgians built—with Congolese muscle and sweat—one of the most modern and extensive transport systems in Africa.

This network served the colony's growing number of textile factories and cotton, coffee, and palm oil plantations, as well as the mining industry burgeoning in the Katanga Plateau and other regions. The discovery of copper, diamonds, gold, and uranium—some of which would give Little Boy, the atomic bomb the United States dropped on Hiroshima in 1945, its explosive power—replaced ivory and rubber sap as fuel for the colonial economy. Rather than struggle along porter paths, workers could ride railroads down to the mines. Or they could climb aboard trucks that plied hard, smooth highways. Belgian executives, meanwhile, could fly directly from Brussels to Leopoldville, then onward by plane to most of Congo's cities and major towns. By the time independence arrived the Belgians had built eighty-five thousand miles of roadways, nearly five thousand miles of rail, and seventy airstrips, six of which could handle international flights.[2] These new networks soon would provide the means for HIV to finally escape Leopoldville.

On December 28, 1958, a lean, fiery Congolese man with a goatee, black-rimmed glasses, and a part shaved onto the left side of his head stepped decisively onto history's stage. Patrice Lumumba, then just thirty-three, was a former postal clerk, an *évolué* in the lexicon of the Belgian Congo. And he had just returned from the continent's first liberation conference, which had been held in Africa's first postcolonial state, Ghana. The Belgians had been hinting at eventual freedom for Congo—one prominent Belgian academic a few years earlier had unveiled a thirty-year plan for independence—but Lumumba, like many Africans, did not want to wait. To a crowd of several thousand people in Leopoldville, Lumumba declared "Independence is not a gift, but a fundamental right of the Congolese."

A week after Lumumba's speech, a riot broke out when members of the jittery colonial army opened fire on a crowd of demonstrators, killing at least forty-nine Congolese and injuring hundreds of others. Over the next three days mobs hurled rocks at white motorists, tore down streetlights, set shops ablaze, and attacked churches, hospitals, and government facilities of all sorts. Belgium's panicked King Baudouin soon announced that there would be an orderly end to Belgian control of Congo. "It is our firm intention, without undesirable procrastination but also without undue haste, to lead the Congolese population forward towards independence in prosperity and peace," the king announced on January 13, 1959, just two weeks after Lumumba's electrifying speech.

Americans tend to remember the decade that followed for the Beatles and Woodstock, the Pill and Vietnam, John F. Kennedy and Martin Luther King Jr. But the 1960s had a dramatically different feel in the Congo, and across most of Africa. This was the era of independence, of hasty pullouts by humiliated colonial powers, of big dreams and even bigger disappointments for the Africans left behind. In this latter-day scramble, Congo's story was among the worst. Lumumba became prime minister in June 1960, and for the nation's independence ceremony gave another famously defiant speech that was greeted by a raucous standing ovation from a crowd of Congolese political figures. But, just a few days later, members of Congo's army went on a rebellious rampage to protest the high-handed ways of their top general, an unreconstructed Belgian colonialist who provocatively scrawled on a blackboard "After independence = Before independence." Amid this turmoil the Western world's patience with experiments in African autonomy quickly wore thin.

The world of Cold War geopolitics was a binary one, sorted ruthlessly into supporters and enemies of American-style capitalism. Lumumba first sought aid from the United States in hopes of steadying his fractured, impoverished new nation. But when that failed, he began publicly mulling the idea of taking aid instead from the Soviet Union, whose anticolonial rhetoric appealed to many African leaders of his generation. Lumumba soon was a marked man. U.S. president Dwight Eisenhower approved his assassination less than two months after Lumumba took office, and CIA director Allen Dulles sent a cable to his station chief in Leopoldville that said, "In high quarters here it is the clear-cut conclusion that if [Lumumba] continues to hold high office, the inevitable result will at best be chaos and at worst pave the way to Communist takeover. . . . His removal must be an urgent and prime objective."[3]

The CIA dispatched an agent with a vial of poison with the intention of somehow infusing it onto Lumumba's toothbrush.[4] But Congolese political rivals got to him first. He was arrested, then turned over to a secessionist leader from the mineral-rich Katanga region, where a breakaway group was working in concert with the Belgians to undermine the new national government. There, he was beaten and tortured just two years and a few weeks after Lumumba declared independence "a fundamental right of the Congolese." Belgian officials oversaw Lumumba's assassination by firing squad and the destruction of his mortal remains, which were hacked into pieces and boiled away in sulfuric acid. The nation, meanwhile, descended into a period of political and economic chaos that became known as the "Congo Crisis." It ended in 1965 when one of Lumumba's former captors, Army Chief of Staff Joseph-Désiré Mobutu, took control of the nation and began a dictatorship that, for nearly three decades, received the staunch backing of the United States and other Western powers. Most Congolese, like many others across Africa, watched helplessly as cruel, despotic, authoritarian European rule was replaced by cruel, despotic, authoritarian African rule.

As those shifts were underway the AIDS epidemic burst loose from its relatively confined existence in Leopoldville. Edward Hooper and others seeking to explain the growth of HIV after independence have pointed to the proliferation of mass inoculation campaigns in the final years of colonial rule. The regular reuse of syringes may have accelerated the spread of the virus.[5] But far more consequential was the end of colonial controls on human movement. The Belgians had restricted where Africans could work and live in a system that resembled what the South African government had built under apartheid. The arrival of independence in Congo included the right to travel across a human landscape far more interconnected than ever before. More infections—of HIV and other diseases—were one unintended consequence. That made the end of colonialism as much a turning point in the story of the AIDS epidemic as the beginning had been. In the 1960s the various subtypes of HIV-1 group M began moving along the roads and railways of which the Belgians had been so proud.

In the end, Lumumba was right. Independence was no "gift" from the Belgians. But this transport network was something new and tangible. So too was the freewheeling sexual ethos of Leopoldville and the other cities along Congo's highways, railroads, and river routes. With these developments some of the ingredients for creating a hyperepidemic of HIV—the waves of sickness and death so massive that ultimately nobody could miss them—were beginning to fall into place. This was the real parting gift of the Belgians: an AIDS epidemic ready to explode.

THE BIG BANG

The Belgian general who scrawled "After independence = Before independence" had a point. The Europeans could take down their flags. Westerners by the thousands could flee. But much of the underlying architecture of colonialism remained. Even as Congolese took political power, Belgians and other Westerners remained key players in nearly every major institution in Congo—churches, schools, mines, hospitals. When independence arrived in 1960 there were only about thirty college graduates in a nation of fifteen million people, and in a bureaucracy that had thousands of senior civil service jobs, Africans held only three.[1] That shortage of education and expertise continued to undermine efforts to instill a deeper sense of African ownership over Congo, which Mobutu rechristened Zaire in 1971 in hopes of distancing his country's future from its dreadful past. He was more effective at such grand gestures than at the basics of delivering government services. Mobutu systematically neglected roads and other infrastructure as he built palaces and siphoned off billions of dollars into personal bank accounts. In the hinterlands far from Leopoldville—now called Kinshasa— the fading remnants of old colonial institutions often were the only ones left functioning. And so it was in an interior town called Yambuku, more than seven hundred miles up the Congo River, where the first signs of a strange

new sickness appeared at a hospital founded by Catholic missionaries and
still run by a group of Belgian nuns.

Yambuku Mission Hospital had no doctor, no certified nurses, and only
the most rudimentary of facilities. Yet with 120 beds and a team of over-
worked nuns, it was the closest thing to a modern medical facility for a health
district of sixty thousand people. When a forty-four-year-old teacher at the
mission's school returned from a road trip with an unexplained fever in August 1976, he went to the hospital for help. One of the nuns, figuring the
teacher had malaria, gave him an injection of quinine and sent him home.
Two days later another man appeared at the hospital with diarrhea. Then
many other patients began arriving, often with troubling combinations of
symptoms: vomiting, dehydration, headaches, sunken eyes, pallid skin. Soon
some of the nuns themselves started taking ill in equally gruesome ways.
The signature feature of this epidemic was blood. It flowed from noses, from
gums, from eyes. It turned up in diarrhea and in vomit. Women in labor bled
to death on the hospital's cramped beds, along with their doomed babies.

Rising panic soon infected those who were not already ill with the mysterious disease, according to journalist Laurie Garrett's 1994 bestseller, *The
Coming Plague*.[2] Villagers burned down the huts of the stricken. Markets and
roadways were closed. Eventually people fled for their lives, putting as much
distance as possible between themselves and Yambuku. Within the hospital
itself, meanwhile, patients kept bleeding as the nuns desperately radioed for
help and prayed to God for relief. When word reached the outside world a
few weeks later, international medical authorities mounted a high-tech, high-
dollar effort that sought to overwhelm the strange new disease with the sheer
might of Western brains and resources. This was an era when belief in the
power of science was running especially high. Men had golfed on the moon.
Global health officials were stamping out smallpox across the earth. Medical
school professors were telling their students that drugs and vaccines might
soon eliminate infectious diseases altogether.

In Congo, international medical teams swarmed the area around Yambuku, and the militaries of eight nations eventually joined the fight with
personnel and machinery. The U.S. government mobilized one of NASA's

old space capsules as a possible isolation chamber.[3] Many of the medical workers who flew into the heart of the epidemic, meanwhile, climbed into head-to-toe protective outfits that inevitably were dubbed "space suits." One European doctor wrote in his diary that the outbreak was "one of the greatest events in contemporary epidemiology. . . . No one of us would pass up such an opportunity for passionate study. Personally, I am delighted to be in this place, and to participate in such an adventure."[4]

By the end of the year this "adventure" had killed nearly three hundred people, including eleven of the seventeen workers at the mission hospital. A final report by the World Health Organization said the response involved hundreds of people and resembled "a small war."[5]

The cause of this outbreak was not HIV. It was Ebola, named for a Congo River tributary that flowed near Yambuku. But in the span of less than a decade, both of these previously unknown viruses would test the remarkable reach of Western medicine, especially when practiced in the unfamiliar environs of Africa. Although both HIV and Ebola are prolific killers, they are nearly opposites in how they attack human communities. HIV is fragile and slow, capable in certain circumstances of building an epidemic gradually, over decades. Ebola is robust and fast, capable of causing hundreds of deaths in a few weeks, but also prone to extinguishing itself in the process. Yet each pathogen has, in its own way, outwitted many of the world's smartest, bravest, most determined doctors. The international medical community that mobilized during the Yambuku outbreak was not, as some had predicted, close to driving infectious disease to extinction. Instead, a new era was dawning that would demand approaches to illness that couldn't necessarily be conjured up in laboratories.

Investigators never found the source of that seminal Ebola outbreak. But they discovered that some of the mission hospital's practices had accelerated it. Sterilization procedures at Yambuku Mission Hospital often involved nothing more than rinsing hypodermic needles in pans of warm water.[6] By regularly reusing these needles, the nuns passed viruses from victim to victim, spreading the infection that first appeared in the teacher much farther than it would have managed otherwise. In Yambuku, where a hospital the Belgians founded in 1935 had barely been updated in four decades, Western medical science had gone awry. Those likeliest to die from Ebola were the hospital's own staff and patients seeking care. And what finally stopped the disease was not the international teams of doctors in their space suits or the scientists with their electron microscopes in an array of First World labs. The most important action was the decision—by a Congolese doctor—to close the mission hospital.[7] The health minister quarantined the entire area three days later. By the time the global SWAT team arrived two weeks after that, the outbreak already was ebbing.

The quarantine certainly helped limit the spread of the disease. But so too did the vastly simpler, more direct methods deployed by the Congolese villagers, who had the good sense to get away from people who were dying from something nobody understood. This instinctive response, born of sheer terror, worked. Ebola soon couldn't find new victims fast enough to sustain the outbreak. "The scientists humbly agreed," Garrett wrote, "that their scientific expertise had not been necessary to arrest the epidemic."[8]

There was another terrible twist, one that was possible to discover only in retrospect. As medical authorities focused on the Ebola outbreak, HIV already was lurking in Yambuku. In the mid-1980s, after the AIDS epidemic overtook Ebola as a matter of urgent international concern, scientists went back and tested 659 blood samples collected during the 1976 outbreak. Five were positive for HIV, for an infection rate of nearly 1 percent. The virus born in the jungles of Cameroon and raised to maturity in Kinshasa had broken free. And it wasn't moving in just one direction. Different subtypes had been radiating in a starburst pattern throughout the 1960s and 1970s. The Big Bang of the AIDS epidemic was well underway.

This metaphor suggests an explosion that was sudden and violent, with sparks of new infection blasting with equal force in every direction. But what genetic researchers reconstructing this history would later find was more subtle, and puzzling. They isolated eleven different subtypes of HIV—labeled with letters A through K—that had dispersed across the planet. All shared the essential DNA of the HIV-1 group M strain, meaning they all were de-

scendants of the original infection that was born of the chimpanzee popula-
tion of southeastern Cameroon and had traveled downriver to colonial
Leopoldville. But the subtypes followed sharply different paths after that.
While subtype A moved slowly, haltingly up the Atlantic Coast and into West
Africa, subtype B found its way to the opposite end of the earth, in San
Francisco.[9] Subtype C rode the postcolonial rail and trucking routes south,
through the Katanga Plateau and into Zambia's Copper Belt region before
finding its way into Zimbabwe, South Africa, and, eventually, India and
Southeast Asia. Subtype D migrated due east, to Rwanda, Uganda, and
Kenya. Several others, meanwhile, never made it far from Kinshasa at all.
This scattershot pattern is a reminder that nothing about the spread of the
AIDS epidemic was foreordained. HIV followed the networks humans built:
the porter paths, the steamship lines, the concrete highways, the air links.
And while the virus eventually found its way to nearly every corner of a
shrinking world, it sparked widespread outbreaks in only some of them.

Florida's Jackson Memorial Hospital, a massive public facility in Miami, was
thousands of miles and an ocean away from Kinshasa. But the Big Bang had
delivered HIV there by 1979, when Haitian immigrants started appearing
with strange cases of tuberculosis that featured swollen lymph nodes. The
illnesses puzzled the doctors, including an inquisitive pulmonologist named
Arthur Pitchenik. A battery of medical interventions—in one case involving
voodoo rituals by a Haitian healer—appeared to bring the symptoms under
control, but in most cases the patients returned later with new troubles. Then,
mysteriously, they died. None of the doctors could figure out why, though
one autopsy provided a clue: The victim had a toxoplasmosis infection in his
central nervous system. This was a common affliction but one that typically
produced serious illness only in people with weakened immune systems. "I've
seen a lot of tuberculosis," Pitchenik thought to himself. "I've never seen it
followed with diseases like this."[10]

   Pitchenik would have been even more surprised if he knew the chain of
infection that spawned these uncommonly aggressive cases of tuberculosis.
For of all of the journeys HIV took across the planet, perhaps none was more

improbable than that of HIV-1 group M, subtype B. This is the one that Worobey's later research—done with the help of Pitchenik, who had saved specimens from some of his Haitian patients—would show moved from Kinshasa to Haiti, then onward to Miami, New York, San Francisco, London, Paris. Nearly every person to get AIDS in the Americas and Europe contracted a descendant of the virus that made this original trans-Atlantic hop sometime during the 1960s.[11]

This crucial moment of transmission appears to have resulted from the turbulence that followed the formal end of colonial rule in Congo. The crisis there caused such alarm that the United Nations, whose secretary general at the time was a Swede named Dag Hammarskjöld, tried to stabilize the situation with an infusion of resources. To address shortages of educated, experienced workers, Hammarskjöld created a program to bring skilled professionals from other French-speaking nations to Congo. Many came from Haiti, which happened to be enduring one of its periodic moments of political unrest. In 1961, Haiti's president, François "Papa Doc" Duvalier, had engineered a sham election in which he claimed 100 percent of the vote. This descent into dictatorship encouraged many Haitians to seek opportunities elsewhere as the relief effort to Congo got rolling. Duvalier was reportedly happy to see them go, figuring that exporting Haiti's most educated citizens meant fewer potential leaders to challenge his rule. And some portion of the money earned overseas would flow back into Haiti, in the form of remittances to family and friends. By some estimates 80 percent of that nation's professionals were living abroad by the middle of the 1960s.

It wouldn't last long. But for a time there was a thriving Haitian expatriate community in Kinshasa, where they filled jobs as engineers, teachers, lawyers, doctors, nurses, and technocrats. Estimates of the size of the community ranged as high as ten thousand before the program faded away in the 1970s. Even though many were fleeing trouble at home, some of the Haitians also felt a pull toward their African roots. A former CIA official in Congo, Larry Devlin, recounts a story in his memoir about a Haitian teacher who came to the interior Kasai region in the 1960s. When a group of local government ministers expressed their gratitude to him for working so far from home, the Haitian man replied proudly that he felt like he was returning to his roots, because his ancestors had come from a tribe based near the area.

Overall, the Haitians in Congo thrived, earning salaries decent enough for them to fly home for occasional vacations. The evidence suggests that on one of these visits an expat based in Kinshasa carried HIV to Haiti, where it infected new victims many thousands of miles from the epidemic's birthplace. It would be wrong to suggest that HIV had no other way to arrive in the New World. The virus had already spread widely enough that it would have found its way to the Americas through some kind of migration, at some point. The world had gotten too small, and the AIDS epidemic too big, to contain it indefinitely in Africa. But the path it actually followed, as Worobey's research makes clear, relied on an extraordinary moment in history, an instance when good intentions went awry.

From that first infection, HIV spread widely in Haiti. And while Haitian immigrants clearly brought some infections to the United States, the essential connection may have been different. Blood products once were routinely imported from Haiti to the United States; some of them could easily have carried HIV.[12] And during the 1970s, Haiti was a popular tourist destination for gay American men, some of whom hired the inexpensive, easily available male prostitutes catering to the tourists.[13] However HIV made this journey to the United States, it gradually spread to millions of others in a new country, a new continent, a New World.

BOOK II

AN EPIDEMIC OF POLITICS

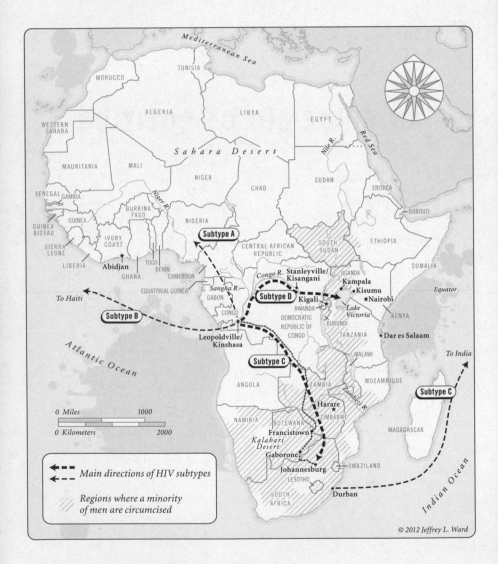

© 2012 Jeffrey L. Ward

[9]

AMERICANIZING AIDS

O n June 5, 1981, the *Morbidity and Mortality Weekly Report* newsletter from the U.S. Centers for Disease Control reported on a cluster of a rare type of pneumonia, *Pneumocystis carinii*. The victims were gay men living in Los Angeles. In a note accompanying the report the editors wrote:

> *Pneumocystis* pneumonia in the United States is almost exclusively limited to severely immunosuppressed patients. The occurrence of pneumocystosis in these 5 previously healthy individuals without a clinically apparent underlying immunodeficiency is unusual. The fact that these patients were all homosexuals suggests an association between some aspect of a homosexual lifestyle or disease acquired through sexual contact and *Pneumocystis* pneumonia in this population.

With that notice, decades of silent spread were over for HIV. And though it would take a few more years to identify the virus as the cause of AIDS, it already had spread throughout the Congo River Basin. It had climbed over the mountains into East Africa. It had ridden the railroad lines into southern

Africa. It had flown to Port-au-Prince, New York, San Francisco, Amsterdam, Montreal, and Sydney, and soon would be found in Moscow, Rio de Janeiro, Mumbai, and Bangkok as well. It had infiltrated blood supplies and the medicines used to keep hemophiliacs from bleeding to death. It had slipped into the syringes of drug addicts. And it had found the all-night dance parties and gay bathhouses in cities across North America and Europe. The rash of discoveries in the early 1980s gave the illusion of an explosive new epidemic, but it was in fact medical science discovering an outbreak already decades old and now in an aggressive new phase.

This was experienced in a way that made AIDS seem to be fundamentally an affliction of gay American men, and this persistent idea would shape impressions of the epidemic, in the United States and elsewhere, for decades to come. Soon after that historic report from Los Angeles similar cases turned up in other North American cities. The puzzling symptoms included swollen lymph nodes, severe diarrhea, thrush, and an unusual type of skin cancer, Kaposi's sarcoma. The medical response—though initially less aggressive than if the disease were hitting less marginalized victims—gradually took on some of the same frenzied quality as the Ebola outbreak in Yambuku.[1] Doctors shuttled around to meetings to swap news and plot their attack on the strange new disease. The CDC and the National Institutes of Health, along with their counterparts in other Western countries, generated a steady flow of discoveries. Even as the scale of the disaster became clearer, official pronouncements remained consistently upbeat about the prospects for a cure, or a vaccine, or some other medical solution capable of saving the world from the epidemic.

But for those watching from the outside trying to make sense of shifting reports AIDS was simply terrifying. Scientists publicly mulled whether the disease might be caused by an accumulation of sexually transmitted infections, or maybe even from the radiation emanating from the tanning beds frequented by gay men. Others suggested it was spread by mosquitoes or bedbugs. Each twist generated its own troubling headlines. Newspapers warned that nobody was safe. And once the disease took hold there was no way to stop it. Death came with a slow, awful inevitability. Early reports about risk groups often spoke of the four Hs: homosexuals, hemophiliacs, heroin addicts, and Haitians. But gay men, so long confined to the fringes of society, were the most visible victims of a disease that scientists initially called "gay-related immune deficiency," or GRID. Soon another name took hold in the popular imagination: The Gay Plague. Conservative newspaper columnist Patrick Buchanan wrote in 1983, "The poor homosexuals—they have declared war upon nature, and now nature is exacting an awful retribution."[2] Moral Majority leader Jerry Falwell followed a few months later with his infamous declaration that AIDS and other sexually transmitted diseases were "a definite form of the judgment of God upon a society."

> If the Reagan administration does not put its full weight against this, what is now a gay plague in this country, I feel that a year from now, President Ronald Reagan personally will be blamed for allowing this awful disease to break out among the innocent American public.[3]

And so AIDS, in the span of a few years, traveled from obscurity to the world's biggest, brashest, loudest stage in the middle of an intensifying, post-1960s hangover of a culture war. This was a nation capable of grooving to the Village People's raunchy "Macho Man" and "YMCA" even as the gay sexuality celebrated by the lyrics remained criminal in most states. In this electrified setting, AIDS became more than just a disease. It became a political cause, a call to arms for opposing sides locked in passionate debate about the direction of their society. Something subtler happened too. An epidemic born thousands of miles away in Central Africa became seen as essentially Western and, in particular, American. Few understood that what they saw in the United States was just one tentacle of a monster already entrenched in parts of the world where nobody knew about Fire Island beach parties, gay pride marches, or Pat Buchanan.

In the same year Falwell pronounced that the disease represented an angry god's wrath, gay activist and novelist Larry Kramer penned a famous polemic against what he considered complacency in the fight against AIDS, both by government authorities as well as by gay men reluctant to change behaviors that were helping spread the epidemic. "If this article doesn't scare the shit out of you, we're in real trouble," Kramer wrote in the *New York Native*, a paper popular among the city's gay community. "If this article doesn't rouse you to anger, fury, rage, and action, gay men may have no future on this earth. Our continued existence depends on just how angry you can get." The article was called "1,112 and counting," for the number of Americans known to be dead so far from AIDS. There was no way for Kramer or anyone else to know it then, but when his article appeared in 1983 the death toll from AIDS in Africa had already reached into the hundreds of thousands.[4] Soon it would be millions.

The AIDS epidemics in the developed world would never approach those in Africa for sheer numbers of deaths, but for concentrated devastation, few places on earth would suffer like San Francisco's Castro Street, Miami's South Beach, or New York's Greenwich Village. A team of San Francisco researchers who collected blood samples from gay men for a study on the spread of hepatitis B later went back and tested the samples for HIV. The results showed that in 1978, the first year of the study, 6 percent of the men had the virus.[5] Just two years later the infection rate was more than five times higher, at 33 percent. By 1982 more than half of the original cohort had HIV.

Even as panic spread throughout many Western societies, AIDS in North America, Europe, and Australia remained mostly a disease of men who had sex with men and of intravenous drug users, and occasionally of their female sex partners.[6] That was especially true after the 1980s, when improved safety procedures removed HIV from most blood supplies and the clotting medicine used by hemophiliacs. At the first International AIDS Conference, held in Atlanta in 1985, federal officials reported that 90 percent of all HIV infections could be traced to male homosexual contact or drug use.[7] Within these communities, HIV had overcome its naturally slow, fragile nature. Blood carried the virus more easily than any other fluid, making transfusions and the sharing of unclean needles the easiest ways to spread HIV.[8] Anal intercourse was also relatively efficient, because torn rectal tissue creates pathways for the virus. An infected man is ten to twenty times more likely to transmit HIV to a receptive partner, of either gender, in anal sex than during vaginal sex.[9] This risk was multiplied over time when men frequented bars and bathhouses where it was easy to have scores or even hundreds of sexual partners a year.

For gay men in the 1980s, this time felt something like Armageddon. Many gradually lost count of the number of friends and lovers lost to AIDS, typically after months of harrowing sickness. And in the years before a reliable blood test arrived, no one was sure who would fall ill next. No one knew when something as seemingly harmless as a new cough or a bout of diarrhea might signal the onset of an irreversible, fatal disease. The rising number of deaths—coupled with the realization that many Americans, including President Reagan, appeared to care little about dying gay men—spawned one of the most powerful political movements of late twentieth century America. Gay liberation had begun explosively with New York City's Stonewall Riots of 1969, but the arrival of AIDS supercharged it. Activists rallying to the slogan SILENCE = DEATH were determined to bring attention to the cause and to push back against conservatives who blamed gay men for their own terminal illnesses—even if it meant occasionally pushing the boundaries of taste.

AIDS activists produced many important successes during this era. They fought for basic rights for those suffering from the disease, raised money for their care, and successfully pushed the medical establishment to accelerate the development of drug therapies. By all accounts this drive meant that antiretroviral drugs arrived earlier—perhaps by several years—than would have happened otherwise, and this breakthrough extended tens of thousands of lives. Gay men also were key players in demanding and disseminating information about AIDS that helped keep themselves and others safer from infection. And they kept the epidemic on the political agenda. The medical response in Africa, meanwhile, almost certainly would have been slower and less energetic without the efforts of Western activists who saw the fights there as an extension of their own struggle against the epidemic.

But there were excesses, too. The most militant activists, convinced that the government and medical community were not doing enough, tossed fake blood on scientists and printed postcards featuring the faces of CDC officials in bull's-eyes. Publishers who dared to cross the activists also faced their fury. After one protest *Cosmopolitan* magazine partially retracted an article suggesting that most heterosexual women were at low risk for HIV—a fact that was demonstrably true for those living in the United States and most of the developed world.[10] And when conservative writer Michael Fumento made a similar argument in his 1990 book, *The Myth of Heterosexual AIDS*, it produced such a furious backlash that some stores stopped carrying it. After *Forbes* magazine ran a favorable profile of Fumento, demonstrators picketed the magazine's offices in Manhattan. Publisher Malcolm Forbes Sr. finally repudiated the article publicly, calling it "asinine."

There was a quieter and no less powerful side to the response too. Many gay men changed their sexual practices to protect themselves and their lovers. Condoms became more popular, anal sex with casual partners less so.[11] And also important, gay men as a group started having fewer partners. Epidemiologist Rand Stoneburner, who worked in New York City from 1983 to 1991, watched this transformation in astonishment. Those most in peril from the disease clamored forcefully for information through new groups such as the Gay Men's Health Crisis, which met regularly with city officials and disseminated cutting-edge information about AIDS as fast as scientists could discover it. The overarching question throughout those hectic years was: How can we survive? "It was terror," Stoneburner recalled years later, his wavy dark hair now graying at the temples, his deep voice softened by the faint remnants of an accent from his native South Carolina. "It was absolute terror."

The toll from AIDS among New York's gay men continued to rise throughout the 1980s and into the 1990s, as those who contracted HIV years earlier gradually succumbed to the disease. But Stoneburner began suspecting that a corner of sorts had been turned as early as 1985, when rates of rectal gonorrhea, another disease spread by anal sex, began to fall. Those who witnessed AIDS deaths personally, of a lover or friend, appeared to be the ones most likely to make lasting changes in their own behavior. Condom use rose dramatically, and many men also resorted to monogamy while others confined their relationships to a small circle of friends they knew well. Some cities closed down bathhouses, and those that stayed open saw business plummet. The experience pointed to an important truth: In the face of a deadly, incurable disease spread by sex, many people make different decisions even at the price of diminished pleasure or less sexual freedom. An accumulation of such individual decisions, if they happen on a large enough scale, is capable of reversing the epidemic. "I didn't believe it was possible, but I saw it with my own eyes," Stoneburner said. However much people enjoy sex, he realized, "it doesn't trump survival."

A combination of factors slowed the spread of HIV in the United States and Europe, but just one emerged as the public symbol of the war against AIDS: the condom. For centuries men had covered their penises during sex to prevent pregnancy or disease. The range of materials used—animal intestines, fish bladders, even linen—varied widely in comfort and reliability. But the essential equation for users was the same, as they traded sexual sensation for protection. And with it many gained the ability to have sex beyond the boundaries of what society officially sanctioned: with prostitutes, with girlfriends before marriage, with mistresses after marriage. An effective condom hid the evidence; there would be no embarrassing itch, no suspiciously swelling belly on a neighbor's wife. Even as other forms of birth control became widely available, condoms maintained their illicit aura as they stayed hidden behind pharmacy counters, in truck stop vending machines, in porn shops.

The AIDS epidemic changed that. Condoms moved forcefully into the American mainstream during the 1980s because of their ability to block the passage of HIV between sex partners. They naturally had an important role in battling the spread of the epidemic, both for straight men and gay ones, who traditionally had little use for a product generally seen as contraception. Many lives have been saved by condoms, and many other sexually transmitted infections, such as gonorrhea and syphilis, have been prevented with greater use of condoms.[12] They also had a metaphorical role as condoms came to symbolize the idea that sexuality carried inherent responsibilities—something that had fallen out of fashion. In the age of AIDS, condoms—and by extension, worrying about both your sexual health and that of your partner—became cool. Activists handed condoms out on street corners. Nurses pushed them on college kids. Researchers began trying to design a female condom to give women more options for protecting themselves.

Another group became focused on condoms during these same years. The global health officials at the U.S. Agency for International Development, commonly called by its acronym USAID, were eager for a role to play in fighting AIDS overseas, so they settled on a familiar product. Condoms had long been the poor cousin of family planning programs, less effective over time than longer-term methods such as the Pill or IUDs. And although veteran family planning experts knew that few people in regular relationships used condoms consistently, they did have the ability to block HIV and many other sexually transmitted infections while also preventing unwanted pregnancies.[13] That meant that if only condom usage rates could be raised, especially in the parts of Africa with major HIV epidemics, then potentially many lives could be saved. This bit of logic in later years would become the focus of significant debate, but at the time it was nearly universally accepted among those working on international AIDS prevention programs. They didn't see many other viable options. And from the point of view of building a government program, condoms had the additional advantages of being small, inexpensive, and easy to quantify when writing up bureaucratic reports. Billboards, bumper stickers, and store displays featuring attractive, amorous young couples clutching each other soon appeared throughout much of the continent with the brand name of the condom—Protector, Trust, Shield—hovering nearby. The promotions were explicitly nonjudgmental and often didn't mention AIDS at all. Yet they were, for many years, the cornerstone of the U.S. government's strategy against HIV in Africa. Now it wasn't just the epidemic that had become Americanized; the world's response to it had been too.[14]

IT CAN'T BE HERE ALREADY!

European doctors began spotting AIDS not long after the Americans announced its discovery in 1981. The first victims fit the familiar profile: gay men in Copenhagen, a Haitian couple in Paris, hemophiliacs in Seville. But a group of Belgian doctors soon reported their suspicions about a new risk group. Africans, all wealthy enough to travel to Europe, started turning up at hospitals in Belgium with wasting, fever, rashes, coughs, diarrhea. When doctors tracked down forty cases of what appeared to be AIDS, only two of the victims were gay men; thirty-seven were immigrants from Africa. "All features suggest that AIDS is endemic now in Central Africa, and that the cases seen in Belgium represent only the tip of the iceberg," the doctors reported in *The Annals of the New York Academy of Sciences,* in December 1984.[1]

These and other reports stirred the interest of a young Belgian physician, Peter Piot, who once had been among the daring heroes of the Ebola outbreak and is credited with codiscovering the virus that caused it. He had been just twenty-seven years old, conducting postdoctoral research in virology at a Belgian university, when he was abruptly dispatched to Yambuku during the height of the disease's spread. It was his first trip outside of Europe, and despite some terrifying moments—Piot later described himself as "young, foolish, and fearless" during the Ebola crisis—the experience helped infuse in him a passion for practicing public health in places where the stakes were high and the resources painfully meager.[2] Over the next several years Piot became considerably savvier about Africa, its diseases, and the international health bureaucracy that dealt with them. When reports began suggesting that AIDS also had roots on the continent, Piot wanted to investigate further, and began seeking funding in both Europe and the United States for a research trip. The most intriguing possibility was that U.S. doctors might have dramatically underestimated the number of people put in danger by the new epidemic. There were nearly as many women as men among the African patients seen at Belgian hospitals, and there was no evidence that drug use or anal sex were significant factors in transmitting the disease. If AIDS could spread among heterosexuals who didn't use drugs, weren't exposed to tainted blood, and had never visited Haiti, the slice of humanity at risk might soon measure not in the millions but in the hundreds of millions.

The possibility of widespread heterosexual transmission was not widely accepted in the early years of the epidemic. When Piot and another researcher submitted a paper on the subject to the prestigious *New England Journal of Medicine* it was rejected. The editors, quoting one expert, wrote "It is a well-known fact that AIDS cannot be transmitted from women to men."[3]

Against that backdrop, and with the support of the National Institutes of Health in the United States, Piot in 1983 organized a research trip to Kinshasa. The city whose residents once proudly called it "*Kin la belle*" for "Kinshasa the beautiful," now was crumbling, as a surging mass of humanity overwhelmed the colonial-era roads, rail lines, and utilities. (The new nickname was "*Kin la Poubelle*," for "Kinshasa the dustbin.") As the population rocketed from half a million at independence to three million in the early 1980s, the slums had swelled, becoming even more dense, squalid, and diseased. The city's health systems, once among Africa's finest, had deteriorated as government investment stagnated. Piot had seen the hospitals during the Ebola outbreak and was aghast to see that now, just seven years later, they were overflowing with patients, many of whom appeared to have AIDS.[4] One doctor at the hospital, Bila Kapita, had collected files on fifty patients he suspected had died of the disease. Piot jotted in a notebook that he kept on such trips, "This is going to change my life."

Even in this era before a definitive test for the AIDS virus, there was plenty of evidence of the disease. Piot's team found patients with low counts of T-cells and white blood cells, along with devastating diarrhea, skin rashes, oral thrush, meningitis, and wasting.[5] The visions of stricken men and women, their skeletal bodies writhing in pain, troubled even the most experienced doctors. Piot already had seen some cases of AIDS among gay men at clinics back in Belgium, but the scale of the suffering in Kinshasa reinforced his impression that something new and terrible had broken loose widely across the world. There was no evidence that homosexuality or unsafe injections were major factors. Yet by his rough initial calculations, seventeen out of every one hundred thousand people in Kinshasa were getting sick with AIDS each year. "This is a minimal estimate, and it is comparable with or higher than the rate in San Francisco or New York," Piot soon would write of his findings.[6]

Piot's team identified two apparent clusters of infection. In one, a forty-six-year-old man had died of the illness, along with his wife and two of his other sex partners. A household maid, who may have been another sex partner, had died as well. In the second cluster, a thirty-year-old woman and her husband had died, as had two of her other sex partners. One of these other men, meanwhile, seemingly had infected at least one other woman. With these findings Piot was certain he had proven that AIDS was spreading through heterosexual contact. The team returned from the trip shaken but ready to ring a new alarm bell—one even louder than the one first sounded in 1981. Piot captured the historic moment in the opening paragraph of the article he was preparing: "The findings of this study strongly argue that the situation in central Africa represents a new epidemiological setting for this worldwide disease—that of significant transmission in a large heterosexual population."[7] Piot also noted the key scientific facts about the spread of AIDS: Those dying had several sex partners each, and often had a history of other sexually transmitted diseases. The men reported an average of seven sex partners per year. The women reported three per year.

But when Piot first submitted drafts of the groundbreaking article to major medical journals, they balked. *The Lancet*, which would publish the piece the following year, initially dismissed it as unworthy, he recalled years later, because the importance was merely "local."[8]

The first bursts of news about a strange new immune disorder caught the attention of David Serwadda, a driven young doctor in Kampala, the capital of Uganda, in the early 1980s. The city was on the other side of the continent from Kinshasa, more than twelve hundred miles east by air. Kampala also had a much different history. While the Belgians had dominated nearly every aspect of colonial life in Congo, the British had maintained a lighter hand in Uganda, allowing a modicum of self-rule. In 1922, Kampala opened its first school of higher education, Makerere University, which went on to produce generations of East African doctors, writers, and political leaders at a time when Congo was producing few. Strife and dictatorship still marred Uganda's first quarter century of independence after the British withdrew in 1962, but by the time HIV arrived the nation had developed a resilient core of highly educated Ugandans, ready to lead. Serwadda was among them.

Though Makerere, on the graceful, hilly outskirts of Kampala, was one of the best universities in Africa, it still was on the fringes of the world's information flow. International phone calls were expensive. CNN hadn't arrived yet. And there was no e-mail. So Serwadda—first as a medical student, then as a junior medical officer at the Uganda Cancer Institute—kept up on international scientific developments by working his way through the latest periodicals at the university library. The reports that he noticed were mostly from the United States, and they focused on strange illnesses among openly gay men—something in short supply in Africa, where homosexuality remained punishable by death in parts of the continent. Yet something about the reports seemed familiar. The men with immune disorders often displayed a rare type of skin cancer, called Kaposi's sarcoma, that caused raised, purplish lesions. The cancer traditionally afflicted middle-age, Mediterranean men, and now, according to the emerging literature, people with AIDS. Yet Serwadda knew that Kaposi's was also common in parts of Africa, including

Uganda. The large public hospital affiliated with the university saw several Kaposi's patients a year.

At first a connection seemed unlikely. The variety of Kaposi's sarcoma typically found in East Africa tended to spread slowly, on people's arms and legs, and rarely killed its victims. The variety afflicting the gay American men, by contrast, moved swiftly and dangerously into their mouths and throats. Serwadda felt the first faint flicker of a historic insight, but nothing more. The differences seemed more profound than the similarities, so he returned to his other research. But his suspicions would not go away. In 1983, the same year that Piot's team visited Kinshasa, Serwadda noticed that four peasant farmers had been admitted to the hospital in Kampala with severe cases of Kaposi's sarcoma. The symptoms were spreading fast, and lesions were beginning to appear on the chests and in the mouths of the patients. There was no evidence that these men used drugs or had sex with other men, yet they had a variety of Kaposi's that seemed to resemble what was attacking the American AIDS patients. One other clue suggested something new was happening: All four patients were from southwestern Uganda, a rural region of farmers and fishermen who plied Lake Victoria for Nile perch and tilapia. It also was a byway for the Trans-African Highway, a vital artery for goods and people moving between Africa's Indian Ocean ports and densely populated inland areas hundreds of miles away.

Serwadda's "eureka!" moment came the following year. In April 1984 he read an article in *Time* magazine reporting the discovery of a strange new virus that scientists believed caused AIDS. Initially dubbed HTLV-III, it reportedly was capable of ravaging a victim's immune system to the point that a lethal combination of other maladies took hold. Among them, the *Time* article reported, was Kaposi's sarcoma. "I remember reading it and thinking 'Bam! I have got the cause,'" Serwadda recalled years later.[9]

The first tests for the new virus, which has since been rechristened HIV, became available in American and European laboratories a few months later. Serwadda drew blood samples from the four Ugandan peasants stricken with Kaposi's and sent them off to a lab in Britain. The reports that came back, via a lightweight-paper aerogram, confirmed Serwadda's guess: All four of his patients had HIV. A surge of elation—the thrill of discovery—blasted through Serwadda's body. The buzz was good for a few hours. Then, as night fell in Kampala, the young doctor gradually realized that the news was, in fact, horrific. A fatal, incurable disease had found its way into Uganda. "It is here already! It can't be!" Serwadda thought, tossing in his bed that night. "It can't be here already!"

The editors of the *New England Journal of Medicine* were not the only ones who had trouble believing that AIDS could spread widely among heterosexuals. Early efforts by global health officials to warn African leaders about the rising threat from the epidemic met with resistance and disdain. Many simply rejected the idea that a disease best known for infecting gay American men was spreading on their continent. Yet as the world struggled to accept that AIDS was something more than a gay plague afflicting Americans and other Westerners, the epidemic was shifting into a new, even more terrible phase in Africa. Piot's article was finally published in *The Lancet* in July 1984, and in that same issue a Belgian-Dutch team reported on similar findings from Kigali, the capital of Rwanda, a small East African nation that bordered both Zaire and Uganda.[10] Thirteen of the seventeen sick men in that study acknowledged having several sex partners, and in most cases to sleeping with prostitutes as well. Three of the seven sick women acknowledged that they themselves had sex for money; two others in that group had husbands who routinely visited sex workers. And here was the worst news: The rate of infection in Kigali appeared to be nearly five times higher than it was in Kinshasa.[11] Of the two, the bustling, cosmopolitan center of Kinshasa seemed a more natural candidate for a catastrophic outbreak of a sexually transmitted disease. Kigali was comparatively conservative and remote, yet AIDS was spreading there with a speed never before recorded outside of gay enclaves or clusters of drug injectors in Europe and America.

The twin reports in *The Lancet* marked the Western world's official discovery of the AIDS epidemic in Africa. Perhaps the most startling elements in these early reports were their frank descriptions of an unfamiliar African sexual culture in which monogamy was not always the cultural norm. Europeans and Americans had been noting this with varying degrees of alarm since the first days of colonialism. Yet polygamy and its informal offspring had not lost their power to surprise Westerners. The Dutch and Belgian researchers in Rwanda reached for a familiar but provocative word to describe what they found. A factor driving the AIDS epidemic in Africa, they reported, likely was "promiscuity."

The word conveys more than just maintaining several ongoing sexual relationships, as was common in many parts of Africa. "Promiscuity" suggested indiscriminate sexuality, unfettered, animalistic. The word also had—and has never lost—the unmistakable ring of moral judgment, as if the same Western nations that had colonized Africa, enslaved its people, and laid waste its indigenous cultures were now blaming it for a disastrous new epidemic. Historian Adam Hochschild named his award-winning book on Belgium's horrific colonization of Congo *King Leopold's Ghost*. In the postcolonial era, relations between the West and Africa were also haunted by condescension, brittle pride, anger, suspicion, and mutual misunderstanding. This toxic mixture poisoned dealings between Westerners and Africans as AIDS spread farther and faster—a time when both could have been facing the common enemy of disease.

ATTENTION NA SIDA

At a conference on AIDS held in Brussels in 1985, fifty African participants issued a joint statement saying that available research "did not show any conclusive evidence that AIDS originated in Africa."[1] This was part of a larger backlash against mounting reports that the epicenter of the global AIDS disaster might not be San Francisco or New York but a stretch of Africa's midsection that some researchers had begun calling the "AIDS Belt." African political leaders, meanwhile, were even less inclined to accept the possibility that AIDS had some inherent connection to their continent. In no place was this truer than Kinshasa, at the heart of the region where the epidemic had in fact flamed to life. The dictator Mobutu banned public discussion of AIDS in Zaire and threatened scientists and public health officials who dared to acknowledge its existence.

Amid this vacuum of information Zairians struggled to make sense of the rash of devastating illnesses. Anybody who lost substantial amounts of weight faced accusations of witchcraft, or of violating society's conflicted sexual code, in which proclamations of Church-sanctioned monogamy existed uneasily with behavior that looked more like a latter-day variation of polygamy. Rumors spread of famous people who reportedly had AIDS—movie stars, singers, even Mobutu's own son—and how they got it. A Pakistani prostitute in the southeastern mining city of Lubumbashi supposedly brought the disease to Zaire. Or the CIA did. Or it began with the spectacular "Rumble in the Jungle" in 1974, when an international congregation of stars gathered in Kinshasa to see the heavyweight bout between Muhammad Ali and George Foreman. The key element in most stories was that outsiders were responsible for AIDS.[2] Among the first victims in Kinshasa, it was said, were so-called Londoners who fraternized with the largely white expatriate community of Europeans and Americans. So intense was the denial of AIDS that doctors dared not even tell their patients what was killing them. American social scientist Jane Bertrand visited Kinshasa's massive Mama Yemo Hospital while working for a family planning project in the city in the late 1980s. The goal was to expose its counseling staff to the ravages of AIDS, which were easy to see while touring wards of wasted, dying men and women. But the doctors strictly forbade the visitors from using the word AIDS. "It was very eerie walking around Mama Yemo," Bertrand recalled, with "our knowing they had AIDS, and their not knowing it."[3]

That strange encounter also captured something of the relationship between the West and African AIDS victims in that era. The ill were there to be studied, their bodily fluids tapped and tested in a hundred different ways. Images of these frail wraiths generated shock and eventually donations from richer, more powerful nations. But there was little that could be done for these doomed men and women in their final weeks of life, and little was done to protect the next wave of victims from becoming infected. Not far from Mama Yemo was *Projet SIDA*, one of the world's preeminent AIDS research centers (SIDA is the French acronym for AIDS). Founded in 1984 by Piot in conjunction with the CDC, its staff grew to nearly three hundred and its budget to $4 million a year while generating a steady stream of discoveries through 120 published research papers and more than one thousand abstracts for scientific meetings. But what it never developed was a way to deliver useful insights about AIDS to Zairians themselves. Even revelations about the relationship between HIV and sexual behavior were not packaged for public consumption. Congolese doctor Bila Kapita, who had helped Piot first understand the extent of AIDS at Mama Yemo and then joined him at Projet SIDA, later told the journal *Science* that "I'm sorry to tell you that *Projet SIDA* had very little impact for infected people here. . . . It would have been useful for them to ask us about what was the useful thing for us they could do here."[4]

Missing from this scene was a figure such as Larry Kramer to call on those most in peril of getting AIDS to take ownership of the disease, or an organization such as Gay Men's Health Crisis to spread the best possible information on how infection could be avoided in the first place. Instead, AIDS in Zaire continued to be viewed as a disease of outsiders—or perhaps one that didn't exist at all. A popular nickname for SIDA in Kinshasa was *Syndrome Imaginaire pour Décourager les Amoureux*: "An Imaginary Syndrome to Discourage Lovers."

That consuming sense of denial finally began to ease thanks to a powerful new song by one of Zaire's most famous men, François Luambo Makiadi, known to his millions of fans simply as Franco. He had grown up in colonial Leopoldville's segregated slums playing the harmonica, kicking a soccer ball made of rags, and helping his mother, a widow, run a street-side food stall. To attract customers Franco sometimes plucked out party tunes on a guitar he had built out of an old tin can and some electrical wire. These performances caught the eye of a local musician, who taught Franco the basics of picking a real guitar. Franco was such a natural that he made his professional debut in 1950, at the age of twelve.[5]

Over the next four decades Franco would grow into the dominant figure in a boisterous Zairian music scene. Franco and his O.K. Jazz band built their tremendous popularity not just on potent guitar licks and irresistible dance grooves; their lyrics chronicled the ups and downs of Congolese life, often in epic songs that stretched on many times longer than the three-and-a-half minutes common for pop tunes in Europe or the United States. Some compositions by Franco's band, including ones commissioned by a Volkswagen dealership and political leaders such as Mobutu, were little more than praise songs, churned out for easy profit. But at his best, Franco was the bard of a new Zaire. He and his team of writers crafted sprawling compositions almost operatic in their ambition. For a nation with low literacy rates and few of its own writers, this music amounted to the nation's indigenous literary canon.[6] It was how Zairians talked to each other, shared triumph and pain. Music was so central to life that on many streets Franco's sweetly intricate guitar riffs spilled out of public loudspeakers for those without radios at home.[7] Franco produced more than a thousand songs and 150 albums over his career, while also touring throughout the world.

Franco's massive appetites, and a girth that testified to his prodigious success, were also legendary. His biographer, Graeme Ewens, author of *Congo Colossus*, wrote that Franco could eat twenty-five slabs of cassava dough, a whole chicken, sausages, and some vegetables in a single feast.[8] His love life was equally expansive. He had at least eighteen children by fourteen women, which was reminiscent of precolonial village chiefs for whom big families signaled wealth, power, and responsibility. At his peak Franco supported one hundred families through the revenues of his sprawling orchestra, with its dozens of rotating musicians. He routinely bestowed new cars and other gifts on those who penned a breakout song or otherwise pleased him. And he demanded absolute respect in return. In person he was not called Franco but Grand Maître, for "Grand Master." Not surprisingly, many of his band's most popular songs were about love, lust, treachery, and unrequited desire. Perhaps his most famous creation was the gigolo Mario, whose adventures filled an epic song that spilled over two vinyl records. And the infamous song "Jacky" was so outrageously lewd that police once tossed Franco and several band members in jail on pornography charges.

That history gave Franco's 1987 hit "Attention na SIDA" ("Beware of AIDS") uncommon authority.[9] No one doubted his love of sex or his fondness for having many women. But fans also knew that their hero, once three hundred pounds, had begun to lose weight. Franco never acknowledged having AIDS, but the lyrics fueled the suspicions. He slyly opened the song with a sexy guitar riff lifted straight from the risqué "Jacky."[10] Then he began singing with a booming, deep-voiced lament, "Oh, AIDS. This terrible sickness / Oh, AIDS, a disease which does not pardon / A disease which spares nobody." Other lyrics read more like a sermon than a song, complete with biblical references to Sodom and Gomorrah and the judgment of God. Franco called on leaders of society—doctors, professors, priests—to fight the disease more aggressively. And he spoke painfully about the scorn and stigma felt by those suspected of having AIDS. But at its deepest level, "Attention na SIDA" was a warning, aimed at those not yet sick:

> *Youths, beware, AIDS can attack you*
> *You are the life force of society*
> *If you let it kill you, who will lead the people?*
> *Avoid dangerous sex. Students beware unknown partners*
> *Be careful who you take money from*
> *It could get you in deeper trouble*
> *Avoid casual partners*

And later:

> *Avoid picking up just anybody*
> *Think before you make love*
> *Even if you desire someone, be careful. Think first*
> *You gentlemen, citizens*
> *Beware of prostitutes*
> *Avoid multiple partners*
> *And you, ladies, citizens*
> *Take measures for your own protection*
> *Workers, in workshops, factories and offices*
> *When you are talking together*
> *Do not neglect the subject*

The song became a huge hit as Franco's legendary bulk gradually withered away. In one of his final public images, on an album cover from early 1989, the big man appeared at about half of his old size, with his face looking gaunt and suddenly old. Franco continued denying that his mysterious illness was AIDS, but when he died a few months later, at age fifty-one, few believed him.[11]

Radios are king in rural Africa. They are cheap and portable and can pick up signals outside of cities, well beyond the reach of televisions or newspapers. During the years that "Attention na SIDA" filled Zaire's airwaves Franco's warnings carried an aura of divine instruction. Denial faded. Discussion began. And, evidently, behaviors started to change. All this happened largely beyond the view of the Western officials, who already were taking charge of the war on AIDS in Africa.

When an American medical student, Peter Kilmarx, visited Zaire in 1987, his goal was to spend time in the small town where he had once been a Peace Corps volunteer. But Kilmarx also had ambitions of entering the field of international public health, so he first made a visit to Projet SIDA in Kinshasa and met the American doctor who was the research center's head at the time. When Kilmarx announced that he wanted to study the impact of AIDS in the village, the official seemed underwhelmed. "You're wasting your time," he told Kilmarx.[12] "These guys won't even have heard of it."

Yet when Kilmarx got to the village, along a quiet road between Kinshasa and the Katanga mining region, many people had heard of AIDS. He interviewed seventeen men, asking them each about their medical histories, their knowledge of the disease, and what they were doing to protect themselves. One said that behaviors were already shifting among the village's women, who were avoiding liaisons with workers from the provincial capital, where AIDS was more common. Prostitutes were even stricter, refusing all clients not from the village itself. Kilmarx also found, as many other observers had before, that a kind of informal polygamy remained common. Five of the seventeen men reported having sex in the previous year with women other than their wives—three extramarital partners for each of these men on average. Yet, strikingly, half of those interviewed thought that such behavior would fall out of favor in the age of AIDS. Most also expressed interest in condoms, but only three of the men had ever used one.[13] The most profound observation, however, came from the chief. He reported once having five regular sex partners in addition to his two wives. But that changed one night after he heard "Attention na SIDA" on the radio of a local bar. The message hit so hard that the chief soon decided to leave his girlfriends in favor of being faithful to his wives.

Jane Bertrand, the American social scientist, also did research into Zairian sexual behavior in the aftermath of "Attention na SIDA." Her team interviewed two thousand people in ten cites across the nation. Though the song was not mentioned in their findings, the results strongly suggest that something had resonated—more so than the condom-promotion efforts of USAID and other Western agencies. Most people in the study reported knowing about condoms but less than 20 percent had ever used one, and less than 5 percent used them with their spouses.[14] The rate among single people was higher, but overall, two thirds of men who visited prostitutes or had other relationships outside of marriage said that they had never used condoms. Overall, neither men nor women liked them much or perceived them to be central to fighting AIDS. Many also complained that they decreased sexual pleasure and sometimes tore, and people expressed fear that condoms might get stuck inside of women's vaginas during sex.

These Zairians weren't ignoring AIDS. They just weren't responding in the ways Western experts expected them to. About half of those interviewed said they had changed their behavior to protect themselves, by avoiding prostitutes or reducing their numbers of sex partners. A different study, of nearly 3,500 health workers in Zaire, reported that those having extramarital sex dropped from 54 percent to 40 percent between 1987 and 1988, the year of peak airplay for "Attention na SIDA."[15] These studies were among the first signs of an effective African approach, born of the continent's own cultures and experiences, to controlling the spread of HIV. The model depended on blunt, outspoken demands by leaders—in a manner reminiscent of the days when power and wisdom resided among village chiefs—for people to change their behavior. That's what Franco did, and what a few other influential Africans elsewhere on the continent soon would do.

Franco's warnings did little to alleviate the shunning, stigma, and shame that accompanied AIDS in Zaire and almost everywhere in Africa. People suspected of having the disease frequently died alone in the darkest, dingiest back rooms their families could find. Others weren't allowed to go home at all; some were attacked. Most faced blame and denunciation for contracting the disease that was killing them. Yet for preventing the spread of HIV, few developments would prove more effective than the kind of broad shifts in sexual behavior that Franco's song helped inspire. Each infection that was prevented because a man gave up a girlfriend or two reverberated throughout an entire society. The many Westerners gathered in Kinshasa to study AIDS appeared to have missed these crucial changes. Even Bertrand, whose research most authoritatively documented them, did not fully understand the power of what she found. Instead, she expected, as did most experts, that the nation was fated to a future of soaring HIV rates and mass death. This was a view that only hardened when Zaire erupted into a civil war that was disastrous even by the grim standards of African conflicts. When the fighting reached the streets of Kinshasa in 1991, the expat leaders of the project fled with the help of Belgian paratroopers. The following year, American government officials quietly flew in a scientist to retrieve most of the computer data and serum samples from what was left of Projet SIDA.[16] Then they shut off the funding. It never was restored. The abrupt end of the research project heightened the sense of doom about Zaire, which the anti-Mobutu rebels eventually would rechristen the Democratic Republic of Congo. "I expected [AIDS in] Congo to take off," Bertrand recalled. "Poor Congo."[17]

But measured in new HIV infections, the epidemic had already peaked there and was beginning to ease.[18] For once, this haunted land had caught a break. The social change that happened amid Franco's powerful warning was almost certainly one reason. The other would soon be discovered.

YOU WON'T BELIEVE

The Pumwani slum in the Kenyan capital of Nairobi is a sprawling, scattershot mix of homes and shops made of reddish mud, rusting tin, and, for a few of the grandest structures, durable walls of painted concrete. Communal taps and toilets gradually brought the rudiments of modern living here, but the homes are tiny and dark, with barely enough room for a bed, a small table, a stove. Visitors find Pumwani chaotic, but there is a social and commercial order. Nearly identical pots of boiling stew bubble outside one cluster of shops, stacks of fabric sit outside other clusters, household goods outside still others. Some of the buildings look just like homes, but where the lady of the house prefers to sit outside on a wooden chair. These are shops too. On sale, for less than one dollar a visit, are the sexual services of the ladies.

Few of the women are from Nairobi itself. Instead, most migrated hundreds of miles to escape crushing rural poverty, and to avoid the glares of relatives who might have objected to the way they were seeking to escape it. East Africa's periodic outbreaks of war, and the accompanying hunger, disorder, and mass rape, also sent many of the Pumwani prostitutes on the move. A particularly large contingent came from the Kagera region of Tanzania, a war-torn border area that was among the first places that AIDS took hold as the epidemic moved east from Zaire in the 1970s. Once in Pumwani these sex workers often had dozens of encounters a month with a rotating group of men, who rarely confined their patronage to a single woman. The result was a vast network of sexual interaction that allowed diseases to travel more efficiently here than most places on earth. Infections such as syphilis and gonorrhea were epidemic, along with chancroid, which causes painful genital ulcers that can last for months or years if left untreated. The arrival of AIDS made this grim picture even worse. In 1986, researchers found that the HIV rate among Pumwani prostitutes was a staggering 85 percent— beyond anything seen in Kinshasa or even San Francisco.[1] Here was a true tinderbox.

It was also an ideal laboratory for studying how sexually transmitted infections moved. A team of Canadian and Kenyan researchers, led by University of Manitoba microbiologists Allan Ronald and Francis Plummer, began examining the spread of chancroid in the early 1980s. As it became clear that AIDS was an even bigger problem, the team began studying the factors that contributed to such an outrageous pace of HIV infection. Ronald was the elder of the team, one of Canada's foremost experts in infectious disease and head of the University of Manitoba's Department of Internal Medicine. Plummer was a rising star in the medical school. Together they turned their attention to a key question: If nearly all of Pumwani's prostitutes had HIV, why didn't most of their customers also have the virus? These men had fairly high rates of infection, but many repeatedly bought sex in the slum without getting HIV. Did some identifiable factor protect them? Or was it simply a matter of luck?

Ronald and Plummer began the study by recruiting hundreds of men who regularly visited the Pumwani prostitutes but did not have HIV. Each man was quizzed extensively about his age, employment, marital status, history of sexually transmitted infections, number of lifetime sex partners, experience with medical injections—anything that could potentially affect risk for infection. The researchers also did physical exams, noting whether the men were circumcised, as was common for most of Kenya's major ethnic groups, or had evidence of the ritual facial scarring still routine in some areas.[2] Then the researchers followed the men over the course of two years, giving them HIV tests once a month as well as regular counseling on using condoms and treating sexually transmitted infections. Each new case of HIV, meanwhile, was tracked.

The next part of the story happened not in Nairobi, with its uncertain electrical supply and slim inventory of computer equipment, but eight thousand miles away, on the leafy campus of the University of Manitoba, in the Canadian city of Winnipeg. As Ronald, Plummer, and the other researchers sorted through the data, they found that men who had genital ulcers—widely suspected as a factor that encouraged infection—had more HIV. But most of the other data was murkier. Married or unmarried, employed or jobless, relatively well educated or not, none of these factors seemed to make much difference in the speed with which the men became infected. It was beginning to look as though, within a highly infectious setting such as Pumwani, HIV was mostly a grand game of Russian roulette. But that picture changed abruptly one night, when Plummer ran the data on the physical characteristics they had recorded about their subjects. Of the 214 circumcised men, 6 got HIV, for a rate of new infections of less than 3 percent. Of the 79 men who were not circumcised, 18 got HIV, for a rate of 23 percent. Over the course of two years the men had effectively sorted themselves into different tiers of vulnerability based on that one factor alone. In a setting where exposure to the virus was routine, circumcised men were one-eighth as likely to get infected than those with intact foreskins. Here was an impact far beyond that of any HIV vaccine yet imagined.[3]

In a world where researchers often have to slice and reslice data dozens of different ways to find a statistically significant result, the circumcision numbers all but leaped off the computer screen. Plummer was so astonished that he immediately placed a call. "Allan!" he said excitedly into the phone, "you won't believe what we found!"

Hunter-gatherers once roamed the great swaths of Africa where AIDS is now worst. They were the closest thing to direct descendants of the earth's first people, and for tens of thousands of years they lived off animals they could kill, and roots and berries they could collect. They were vulnerable to any number of maladies, but AIDS was not one of them, and probably never could have been under such conditions. The hunter-gatherers were too few, too spread out, and too limited in their mobility to incubate a sexually transmitted epidemic with a virus as fragile as HIV. Had the outbreak somehow started in this era, only a handful of people might have gotten sick before the virus itself died out. Ancient Africa was inherently resistant to an AIDS epidemic.

The shape of human settlement on the continent began changing more than one thousand years ago, with a series of historic migrations. The most powerful was the journey of Bantu-speaking peoples out of West Africa, where they had an ancestral home near today's border between Nigeria and Cameroon. The Bantus were farmers who favored the settled life of villages, where they could grow such staple crops as yam, oil palms, sorghum, and, eventually, millet and corn. Later they also took to raising cattle and other livestock. As the Bantu influence spread east and south, this language group gradually populated most of the continent from the Sahara Desert down to the Cape of Good Hope at Africa's southern tip. And this pushed the hunter-gatherers into increasingly marginal areas, the densest of equatorial forests, the driest of deserts.[4]

Throughout the continent, a defining moment of Bantu life was a coming-of-age ceremony whose importance was rivaled only by weddings and funerals. In this initiation ritual a village elder would grasp the penis of an adolescent boy and cut away the foreskin, exposing the previously sheathed head within. The boy, though only a few painful moments older, was now considered to be a man, free to marry, own property, join discussions of village affairs. Anthropologists and linguists date this tradition back millennia in Africa. And for that reason, the basics of it were remarkably similar across many Bantu-speaking groups living thousands of miles apart. The boys, at puberty or shortly before, moved to an area outside of their main village, away from their mothers and other women, and were put under the care of men. The initiates shed even their clothes as part of the break from their childish pasts. The cutting of the foreskin happened fairly quickly, often on the first day of the circumcision school, and generally was followed by weeks or months of healing and rituals. The boys would sing songs, dance, and cover their bodies in clay or ash to help signify the transformation they were experiencing. Elders spoke to them about the responsibilities of manhood and the importance of facing life's trials with stoicism.

In Nelson Mandela's autobiography, *Long Walk to Freedom*, he wrote movingly about how he gathered before relatives with a group of two dozen other boys at a ritual spot near the Mbashe River. When the *ingcibi*, a circumcision expert, sliced their foreskins off with a spear each boy was supposed to respond not with tears or screams of pain but with the proud declaration *Ndiyindoda!*: "I am a man!" Mandela's momentary hesitation before shouting the word—he was stunned with pain and could barely think for several seconds—embarrassed him.

> I was distressed that I had been disabled, however briefly, by the pain, and I did my best to hide my agony. A boy may cry; a man conceals his pain.
>
> I had now taken the essential step in the life of every Xhosa man. Now, I might marry, set up my own home, and plow my own field. I could now be admitted to the councils of the community; my words would be taken seriously.

Bantu circumcising communities often regard uncircumcised men, no matter what age, as trapped in perpetual boyhood. And these communities regard groups of men who were circumcised in the same ceremony as a cohesive social unit, as close as brothers, creating relationships that reached beyond individual clans and villages. These bonds were among the crucial building blocks of Bantu society wherever this language group's influence reached. In many places, the procedure itself later moved into sterile hospitals and clinics, and increasingly it was newborns rather than adolescent boys who were cut, but the tradition survived. The near universality of male circumcision in the Congo River Basin was the mystery factor that helped keep HIV from developing into a hyperepidemic during all those years of the twentieth century that it was spreading steadily but slowly. Without foreskins to infect, vaginal sex couldn't spread the virus fast enough to create outbreaks in which 10 percent or more of all adults had the virus. That held Kinshasa's adult infection rates in the single digits for many decades—even before Franco started singing "Attention na SIDA."[5]

The reason is simple—if rather graphic—anatomy. The skin on the shaft of a man's penis is, like that on most of his body, relatively thick and tough, allowing it to serve as a natural barrier against infection. But the foreskin of an uncircumcised man is unusually vulnerable, because it is soft, thin, and a bit moist, making it easier for pathogens to penetrate. HIV targets certain types of immune cells that are close to the surface in foreskin tissue.[6] During erection the foreskin is stretched back down to cover the upper part of the penis shaft, turning this most vulnerable skin outward, where it can come into contact with fluids that may contain the virus. The penis of a circumcised man, by contrast, presents a daunting challenge to HIV. There is no inner foreskin to turn outward and fewer easily accessible immune cells that the virus could infect. The man is much safer, and so are his future sex partners. With fewer infections, the overall community is safer too.

The reverse is true for communities that no longer circumcise, or never did. As Bantu speakers migrated into East Africa they ran into two other ethnic groups moving down from the north. The Cushitic people from Ethiopia had their own tradition of circumcision that may date back even further than that of the Bantu speakers, by some estimates up to nine thousand years.[7] (Their ethnic cousins, the ancient Egyptians, scrawled images of circumcision rituals on stone tomb walls that survive to this day.) But the second group, the Nilotic peoples from Sudan, did not circumcise. As these three groups mingled and jostled for land in East Africa, their cultures blurred at the edges. An eastern branch of the Nilotic people began circumcising, apparently after contact with some Cushitic groups. Some nearby Bantu-speaking groups, meanwhile, abandoned the ritual.

The story in southern Africa is even more complex. The dominant ethnic group in Zimbabwe, the Shona, are Bantu speakers but do not circumcise, possibly because of some historic contact with Nilotic groups to the north. But nearly every other major ethnic group in southern Africa did circumcise their boys, as recently as the beginning of the 1800s. Historians say the decline coincided with the rise of European influence, first along the coast and later deep into the interior as well. Christian missionaries discouraged traditional Bantu ceremonies, including the coming-of-age rituals that involved male circumcision, as unacceptable relics of a heathen past. This was also a time of rising tribal warfare in southern Africa. Some historians believe that efforts to reorganize society on militarized lines—with fighting regiments taking the role once filled by brotherhoods of men circumcised together—caused the ritual to decline in importance.[8] And some tribal leaders may simply have wanted to make sure their young warriors were ready for battle, not off in the bush healing from their circumcision wounds.

Whatever combination of causes proved decisive, the Zulus and Swazis had abandoned circumcision by the mid-1800s, and it declined among several other Bantu-speaking groups, as Africans gradually moved out of traditional rural areas and into cities, where tastes and values were rapidly Westernizing. This contributed to the creation of a large, contiguous area where male circumcision was far from universal. It ran from East Africa's Great Lakes Region, south past Harare and Johannesburg, and down through the bustling port cities of Maputo and Durban. Within this swath of the continent, about five hundred miles wide and fifteen hundred miles long, the AIDS epidemic finally revealed its disastrous potential.

By the time Plummer and his colleagues finished analyzing the data and writing it up into a formal report, a discussion over possible connections between HIV and male circumcision had already begun in some medical journals. A physician in suburban New York, Valiere Alcena, first publicly raised the issue in a letter to the small *New York State Journal of Medicine* in August 1986, about five months after the Manitoba-Nairobi team started enrolling subjects in their study. A bigger splash came in October that same year, when a letter from California urologist Aaron Fink was published in *The New England Journal of Medicine* noting the higher rates of sexually transmitted infections among men who had not been circumcised. He suggested that something similar may be happening with HIV. "The likelihood that the cervical secretions of a woman with AIDS can be transferred by similar means is greater when the skin surface is a delicate, easily abraded penile lining, such as the mucosal inner layer of the foreskin, than when the foreskin is absent," Fink wrote.

Soon after, Priscilla Reining, an American anthropologist working in the Kagera region of northwestern Tanzania, began noticing that the Haya people were dying of AIDS at much higher rates than other ethnic groups. She wondered whether circumcision—the Haya didn't perform the ritual on their boys but most groups in the region did—was the reason. Reining enlisted demographer John Bongaarts and two other colleagues to study the issue more broadly. The resulting paper, published in the journal *AIDS* in 1989, found striking geographical relationships between high rates of HIV and low rates of circumcision among 409 ethnic groups in Africa.[9]

By the time the Pumwani findings appeared in *The Lancet* that same year there already was a way to explain how HIV might more easily infect uncircumcised men—what scientists call "biological plausibility"—and the first evidence of higher infection rates among groups that don't circumcise—what scientists call "ecological evidence." The Pumwani study, then, was offering confirmation of that connection by reporting on new cases of HIV as they occurred. This emerging science had the additional benefit of a potentially powerful advocate. One of the eleven authors on the Pumwani study was Peter Piot. Though junior to Ronald and Plummer on the research team, Piot already was emerging as an unusually energetic and politically adept AIDS scientist.

As his star rose—and no one would become more influential in the global AIDS fight over the next two decades—Piot moved into positions that could have allowed him to publicly raise the issue of circumcision and push for further research. An essential piece of the epidemic's spread, and one that offered the prospect of effective interventions, could have fallen into place. Global health officials could then have begun alerting tens of millions of uncircumcised African men—and their tens of millions of wives and girlfriends, who also would have benefited from having sex partners who were not infected—that a simple surgical procedure could help save their lives amid an incurable, lethal plague.[10]

FEAR WORKED

A mong the African nations with low rates of male circumcision was Uganda, home to scientist David Serwadda and the strange cluster of Kaposi's sarcoma cases he discovered. The nation is shaped like the head of a hatchet, with its rounded blade aimed eastward toward Kenya. Rivers and lakes define much of this geography, but none so powerfully as Lake Victoria, named for the British queen who oversaw her nation's most ambitious period of colonial expansion into Africa. This massive inland sea is the centerpiece of East Africa's Great Lakes Region and also the fabled origin of the Nile River. If the Congo River and its tributaries formed the geographic center of the early AIDS epidemic in Africa, the next phase belonged to Lake Victoria and the other great lakes—with their vast, vulnerable populations of uncircumcised men living nearby, along with their sex partners. Here the disease began killing on a scale not seen before, as if HIV somehow shifted into a higher gear as it climbed past the mountains of Africa's Western Rift and into East Africa.

Two of the hardest-hit areas in Uganda were Masaka and Rakai, lowland provinces inhabited mainly by subsistence farmers and fishermen living near Lake Victoria. Serwadda knew that all four of the men with Kaposi's sarcoma came from that region. So to investigate further, in January of 1985 he and two other doctors began driving south from Kampala, the capital, on a battered highway running parallel to the shores of the lake. When they stopped a few hours later, it was clear they had found a public health crisis in full bloom: There were shrunken babies, terrified widows, hospital beds spilling over with dying patients. Several members of a single family often had the disease. The survival of entire villages seemed imperiled as the epidemic struck adults who should have been at the peak of their power to work and raise families.

As before, the purplish skin welts of Kaposi's signaled underlying immune problems. But in Masaka and Rakai, Serwadda's team discovered the rattling cough and profound wasting that would become the awful signature of the AIDS epidemic there. The cause was tuberculosis, a vicious airborne disease that had grown rare in much of the world, but not in Africa, where it afflicted miners, slum dwellers and others forced to breathe fetid, infected air. The arrival of AIDS supercharged tuberculosis and spread the disease more widely than ever before. As HIV weakened the immune systems of its victims, tuberculosis destroyed their lungs and consumed their flesh. The Ugandan peasants had a name for what they were seeing: Slim. "It was unbelievable. You could make the diagnosis from the weight loss," Serwadda recalled years later, his voice low and grave.[1]

Long before Serwadda and his colleagues made their visit, an elaborate mythology had developed around Slim in this corner of Uganda. Sickness and death in many African villages were not typically blamed on tiny germs visible only to white-coated scientists peering through microscopes. Instead, disease was seen as the physical manifestation of forces much more apparent. A shunned neighbor, angry that food from a recent harvest wasn't shared fairly, might have put a spell on the victim. Or the community's ancestral elders, still regarded as present and easily angered, could punish the living for failing to slaughter a goat at a time when tradition demanded such ritual sacrifice. And those who thrived financially but failed to share their bounty were seen as particularly ripe targets for witchcraft and disease. So when the first victims of Slim appeared to be prosperous traders who made frequent runs over the border into nearby Tanzania, their illnesses fit neatly within the popular understanding of how the vengeful new sickness worked. Many of the next wave of victims were fashion-conscious young women who were known to push the limits of village social standards and hence were also seen as natural victims. Slim soon acquired a second nickname, Juliana, for a popular line of clothes the young women preferred.[2]

Serwadda and the other doctors from the capital believed not in witchcraft but in medical science, yet there wasn't much they could do about Slim other than document its destruction. The team drew blood, took swabs, and administered basic care for the symptoms that could be treated. All of the twenty-nine patients the medical team identified as suffering from Slim later tested positive for HIV.

When Serwadda and his colleagues returned to Kampala and conducted lab work on the samples they discovered something even more alarming. As is routine in investigations of new outbreaks, they had also taken blood samples from a so-called control group, hospital patients who did not display obvious symptoms of AIDS. That was supposed to allow for a higher degree of confidence that HIV was causing the symptoms among those visibly ill. Yet of the thirty people Serwadda's team tested as part of their control group, five had the virus, for the equivalent of an infection rate of 17 percent. In another troubling sign, ten people among a group of fifteen cross-border traders had HIV. The team also tested five apparently healthy women who were married to men who had Slim; all five of these women had HIV too.

These results made clear that just beyond the visible wave of AIDS deaths lay another wave, and perhaps another and another and another. Not only could the doctors not stop this; there was no practical way for them to know how large those next waves would be, or when they would crash. Serwadda's report, published in *The Lancet* in October 1985, was striking as well because it was among the first reports of rampant AIDS outside of a major city.[3] The once tiny outbreak had grown into a voracious epidemic capable of killing across vast areas of entire nations across almost all parts of a society. Men and women, young and old, farmers and fishermen, teachers and technocrats—nobody seemed safe.

. . .

In January 1986, three months after Serwadda's piece appeared in *The Lancet*, a leftist guerrilla leader blasted his way into Kampala. The victory by Yoweri Museveni—a balding intellectual with a round face, a mustache, and a knack for evocative metaphors—came after twenty disastrous years of dictatorship and civil war that had brought Uganda to ruin. Museveni was not much of a democrat himself, and many Ugandans would eventually come to see him as a despot. But his record on fighting AIDS was remarkable for a degree of boldness and clarity achieved by few other leaders in Africa, or anywhere else. This began after Museveni got an unwelcome warning that the burgeoning HIV epidemic threatened not just the nation's recovery but also his hold on power. Uganda's military had sent sixty top officers to Cuba for training. In routine medical screening there eighteen of the Ugandan officers tested positive. Several months later, at a conference of nonaligned nations held in Zimbabwe's capital of Harare, Cuban president Fidel Castro told Museveni, "Brother, you have a problem."[4]

Museveni soon convened his top health advisers. And through a haze of contradictory information that clouded the decision making about AIDS almost everywhere else, they focused on the basics: HIV was incurable. It was fatal. It was spread by sex. And if left unchecked millions of Ugandans—and perhaps the nation itself—were doomed. From this basic calculation the Ugandan government became the first in Africa to fight AIDS with the necessary urgency. The nation had little financial or institutional capacity after years of war, and its friendship with Cuba made it an unlikely recipient of major investments from Western donors. But the Ugandans developed a plan of their own—one that emphasized plain talk about a disease that carried profound stigma on the continent. "When we went to conferences and talked about AIDS, fellow Africans would say, 'Don't embarrass us. We don't have that thing in Africa,'" recalled Ugandan epidemiologist Fred Wabwire-Mangen, one of Museveni's early advisers.[5]

The Ugandans saw themselves as anything but un-African for talking about a disease that was clearly in their midst and putting their communities in such danger. It was at the essence of their culture to regard threats not just individually but collectively. Said Wabwire-Mangen: "In the African tradition, when a snake comes in your hut, you bang your drum and call all your neighbors."

Museveni also had been a born-again Christian.[6] And at times there was a moral—occasionally even moralistic—element to Uganda's early AIDS campaigns. But Museveni and his advisers, which included several prominent physicians, were most focused on the public health imperative of preventing more infections however they could. Museveni and his team worked closely with religious and community leaders to hone a message that could be delivered across the nation, anywhere and everywhere, at any time. This included in rural areas that remained largely polygamous and where any message that advocated strict monogamy could provoke a backlash. And yet Museveni did not shy away from focusing his nation on the relationship between sexual culture and disease. The available data made clear that anybody who had several sex partners had a high probability of contracting HIV. One study from that era found that the average African who died of AIDS reported having sex with thirty-two different people.[7] That likely was an overestimate, especially after HIV moved more widely across the region, but truckers and some other men who traveled regularly were having many partners—and, as a result, a powerful role in spreading the virus. A Ugandan businessman named Akokoro told *The Washington Post* in 1986 that he contracted Slim after having sex with about one hundred women a year in Uganda and neighboring countries. "He says he has never had a homosexual encounter or received a blood transfusion," the *Post* reported. "Asked what kinds of women he had sex with, Akokoro said, 'All.'"[8]

Most Africans suffering from the disease had far fewer partners than Akokoro—and in many cases women got HIV after being faithful to husbands who did not return the favor. But Museveni and his team stayed focused on the perils of sex outside of committed relationships. The distinction they made here was key. Traditional polygamy was appropriate to Ugandan culture and, if practiced strictly, not especially risky, because the man and his wives formed a closed sexual circuit that kept out new infections. But the modern variant of polygamy that had become common during colonialism and its aftermath—when multiple wives were replaced by a succession of informal, often overlapping relationships—was portrayed as alien, decadent, and deadly. The only way to fight back, Museveni and his advisers decided, was to use one of the few weapons more powerful than the human sex drive: fear.

Fear has a bad name among many public health campaigners, especially in the AIDS world. Positive, uplifting messages are supposedly the key to beating the disease. And messages that inspire fear can heighten stigma against those already sick while making everyone else feel overwhelmed, fatalistic, and perhaps more prone to reckless behavior. In some places the terror that accompanied the arrival of AIDS provoked awful consequences, including attacks on those who were already struggling with a fatal disease. Uganda's leaders had lived much of their lives amid the fear of war and dictatorship, and in fighting AIDS, they tried to harness its power to change the behavior that was helping spread HIV. "Fear worked," Sam Okware, a key health official in Uganda's original anti-AIDS campaigns, said years later. "That was the weapon we used."[9]

But it was not fear alone, according to Jesse Kagimba, the longtime personal physician to Museveni and another of his top advisers on AIDS. The Ugandans drew upon the natural fear that the epidemic created and focused it on a practical goal. Western social scientists eventually would rediscover this approach and dub it "fear plus an efficacious solution."[10] Kagimba, who has a bald head, narrow face, and a fierce gaze, once explained it to me more succinctly than any journal article could: "You change because of fear. And you change because of love," he said. "Fear is stronger than love."

In addition to being a doctor, an AIDS theorist, and an amateur psychologist, Kagimba was a habitual doodler. During an interview at a hotel bar in Kampala one evening, many years after the peak of the epidemic in Uganda, he reached over and grabbed my notebook. In it he drew an enclosed, circular structure, almost like a circus ring with raised walls. Inside he drew a cat that was hungrily eyeing a mouse. The mouse meanwhile was looking back at the cat, no doubt contemplating its extremely limited range of options. Kagimba drew a line to this mouse, and at the top of the line, he wrote FEAR. Kagimba then lifted his pen just long enough to explain the obvious: The mouse was doomed. It could run around in the circle, but it couldn't escape. Death would soon arrive, if not from the cat's claws, then from the heart attack to which the terrified mouse soon would succumb. But then Kagimba drew another circular structure below the first. In this second ring the cat gazed not at the mouse but straight ahead, at me, looking puzzled. The mouse in this second image was not looking at the cat but off to the side. A line still connected the mouse to the word fear, but it was a different kind of fear, because now Kagimba had drawn several holes to the right of the mouse in the base of the circular structure. These were the ways the mouse could escape. And it looked like that mouse was about to run for its life, straight for the hole that Kagimba had labeled BE FAITHFUL.

"This is what Museveni did," Kagimba said solemnly, as he finally raised his eyes from my notebook. He then explained how Museveni's team had attempted to reverse a trend in Ugandan sexual behavior that he dated to the arrival, decades earlier, of widely available contraceptives and antibiotics to treat sexually transmitted infections. Shame surrounding out-of-wedlock births had declined at around the same time, he said. So the Ugandan government's anti-AIDS campaigns did not rely on promoting behavior that was culturally unfamiliar, as were approaches like condoms and HIV testing. Instead Museveni and others called on their fellow Ugandans to restore a moral order they said had been lost, to revert to an era when it was acceptable to have three wives but not to have three girlfriends (or boyfriends).[11]

This message was most famously crystallized in the slogan ZERO GRAZING, an evocative term known to the vast majority of Ugandans, who had grown up on farms. A goat tied to a pole, grazing in a single spot, left an unmistakable, zero-shaped pattern in the grass. The pole, of course, represented a man's homestead, whether it included one wife or several. The area of chewed-up grass was his sexual activity, which in the metaphor was portrayed as entirely natural and proper. The only issue was where that sexual activity happened: within the homestead or beyond it. There were other versions of this message of sexual caution. Billboards urged Ugandans to LOVE FAITHFULLY. At public appear-

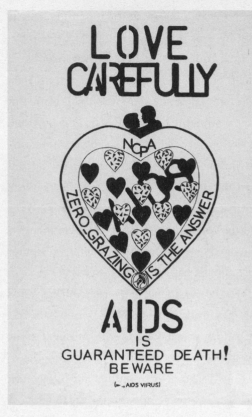

During the Zero Grazing era in Uganda, posters like this one from the late 1980s were used to promote changes in sexual behavior.

ances Museveni often asked crowds whether they were familiar with the elaborate mounds of dirt that termites build in fields across much of Uganda. When many in the audience nodded Museveni pointed out that these mounds often had holes. "If you put your finger in enough holes," Museveni said to embarrassed laughs from the crowd, "sooner or later you will get bitten by a snake."[12]

These messages were broad enough, and delivered frequently enough, to reach across a diverse nation that did not necessarily have uniform views on what amounted to appropriate sexual behavior. The Christian churches, which for generations had been attempting to banish polygamy from Uganda, were at first slow to embrace any message that didn't explicitly forbid sexual relationships outside of monogamous marriages. But the epidemic produced such a toll that Christian leaders soon adopted the "Zero Grazing" message. One of that effort's leaders was Anglican reverend D. Zac Niringiye, who in 1987 discovered that his own brother was dying of AIDS. Soon after one of Niringiye's young theology students fell ill as well. This terrible news made it impossible for Niringiye to simply regard the sick as sinners paying the price for misdeeds. "I can no longer say this [AIDS] is for nonbelievers," he recalled thinking. "This is for them. This is for us."[13] Religious leaders became among the most fervent and influential leaders in the Zero Grazing campaign. Some delivered their warnings about the dangers of AIDS not only in their services but at funerals as well, paired with blunt language about the danger of casual sex. Long past midnight, after hours of grieving for a loved one lost to the disease, the minister or priest would say, "You know why this man died. Who will be the next to go?"[14]

Such pointed messages became hard for Ugandans to avoid in this era. National television carried before-and-after pictures of a young woman with AIDS, showing her beautiful body wasted away. Museveni also required that every government official warn about AIDS in every public appearance. Each morning the national radio stations carried the sounds of drums pounding out the signature cadence of war *BOOM-BOOM ba-BOOM-BOOM*. As the drums subsided a young girl would plead, "My father, I'm still too young. Please don't die. Be faithful." Meanwhile, grim AIDS posters started appearing all over Uganda. A typical one from the era showed a massive skull and crossbones perched atop a coffin. The tiny image of a couple stood hand-in-hand off to the left. "I wish I had said NO to AIDS," the poster said. "My quick pleasure led to a slow, painful death."[15]

Newspaper coverage at the time captured the changing attitudes. The *New Vision*, the nation's biggest paper, headlined an article in October 1987 SLIM IS FORCING PEOPLE TO CHANGE SOCIAL HABITS. It reported on how more men were staying home at night, and those who went out to bars were avoiding casual hookups with waitresses out of fear of getting HIV. Some men, the article reported, resorted to long walks in their neighborhoods to work off the physical tension created by having less sex. One woman complained about the rising sexual demands of her husband at the height of the Zero Grazing campaign. But others saw it as improving the intimacy of their marriages. "In Bugolobi, a young housewife with three children, declared with a gleam in her eye, 'There has been a positive change in our marriage. My husband stays at home much more. And I encourage him to do so by enthusiastically keeping him informed of the latest gossip about Slim victims.'"[16] Other coverage at the time reported on bar and disco owners lamenting their loss of business; sex workers complained too, as men stayed away. Prices for sex fell, and many prostitutes moved back home to their villages and found other ways to eke out an existence.

These changes began making a difference in the spread of HIV. But even all of that may not have been enough to produce the shifts in sexual behavior as swift and serious as were soon recorded in Uganda. For that it took a man universally admired for his talents and magnetism to die a very painful, very public death. It took Philly Bongoley Lutaaya.

[14]

BORN IN AFRICA

IDS hit Africa's boisterous music industry with fury. Not only did musicians have access to the easy sex available to celebrities the world over, they also were having their encounters within a population where HIV was rising fast, far beyond anything seen among heterosexuals in the Americas, Europe, or Asia. AIDS is widely presumed to have killed Franco, as well as Nigeria's Fela Kuti, whose raucous rhythms and blistering lyrics captured the spirit of a restless, rising generation of postcolonial Africans. Franco and Fela were all but secular gods for their respective nations.[1] But crucially, both singers denied they had AIDS even as their bodies wasted away. In societies where illness was seen as a sign of weakness and shame, acknowledging AIDS may have been too much to ask of these men. But one star of their generation, Uganda's lithe and lyrical Philly Lutaaya, accepted his fate in a way that dramatically shaped his nation's response to AIDS. In part because of Lutaaya's example, Uganda became the first—and among the only—African nation to curb the spread of HIV while also increasing respect and sympathy for those already infected.[2]

Lutaaya had his first hits in the late 1960s, and by the 1980s he was living comfortably much of the year in Sweden, away from the chaos and destruction in his homeland. In Europe he had access to First World medical care and all the comforts that a wealthy, dying man might want. But instead of spending his last days in the ease of his second home, Lutaaya traveled back to Uganda on a mission. He had decided to make his last act as a performer his most important one. He would die fighting AIDS.

Newspapers first carried reports that Lutaaya had AIDS in April 1989. This was more than two years into Uganda's aggressive campaign against the epidemic. Yet skepticism remained high. Many Ugandans, and especially young men and women at the primes of their sexual lives, at first regarded the news about Lutaaya having HIV as mere propaganda crafted by elders bent on dampening youthful fun. But as the months went by, and Lutaaya took his campaign around the country, Ugandans came to believe. A documentary released shortly after Lutaaya's death captures this transition, as rapt audiences turned out to see one of Uganda's most famous men warn about AIDS. Lutaaya was painfully gaunt. His long, dark braids were gone, leaving behind sad gray wisps of hair. His skin was splotchy, and even his signature goatee had fallen away. Lutaaya's biggest hit during his career had been the assertive, upbeat "Born in Africa." But during these final months Lutaaya found time to record a final song that was closer to a national cry for help. The lyrics of "Alone and Frightened" carried none of Franco's Old Testament admonitions about sex and sin. Instead Lutaaya focused on being a public example that others could not dismiss.

> Today it's me, tomorrow someone else.
> It's me and you, we've got to stand up and fight.
> We'll shine a light in the fight against AIDS.
> Let's come on out, let's stand together and fight AIDS.

In his public appearances Lutaaya offered a message that was even sharper, and more in line with what Museveni and church leaders were saying. "Changes must be made in our sexual behavior," Lutaaya told one crowd. "In past years we have been so free in our sexual behavior. Adultery is a serious threat to all of us and must stop. Adultery by either sex presents a danger to the other partner, and ultimately to the whole family. If we don't work hard, the human race is going to die."

There was a jarring harshness about such messages. They were blunt. They were scary. And for people who had AIDS, they must have been hard to stomach. The documentary on Lutaaya shows him turning his head away and wincing during a rally where village women sing a song about the disease:

AIDS was inflicted upon the rebellious,
The promiscuous and the criminals.
It's terrible now because it strikes children
Who know nothing about the world.
How should we pray?
Help us, Father. We are perishing.

The Ugandan approach to fighting AIDS relied on Western discoveries about the cause and nature of the disease, but it emphasized changing the nation's own social norms rather than counting on condoms or other biomedical tools to arrive from the outside world. Perhaps this was a hardheaded judgment that, even if scientists discovered some medicine to cure or control the disease, it would not reach poor Africans any time soon. But it also grew from the more communal, less individualistic nature of most African societies. They tended to treat AIDS as a malevolent outside force—in the image offered by epidemiologist Fred Wabwire-Mangen, as a deadly snake threatening the village—that should be fought off. Whoever got bitten might die, but the survival of the village was paramount. In many places this contributed to shunning and stigma that made the final months of the illness even more horrendous. But Lutaaya's example made this less of a problem in Uganda, because the openness of the conversation demystified the disease. Government, community, and religious leaders generally paired their warnings about dangerous sexual behavior with equally strong admonitions against judging those already infected.[3]

Africans also made a range of practical adjustments that—as was the case with Ebola in Yambuku—often were more direct and effective than most imported Western strategies. In the Kagera region of northwestern Tanzania, which had an approach similar to that of neighboring Uganda and saw HIV decline by a similarly dramatic rate, Christian churches began requiring HIV testing before performing marriage ceremonies, and there were broad shifts away from casual sex. "In the past, during wedding ceremonies, men spent all night partying and drinking," a woman told Tanzanian researcher Joe Lugalla. "They would move from one party/ceremony to another sleeping with different women. . . . Thank God that this is no longer the case today."[4]

In many places the practice of widow inheritance, in which custody of a dead man's wife switched to a brother or other male relative, ended because of the obvious risk of keeping the virus circulating through a family. And, more generally, regular sex partners started keeping closer eyes on each other; even co-wives teamed up to monitor the travels of their shared husbands.[5] In Uganda, the sexually transmitted disease clinic at Makerere University's hospital once had been known colloquially as the "Hall of Heroes," but during the Zero Grazing era having syphilis or gonorrhea suddenly became a source of embarrassment and even shame. Some men still spoke nostalgically about the carefree time when even a chance encounter on the street could lead to an impromptu conversation and, eventually, sex. That changed as the reality of the AIDS epidemic sank in. Several men said that now when they saw appealing young women approach, they crossed the street. "I saw this girl in a miniskirt walk by," one man said. "In earlier times I would get excited. Now she looked like death."[6]

The architects of Zero Grazing portrayed the behavior spreading HIV as aberrant, a bastardization of traditional village mores caused by colonialism and its aftermath. That belief—even if it was a product of some degree of mythologizing of the past—meant that there were core cultural values that could be invoked in the battle against AIDS. Museveni and some other Ugandan leaders, meanwhile, tended to see condoms as inherently foreign.[7] Their use in the Western world dated from at least Shakespearean times, but they had been all but unknown in most of Africa. Traditionally many Africans believed that sex in which a man's semen fails to enter a woman's womb was not sex at all. The act was incomplete. To the extent condoms were used at all in Africa, it was mainly by prostitutes and soldiers, or for hookups in bars. Some early, targeted programs in Uganda did attempt to get condoms to these groups, but the main focus was to discourage extramarital sex. Museveni saw enough potential conflict between the Zero Grazing message and some aggressive condom campaigns that he banned mass-media condom promotion outright in 1991.[8]

Museveni's stance against condoms generated intense rebuke from Western health officials, who leaned hard on Museveni to relent, which he eventually did. But during the period when Museveni was resisting this pressure, he gave a speech at the Seventh International AIDS Conference in Florence, Italy, in which he pushed back against these critics by casting the epidemic as part of a broader story of eroding values and "permissive sexuality" caused by Western influence. "Traditional checks based on morality and self-control were thrown aside," he said.

> I have been emphasizing a return to our time-tested cultural practices, which emphasized fidelity and condemnation of premarital or extramarital sex.
>
> I believe that the best response to the threat posed by AIDS and other sexually transmitted diseases is to reaffirm publicly and forthrightly the reverence, respect, and responsibility every person owes to his or her neighbor. Young people must be taught the virtues of abstinence, self-control, and postponement of pleasure and sometimes sacrifice. Just as we were offered the magic bullet of penicillin from the early 1940s, our public health figures now are offering us the condom and "safe sex." In countries like ours, where a mother often has to walk twenty miles to get an aspirin for her sick child or five miles to get any water at all, the practical questions of getting a constant supply of condoms or using them properly may never be resolved.
>
> Meantime, we are being told that only a thin piece of rubber

stands between us and the death of our continent. I feel condoms have a role to play as a means of contraception, especially
in couples who are HIV-positive, but condoms cannot be the
main means of stemming the tide of AIDS.

In this speech, which was met by disdain by some in the audience, were
echoes of Franco, with his melodic warnings to Congolese to "[b]eware of
prostitutes, avoid multiple partners," and also of Philly Lutaaya as his death
neared.[9] They believed that with the right prodding their fellow Africans
could make different sexual choices, and that this might save lives.

Among those watching Lutaaya deteriorate was his physician, Elly
Katabira, and he was astonished that despite the singer's crusade many Ugandans still didn't believe that he had AIDS. "People were saying, 'He's lying,'"
Katabira later recalled. "It was only after he died that people said, 'Ah, maybe
the man was right.'"[10]

Lutaaya's final public appearance came at a large soccer stadium in downtown Kampala, thronged with cheering fans. Music played throughout the
evening, and as dusk settled, the crowd lit thousands of candles, filling the
stadium with a solemn glow. Lutaaya appeared in a white tracksuit that fit
so loosely it was painful to imagine what his body looked like underneath.
Though only thirty-eight years old, he walked like an old man, ambling
awkwardly up onstage while waving to tens of thousands of screaming fans.
"Hello, hello ladies and gentlemen," Lutaaya said weakly, as the band geared
up to play "Born in Africa." "I'm giving you one last song. I'd like to tell you
that I'm going back to Europe on Saturday, but I'll be back in Uganda in
December. And if God wishes I shall stage a concert at Lugogo Stadium.
Thank you very much."

Lutaaya never returned for that final concert. He died soon after, on
December 15, 1989, in Sweden. But the images of his deterioration over
those final months burned into Uganda's national consciousness. Sociologist
Swizen Kyomuhendo, who was a student at Makerere University at the time,
later recalled Lutaaya's public sacrifice as crucial to reversing what until then
appeared an irreversible plague. "We saw him," Kyomuhendo recalled. "We

saw him die. We abandoned the girlfriends." Many others did too. Data analyzed years later showed that the pace of new HIV infections in Uganda had climbed steadily between the late 1970s and the late 1980s, the height of the Zero Grazing era. Then it tumbled sharply downward and stayed low for years to come.[11]

[15]

THE CONDOM CODE

S uch a profound reversal of almost any other disease would have generated immediate, universal acclaim. When Ebola began its retreat after Yambuku's mission hospital finally closed, it was clear something good had happened, because people soon stopped falling sick. The same would have been true for outbreaks of cholera, salmonella, flu. Their spread caused obvious symptoms, and their ebb obvious relief. AIDS was the opposite. Because nearly a decade on average passed between the start of infection and visible illness, there were no clear signs when HIV began to spread widely in a society, and also no clear signs when the rate of new infections eventually slowed. This disconnect bedeviled scientists nearly as much as ordinary people. Again and again seemingly shrewd, reasonable judgments about the epidemic's course turned out later to be wrong.

In the late 1980s and early 1990s, during Uganda's historic decline in new HIV infections, the conventional wisdom held that the nation's homegrown approach to AIDS prevention was off track.[1] And it wasn't only Western experts who doubted the effectiveness of Zero Grazing; many Ugandans did too. In hindsight it's easy to see why. Even as the spread of HIV slowed, it would have been nearly impossible to notice, because the overall numbers of people with the virus continued to rise. This would remain true so long as people were getting HIV faster than they were dying from it.[2] Most of those who got infected in, say, 1985, would not appear visibly ill until they approached death about ten years later. So the slowing of HIV's spread in Uganda was entirely missed when it was happening, and it has been poorly understood ever since—even by many who lived through it. Amid all this illness and death it was hard to imagine that anything was working, and it would take several more years of analyzing data before scientists would be convinced that HIV had actually declined. The horrific waves of death during these years also obscured the picture. Researchers later would debate whether HIV fell because the pace of new infections had slowed, because sick people were dying so fast, or some combination of both factors.[3]

Among the first to spot signs of Uganda's reversal was American anthropologist Edward Green. During a trip to the country in 1993, while he was working as a consultant for USAID, Green noticed that rates of other sexually transmitted diseases were falling and deduced that new cases of HIV were as well. He wrote in his report to USAID, "If a high AIDS-prevalence country like Uganda shows a significant decline in STDs . . . *in the absence of a male condom prevalence rate over 5%*, it might suggest that *other* types of behavior change (premarital chastity, "Zero Grazing," marital fidelity, abstinence, non-penetrative and other safer sexual practices) can significantly affect STD incidence if not HIV incidence."[4]

This declaration gained no traction within the U.S. government, or anywhere else. But two years later, in 1995, it appeared that Green's analysis was on the right track. Rand Stoneburner, the epidemiologist who had witnessed the decline in HIV rates in New York City, now was working for the WHO. His job included monitoring the spread of the virus. Within Uganda, Stoneburner noticed encouraging signs from two Kampala hospitals that had been regularly conducting HIV tests of pregnant women for several years. At Nsambya Hospital the infection rate of nearly 30 percent in 1992 fell to 21 percent in 1994. At Rubaga Hospital it went from 29 percent to 17 percent over the same period. Uganda still had one of the worst AIDS epidemics on the planet. But at last—and for one of the first times anywhere in Africa—the trend lines pointed in the right direction. In a memo to a senior WHO official on February 17, 1995, Stoneburner plotted the infection trends at both hospitals. Then he took a leap of logic, writing, "The interaction of behavioral modifications and the natural HIV infection dynamics shows that preventing further spread of HIV may not be completely beyond simple and readily accessible interventions. If this is true, then it is good news for Uganda and elsewhere."

The statement was as much a product of scientific intuition as anything yet confirmed in the data. But the consequences were profound. Stoneburner didn't know it yet, but the AIDS world—made up of scientists, activists, and officials for the big international donors—already was beginning to cleave along a fault line that would define the next decade of the global response against the epidemic. On one side were a few like Stoneburner and Green who believed that social shifts could either accelerate the spread of HIV or reverse it. Embedded within that logic was the assumption that people could change, could shape their own fates and those of their communities. That's what made Stoneburner optimistic about "simple and readily accessible interventions." On the other side of the divide were those who were skeptical that affected communities could do much to slow HIV. Sex was too elemental, too deeply rooted in cultures that were inherently slow to change. That made these skeptics invest most heavily in potential biomedical fixes—pills, shots, condoms—that didn't rely on more fundamental changes in sexual behavior to work.

Faith in such technological solutions was rooted in the deepest folds of the Western brain, and especially among those schooled in medicine and most other branches of the hard sciences. Yet watching the epidemic turn around in New York, and now apparently in Uganda, caused Stoneburner, himself trained as a physician, to break with this orthodoxy. He believed that the Ugandan approach—blunt, intentionally scary, and at times even judgmental, all things the Westerners in charge of the global AIDS response resisted—might well work. Stoneburner later would conclude, after years of professional setbacks pushed him to the fringes of debate over the epidemic, that the optimism expressed in that seminal memo "pretty much cooked my goose forever."

• • •

At the same time that Museveni was resisting condom promotion officials thousands of miles away in the Southeast Asian nation of Thailand were embracing it with remarkable success. Thailand had an estimated three hundred thousand prostitutes, in addition to substantial populations of intravenous drug users and homosexual men.[5] Large percentages of Thai men had their first sexual experiences at brothels and remained regular visitors even after they married. The most visible proponent of Thailand's condom campaigns was a government health official, Mechai Viravaidya, who had been a family planning enthusiast before becoming a politician. He founded a chain of restaurants called Cabbages and Condoms that had a "vasectomy bar" inside and an abortion clinic next door. Beyond the restaurant he handed out condom key rings that bore the slogan DON'T LEAVE HOME WITHOUT IT and sponsored contests in which people competed at blowing up condoms like balloons.[6] Mechai's name even entered the Thai vernacular as slang for condom. Thailand's government backed the public health campaign with police action, threatening to close brothels if anyone was caught having sex there without a condom. The number of Thai prostitutes who said they always used condoms with their customers soon jumped, from 14 percent in 1989 to more than 90 percent in 1994. There was also a large reduction in the numbers of men frequenting the brothels, to the point where today it is no longer a common practice for Thai men to go there.[7] The adult HIV rate in Thailand, though high by Asian standards, never reached beyond 2 percent, and it declined after the condom campaign was implemented.

Reports of Thailand's reversal inspired similarly effective approaches in other places where HIV was spread mainly by sex industries, including Cambodia, Senegal, the Dominican Republic, and parts of India.[8] Such successes bolstered the argument that condom promotion, done aggressively enough, should be capable of turning back the AIDS epidemic—without raising awkward questions about people's sex lives. It was an appealing notion, especially within an AIDS activist community in which gay men were among the most forceful voices. Gay journalist Gabriel Rotello, author of *Sexual Ecology: AIDS and the Destiny of Gay Men*, called this belief the "condom code." He said it grew from the desire among many men to preserve a lifestyle born of the unbounded sexuality of the early liberation era, between the Stonewell Riots of 1969 and the discovery of AIDS in 1981. Those dozen years saw a profound surge in gay men embracing their sexual identities and fighting for equal rights. The era also saw the development of a fast-lane culture that celebrated casual sex as something akin to a revolutionary act against the mores of a middle-class society that often treated homosexuals as freaks.[9] "For many, the code is the only possible approach to AIDS prevention because it is the only possible response that allows the so-called sex positive aspects of gay male ideology to go essentially unchallenged," Rotello wrote. He added, "the condom code is the only strategy that allows many gay men to pretend that sex without consequences is still a possibility. As such, it simply *has* to work."[10]

That conviction carried through to the international AIDS arena, as many gay men became fierce advocates of fighting the epidemic in Africa as well. To these activists it seemed a natural extension of the domestic battle, because the victims still were people given short shrift by the world. If anything, Africans seemed more vulnerable to mistreatment and less able to advocate successfully for medical care or other resources. Randy Shilts, author of *And the Band Played On*, began predicting that AIDS would gradually become a chronic but manageable disease in the United States while death rates in Africa created a latter-day Holocaust. This infusion of attention and energy helped push the epidemic even more firmly onto the world's political agenda, but also reinforced impressions that HIV was essentially the same everywhere it struck—and therefore would be amenable to the same approaches.

Condoms were a familiar tool to members of a public health community that, in many cases, had gotten into AIDS work after experience in the field of international family planning. One early advocate was Malcolm Potts, the British-born head of Family Health International, an American nonprofit group with years of experience distributing contraceptives in poor countries. Potts had worked extensively in Africa and seen the potential for devastation shortly after the first reports of AIDS on the continent. He was so convinced that condoms were the answer—and that only financial constraints and lingering prudishness were keeping them from being deployed widely enough to defeat the epidemic—that he commissioned the construction of a hot-air balloon in the shape of a thirty-five-foot-tall condom. Local authorities in North Carolina, where Family Health International was headquartered, balked at letting Potts unfurl the giant condom there. But at the Fifth International AIDS Conference in Montreal, in 1989, Canadian officials allowed the condom to be inflated as long as the slogan emblazoned on the side, I SAVE LIVES, appeared in both English and French. Potts wryly recalled, "It was the only condom that required air traffic control to get it up."[11]

His enthusiasm was shared by the major donors, including USAID, which already had settled on condoms as the essential weapon against HIV's spread in Africa. Family Health International and other nonprofit groups received much of their funding from contracts with the government, and they were eager to keep the revenue flowing. Potts would later criticize this business side of the war on the epidemic, dubbing the constellation of government agencies, philanthropies, and nonprofit groups that worked on the issue the "AIDS Industrial Complex." But during his years at Family Health International (Potts retired from the group in 1990), he thought that he could fight HIV successfully while still harnessing the rising funding stream to keep his organization and its payroll growing.

Potts later acknowledged that there was a conflict lurking, especially as HIV work became steadily more lucrative. There were powerful incentives to work within the conventional wisdom so long as success was measured in terms of winning the next round of contracts from USAID or other big funders—and powerful disincentives to consider that something as accepted as condom promotion might not be succeeding in Africa the way it had in Thailand or San Francisco. That was especially true during the procurement process, when USAID issued detailed requests for proposals and scored them on the ability of bidders to fulfill easily measurable goals, such as numbers of condoms delivered, instead of more elusive ones, such as slowing the pace of new HIV infections. "When you're responding, it's like a game of Jeopardy," Potts recalled. "You're trying to come up with the right people and the right things to win the contract."

. . .

Amid this rising focus on condom promotion in Africa, Stoneburner was trying to determine what exactly had pushed Uganda's epidemic into reverse. He looked for answers in a pair of WHO surveys of sexual behavior that, by chance, managed to capture HIV's decline there. One was conducted in 1989, not long after the height of transmission, and the second in 1995, a few years after transmission had fallen steeply. The surveys opened with basic questions about the ages of subjects, their educations and incomes, and whether they were single, married, widowed, or divorced. Then the questionnaires probed more intimate matters. At what age did the survey subjects start having sex? Did they visit prostitutes? Did they use condoms? Did they have sex before marriage? Or outside of their marriages? To sort through the resulting mass of data, Stoneburner had the apparent benefit of a WHO colleague, Belgian anthropologist Michel Carael, who had helped construct the surveys and now was helping analyze the results.

The timing appeared auspicious as well. This all was coming together in December 1995, just as the Ninth International Conference on AIDS and STD in Africa was about to convene in Kampala. The conference brought together two thousand scientists, policy makers, and aid-group officials for four days of reports and debate. One of the final presentations would go to the analysis by Stoneburner and Carael. The turnaround for such a major piece of research was tight. They worked hard for weeks and then, when the presentation was just a day away, toiled through their final night in a hotel in the pleasant inner core of Kampala. Several times during that bleary-eyed session, Stoneburner had the sense of being on the verge of history, of unraveling a mystery at least as important as how gay men in New York City slowed the HIV epidemic through their sexual choices. But there were also signs of trouble.

Stoneburner had brought along his young protégé, the cerebral, Oxford-trained mathematician Daniel Low-Beer, to help with statistical analyses. But as Carael worked on data comparing some of the 1989 WHO survey results against those from 1995, Stoneburner grew frustrated that he and Low-Beer couldn't do the same. They had all the data from the 1995 survey and a summary of the 1989 set as well, but to generate a full range of analysis about the changes in Ugandan sexual behavior, they needed the entirety of both data sets.[12] As that long night wore on, Stoneburner began to suspect that Carael was withholding it. A certain degree of professional rivalry is hardly unknown within the ranks of scientific researchers, but this felt different to Stoneburner. His unease would only grow. "It was Michel Carael's baby," Stoneburner later recalled. "He did the survey. He kept the data."

Carael would later give a different account, saying that he too did not have all of the 1989 data on his laptop that night. Afterward, he said, Stoneburner and other researchers were free to request it. But Carael did not dispute the emerging sense of professional rivalry with Stoneburner, and if the data set was indeed freely available, few outsiders seemed to know how to access it. It would take years before Stoneburner got his own copy and was able to definitively answer the questions he was pondering that long night.

Carael and Stoneburner, it soon became clear, were opposites in nearly every way. Stoneburner was a relative newcomer to Africa, an outsider with a blunt manner and ideas that soon would push him out of the mainstream of thinking within WHO. Carael, meanwhile, was a smooth insider with years of experience both in Africa and in AIDS work. His research into breastfeeding, maternal nutrition, and birth rates in Zaire had brought him into rural areas where ancient rules still governed the social lives of most people he met. He would later recall being "a little bit shocked" at the sex lives of some of his African friends. But among Zaire's rural peasants, Carael said, he saw little evidence of either prostitution or extramarital sex. "Sexual behavior is deeply rooted in culture. And in Africa especially, there's a lot of taboos about sexuality," Carael said in an interview years later.[13]

> It is not easy, and certainly not the role of foreigners, to come out
> with a model of sexual behavior. . . . How can you change that?
> It's not just a question of saying, "Oh, you should stick to one
> partner." So the idea that you can, you know, drastically modify
> sexual culture, or determinants of sexual behavior, was not mine
> from the start.

The morning after Stoneburner and Carael completed their all-nighter in Kampala, the weary scientists emerged from their hotel and made their way to the nearby conference center where the international meeting on sexually transmitted diseases was in its final hours. The tension between Stoneburner and Carael did not break into the open immediately. They took turns making presentations, and they each agreed on several consensus facts: As HIV transmission in Uganda declined, condom use rose. Sex outside of marriage or other steady monogamous relationships fell. And young Ugandans were waiting until a later age to have sex for the first time. But the presentations obscured deepening divisions over which of these three developments mattered most in reversing AIDS in Uganda. Was it rising condom use? Declining numbers of sex partners? The delay in sexual activity among youths? Or was it some combination of the three? The answers would help determine which strategies should be emphasized—and which would get the largest share of what eventually became billions of dollars of AIDS prevention funding—in the years to come. Perhaps more deeply still, these questions would shape how Africans would be portrayed amid this devastating crisis. Had Ugandans made fundamental changes in their lives for the sake of the safety of themselves and their families? Or was dangerous sexual behavior so deeply ingrained that the only hope was, in the words of Museveni, "a thin piece of rubber"?

Carael was convinced that while all three factors played a role, there was clear evidence for the importance of rising condom use. The number of Ugandan men who used condoms, according to the two surveys, rose from 15 percent to 55 percent—a number that on the surface sounds impressive. In a 1997 article based on the same data, Carael and several coauthors described a "sharp increase" in condom use for both sexes, contrasting it with what they called a "slight decrease" in casual sex by young men. But these statistics masked deeper problems with Carael's theory: The 55 percent represented the number of Ugandan men who had *ever* used a condom, even once in their lives.[14] Yet nearly half of Ugandan men, and more than half of Ugandan women, had said they had *never* used a condom, even once in their lives. Condom use rates, meanwhile, were little different than in the many neighboring African nations where HIV's spread had not yet slowed down. Certainly the increase in Uganda's condom usage rates was not nearly as impressive as what had happened in places like Thailand, where the practice had become nearly universal in the brothels, where transmission was concentrated.

Stoneburner, meanwhile, remained skeptical that condom use at the level reported in Uganda could have produced meaningful declines in new infections. He believed instead that Ugandans, like gay New Yorkers, had responded to a dramatic threat with dramatic action: cutting down their overall numbers of sex partners. Stoneburner had detected a roughly 60 percent decline in casual sexual relationships—meaning ones outside of marriage or some other steady relationship—between 1989 and 1995.[15] The numbers of men reporting three or more casual partners fell even more steeply, from 15 to just 3 percent. Here was an analysis entirely in line with what President Museveni and Philly Lutaaya had been trying to do. Ugandans had sharply pared back their numbers of sexual relationships. Zero Grazing had worked.

For those determined to understand what had happened in Uganda, further revelations would eventually sharpen the picture even more. Stoneburner and Low-Beer later calculated that the historic drop in Uganda's HIV transmission happened quickly, between the late 1980s and the early 1990s. Condoms were hard to find in Uganda during those crucial years. Fewer than 1 percent of Ugandan women reported using condoms in a large national survey conducted in late 1988 and early 1989.[16] Museveni had banned condom promotion in 1991, and mass condom campaigns did not begin until several years after that. Even in 1994, after the drop in new infections was nearly complete, only about twenty million condoms were imported into Uganda, enough for every man to use three that year. Compared to this, Stoneburner's 60 percent decline in casual sex looked impressive. But this was a story that many within the growing universe of AIDS experts and activists were not ready to hear.

THE BEAT-UP

n July 1996, thousands of scientists, journalists, and activists gathered in Vancouver, on Canada's Pacific coast. It was the Eleventh International AIDS Conference, and by now such events had taken on their own distinctive rhythms and rituals.[1] Activists hung a giant banner proclaiming GREED = DEATH,[2] shouted at speakers they disliked, and tossed fake money into the air emblazoned with the names of pharmaceutical companies.[3] Scientists rushed from meeting to meeting in hopes of keeping up with the frenetic pace of discoveries reported over the five days of the conference. Journalists, meanwhile, tried to marshal this chaos into stories digestible for audiences that regarded AIDS as just one of many dramas playing out across a troubled planet.

Vancouver provided a coming-out party for a new United Nations agency, UNAIDS, which had formed that year in Geneva at the behest of rich donor nations tired of being pestered by a half dozen or more agencies seeking funding for HIV programs. The job of UNAIDS was to coordinate efforts by WHO, the World Bank, the United Nations Development Program, and other global agencies into a coherent, unified response. Grumbling started almost immediately about the expansive nature of its ambitions. UNAIDS sought to encompass both the scientific mission, of gathering and disseminating the best available information on the epidemic, and the political one, of keeping AIDS in the international spotlight and raising money to fund programs against it. But overall, the creation of the new agency raised hopes of a more energetic global response at a time when falling HIV rates in Thailand and Uganda suggested that prevention programs could work well in very different places. The founding executive director of the agency was Belgian scientist Peter Piot, whose appointment drew no discernible objection from any quarter.

At a time when conversations might have started brewing about alternative prevention strategies—focusing more on curbing numbers of sex partners, for example, or circumcising men—the Vancouver conference was dominated by exciting news coming from the world's laboratories. Activists for years had demanded major infusions of research dollars, and, at last, scientists had delivered a new class of drugs, called antiretrovirals, that appeared to be working when taken in the right combination. The medicines prevented HIV—which within the cosmology of viruses was called a "retrovirus" for the way it used RNA to make copies of itself—from reproducing. When older copies of HIV gradually died off they were not replaced by new ones. As the overall viral load in a victim declined, the immune system began restoring itself.

ARVs, as they soon became known, had limitations. They were unusually toxic by the standards of most medicines. They were wildly expensive. And HIV still managed to survive in its victims by hiding in places, such as lymph nodes, where the medicine was not able to eliminate it. But many hoped that the drug regimens eventually might become sophisticated enough to actually cure AIDS. Even if that never happened, ARVs were nothing short of miraculous compared to the treatments that were available before. Diarrhea eased. Fevers cooled. Weight returned. For many, life began anew. Some people started calling it the "Lazarus effect." Even the august *New York Times* got caught up in the excitement, running a story on its Sunday front page declaring, "The 11th international AIDS meeting opens here on Sunday with leading researchers' hopes as high as the sun-drenched mountains surrounding this Canadian seaport."[4] Amid the giddiness of Vancouver Piot used his public stage to pioneer what would become one of his recurring roles. He was the stern, goateed scold who spoke for the world's poor, the unseen billions for whom such heady progress seemed to be beside the point. Piot reminded the *Times* that 90 percent of those with HIV were too poor for aspirin, much less ARVs.

His statement anticipated the coming fight over whether the new class of medicines ever could reach poor nations with meager health budgets, weak medical systems, and many other serious diseases to combat. But Piot also tried to convince other parts of the world—and especially the ones with the financial and scientific capacity to lead the fight—that AIDS was not merely a battle for the mostly poor countries where most victims lived. Piot declared the epidemic to be a fight for all, not only for humanitarian reasons but also because, he said, the epidemic soon would explode across the earth. Piot began developing that message before Vancouver, but it first achieved lift-off as the conference convened. In a lead editorial in *Science* magazine, Piot warned

> In heavily affected countries in Africa and Asia, where one out of three urban adults may be infected, AIDS deaths among young and middle-aged adults—workers, managers, political leaders, and military personnel—are threatening health systems, economies, and national stability. With the current scale of global travel, the largely invisible, shifting, and expanding global epidemic of HIV makes the planet a more dangerous place for all.[5]

Other global health authorities, along with a growing number of politicians and celebrities, were making similar arguments at the time. But Piot's was more prominent and, arguably, effective. His message soon ricocheted across the world, courtesy of journalists eager to make sense of an epidemic whose complex realities continually challenged the desire for straightforward story lines. And Piot kept building that momentum in the months that followed. He campaigned north and south, east and west. At each stop the message was the same: AIDS is coming for you. The Chinese were in danger. So were the Indians, the Filipinos, the Vietnamese, and the Salvadorans. All needed to put AIDS at or near the top of their public health agendas. For each stop there were statistics—generated in increasingly sophisticated UNAIDS reports—highlighting the danger. In Ukraine there were some Black Sea towns whose HIV rates among heroin addicts in just a few years shot from less than 2 percent to more than 50 percent. In China HIV infection supposedly had risen tenfold in two years. In the Cambodian capital of Phnom Penh, rates among blood donors had gone up one hundredfold in just four years. India appeared to be experiencing such a surge in HIV that many feared it soon would have more infections than any nation in the world. Spain, meanwhile, was the European country with the highest infection rate. Piot demanded international action to make sure that travelers to Spain saw notices at border posts warning about AIDS.[6]

Activists seeking to spur action on other public health issues often have been accused of overstating their numbers. And Piot, while defending the rigor of the HIV estimates the agency generated under his stewardship, would later say, "My job was really to make sure AIDS was taken seriously."[7] Yet what eventually would create trouble for UNAIDS was that its mission was supposed to encompass both advocacy and science. During this era the agency often failed to lend the perspective that a global scientific authority should provide. While in the early 1980s Piot and some of his colleagues were pioneers in identifying the emerging danger of widespread heterosexual HIV outbreaks, by the late 1990s it was increasingly apparent that the epidemic wasn't exploding—and wasn't very likely to—in most regions. When reviewed carefully, what a broad accumulation of statistics from that time showed was that HIV was a total disaster in some places, and especially among the high-risk populations of prostitutes, injecting drug users, and men who had sex with other men.[8] But it was a much smaller problem in most of the world. In only a few countries was AIDS among the top health problems demanding a national government's attention.[9] In most places the problem was confined to particular groups—heroin addicts in Black Sea towns, gay men in Barcelona, or sex workers in Phnom Penh's brothels. These were communities that faced disproportionate risk but that might not notice vague highway billboards.[10] Big national responses that portrayed the danger as equal for everybody tended to obscure messages, not sharpen them.

This would become a problem in several regions where prevention messages grew vague about their target audiences. HIV spreads much faster and more extensively among men who have sex with other men—many of whom do not consider themselves gay, because they also have sex with women—than in the rest of the population. Yet prevention campaigns in much of Latin America and Asia tended to focus overwhelmingly on the risk of heterosexual transmission. Some men in India later told British researcher Jeremy Seabrook that they had anal sex with other men, in part, because they believed they were protecting themselves from HIV—which they viewed as resulting from sex with women.[11]

The statistics UNAIDS produced varied in their underlying elements—sometimes they applied to entire nations, sometimes regions, sometimes an individual city, and sometimes high-risk groups within just one of those geographic frames. But to nearly every country Piot visited he offered numbers that, in the absence of broader context, were alarming. The news coverage followed Piot's lead, with words like "exploding" and "skyrocketing" popping up in stories by news organizations that, on almost any other issue, applied a healthy dose of skepticism when it came to the assertions of public officials. Piot's *Science* editorial before the Vancouver conference was a good example of how he employed statistics. It may have been true, as he wrote, that in some of the most heavily affected African cities—Kampala, Harare, Lusaka—up to one in three adults had HIV.[12] But the same wasn't remotely true for anywhere in Asia.[13] Nor was it true in most of Africa. In vast swaths of even that hardest-hit continent HIV rates were much lower, and in some places all but undetectable. Yet Piot wrote vaguely of the "heavily affected countries in Africa and Asia, where one out of three urban adults may be infected."

The scariest numbers, meanwhile, were broad extrapolations of narrow statistics. Some of the dramatic increases resulted mainly from expanded tracking systems. The estimated HIV rate in India was less than 1 percent of adults in that massive nation. The raw numbers in a country so large still

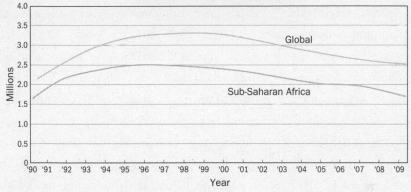

The number of new HIV infections peaked in the mid to late 1990s.

amounted to many people, but dozens of other maladies, many of them also potentially fatal, were much more common. The rate in Spain, at half of 1 percent, never came close to the levels found in most of Africa.

Piot later defended the handling of data analysis at UNAIDS under his tenure, saying the agency relied on the best available science at a time when existing monitoring systems for the epidemic were rudimentary. But there were skeptics, including the main architect of the previous WHO estimates of HIV rates, American epidemiologist James Chin, who was appalled at the extensive reworking of HIV estimates under Piot. Chin would later single out a quote from a 1997 conference in Manila when he recalled Piot saying, "HIV will cut through Asian populations like a hot knife through cold butter!" as evidence that he either didn't understand the way the epidemic spread or was hyping it for political purposes.[14]

The ironic footnote to this first major UNAIDS initiative was clear only in retrospect: As Piot was warning about HIV's looming explosion, the opposite was actually happening. All outbreaks of infectious disease—cholera, SARS, bubonic plague, flu, Ebola—follow what scientists call an "epidemic curve," in which the rate of new infection rises, then falls. That's because the most vulnerable members of a community get sick first, and each subsequent wave of infection has more trouble finding new victims. That's why in epidemics, generally, what goes up must come down—even in the absence of useful human interventions.

Most experts today, including those at UNAIDS, believe that its spread peaked somewhere in the second half of the 1990s.[15] There was no way anyone could have known that then, or for several years after. But it's now clear that when Piot was delivering his warnings, the epidemic curve already was peaking and the pace of new infections was about to turn downward.[16] That didn't mean that AIDS wasn't still a problem of staggering proportions. It was, and still is. AIDS deaths, which lag behind initial infections by about a decade, would continue to rise into the first years of the twenty-first century. But the pace of HIV's spread wasn't rising inexorably everywhere. Instead, it was digging in precisely where it had already caused the most damage— among gay men, intravenous drug users, prostitutes and their sex partners, and Africans living in that mysterious stretch of the continent that some scientists called the AIDS Belt.

Two narratives about the path of the epidemic through the world emerged soon after the discovery of AIDS in Africa. One held that HIV was rampant in certain areas because it had arrived there first, and it would eventually spread to most of the rest of the world in roughly equal measure. Many diseases moved this way: Bubonic plague, the Spanish flu, smallpox. They were so contagious that human density alone was enough to enable them to infect large percentages of nearly any population they encountered. But the other narrative relied on the understanding that many diseases—and most sexually transmitted ones fall into this category—move faster or slower depending on behavior, culture, or other factors. Rates of syphilis and gonorrhea, for example, long had been higher in much of urban Africa than almost anywhere in Europe. If HIV was following a similar pattern, that meant that the factors controlling the spread of HIV theoretically could be identified and addressed. That was a reason for hope.

Researchers trying to understand the spread of a sexually transmitted disease turned first to sexual behavior, and a blunt focus on that was evident in the initial waves of research into HIV's spread. The first CDC notice about the mysterious immune disorder noted that the men were gay, and examinations of sexual behavior were included in the early work of Piot, Serwadda, and some other scientists. But by the 1990s political sensitivities were on the rise, and discussions of variations in this area gradually acquired a taboo status. British researcher John Cleland, who oversaw a massive, groundbreaking survey into sexual behavior worldwide for the WHO in the early part of that decade, remembered that no subject was more radioactive than African sexual culture. "I do remember the tension, the unstated assumption that the one thing you couldn't do is point the finger at African sexual behavior, essentially from political correctness," he recalled in a later interview.[17] Cleland added, "I'm sure [UN] staff would have put red lines through any section that suggested that African behavior is radically different than the rest of the world."

The book Cleland's team produced, *Sexual Behaviour and AIDS in the Developing World*, published in 1995, bears the mark of that sensitivity.[18] It presented powerful data pointing to a fundamental difference in sexual behavior between Africa and the rest of the world, but it seemed to barely notice its own revelations. The research actually undermined long-standing Western suspicions that Africans simply had more sex, or more sex partners, over their lifetimes. On these measures, as well as on those for casual and premarital sex, Africans overall reported behavior that was close to global averages. And the age of the first sexual encounter was higher for Africans than in some other places. But what was striking were the percentages of men and women who reported having more than one regular sex partner at a time—a wife and a girlfriend, for example, or a husband and a boyfriend. These were an order of magnitude beyond anywhere else.

Tellingly, African women knew that they did not have exclusive sexual province over their men, suggesting this was an accepted part of the culture, not an accumulation of individual aberrations. In Ivory Coast nearly half of all women believed that their male partners regularly slept with at least one other woman. (Many men confirmed this practice in their own survey answers.) The data was sketchier for the behavior of the women themselves, but in the Zambian capital of Lusaka, for example, 11 percent of women said they maintained more than one regular sexual relationship.[19] And in the tiny mountain kingdom of Lesotho, a chunk of rugged highlands surrounded on all sides by South Africa, 39 percent of women said they had more than one regular sex partner. Among men there, it was 55 percent. The nation would eventually have the world's third-highest HIV rate, slightly behind only Swaziland and Botswana. The percentages of such multiple relationships elsewhere in the world were small by comparison. In Sri Lanka 2 percent of men and 1 percent of women reported more than one regular sex partner. In Thailand, it was 3 percent of men and fewer than 1 percent of women. Even in legendarily sensual Rio de Janeiro, with its sand, surf, and tiny bikinis, 7 percent of men and fewer than 1 percent of women reported having more than one regular partner at a time. For keeping more than one ongoing relationship, no group came close to Africans in Cleland's WHO surveys. A number of subsequent studies would confirm this general pattern.[20] Members of the original WHO team, a group that included Michel Carael, were startled by their findings. In a later interview he called the differences in multiple relationships "spectacular."[21]

Yet in a chapter Carael authored for the book produced by the WHO team, he moved blandly through various elements of the survey results before concluding: "In this first cross-cultural attempt to examine aspects of sexual lifestyles, it has been consistently shown that broad generalizations about one particular population or region are misleading."[22]

Another strand of research from the era did take crucial new steps toward understanding why such variations in sexual culture proved so powerful in explaining how HIV spread. American social scientist Martina Morris was investigating the power of a kind of behavior called "concurrency"—meaning having more than one sexual relationship at a time.[23]

In one large survey in Rakai, one of the most heavily affected parts of Uganda, a majority of men and women over the age of thirty reported having had at least one concurrent relationship. More than 90 percent of these relationships were long term, with the period of overlap averaging more than two years. These weren't Western-style extramarital flings that flamed briefly,

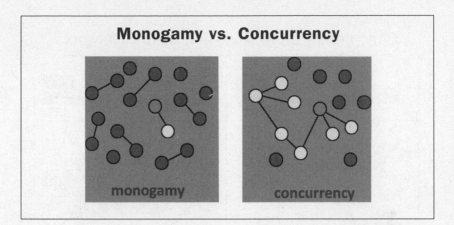

Monogamy vs. Concurrency

monogamy

concurrency

Serial monogamy traps sexually transmitted diseases within a single relationship for a period of time. Concurrency creates a network that can greatly accelerate the spread of HIV.

then died out. This was a modern variation on polygamy operating below a surface pretense of monogamy. And only about 10 percent of the people maintaining such overlapping relationships reported using condoms regularly. "It is not the number of sex partners that appears to be at work here; rather, the network structure that links them together," Morris explained at a conference around that time. "The network structure created by this pattern of concurrent partnerships may help to account for the speed and pervasive spread of HIV and other STDs in this population."[24]

The consequences for understanding the transmission of HIV could hardly have been more profound. Ugandans didn't have to become permanently monogamous, or have less sex overall. The potential for the virus to spread through the community would be dramatically reduced if people had only one partner at a time. Or—and here is the crucial insight that Morris's work made so clear—Ugandans as a whole just needed to average fewer sex partners at a time. Morris sharpened the picture over the years with a series of increasingly sophisticated computer models; they made clear that even small shifts toward fewer concurrent partners, if they happened broadly enough, could make a big difference in HIV's spread. When Morris modeled a society whose members averaged 1.86 sex partners at a time, the resulting web of relationships looked like the map of a major city bustling with traffic, with

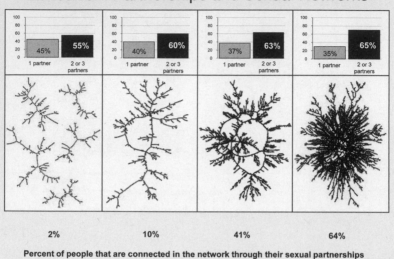

Concurrent Partnerships and Sexual Networks

2%   10%   41%   64%

Percent of people that are connected in the network through their sexual partnerships

Modeling suggests that even modest changes in levels of concurrent partnerships in a society can have large impacts on the sexual networks that spread HIV.

lines darting and swerving in dense tangles of interaction. But when Morris throttled down the average number of ongoing sexual relationships slightly the model showed the tangles gradually breaking apart and becoming far more resistant to HIV. Here, Morris found, was a small change with big consequences for saving lives. As Yoweri Museveni and Philly Lutaaya demanded that fellow Ugandans change their sexual behavior, their nation already was teetering near a tipping point, in which relatively modest changes by millions of people could cause major relief across the society. In *The Invisible Cure*, science writer Helen Epstein compared this dramatic precipice to a "phase transition" in nature in which even a minor temperature shift could cause water to become ice or ice to melt back into water.

While it was clear that variations in sexual behavior were part of the answer, they were, by themselves, not enough to produce the pattern of HIV infection seen across the world, or even across Africa. A husband-wife team of

Australian anthropologists, John and Pat Caldwell, had grown interested in AIDS after spending decades studying sexual behavior and fertility in Africa. Based on that experience, they were particularly puzzled that many places with high rates of sexually transmitted infections, such as Nigeria in West Africa, did not have HIV outbreaks on the scale of Uganda or other parts of East Africa. Given that most sexually transmitted diseases spread through similar behavior, it seemed that HIV should move across the continent in patterns resembling those of syphilis or gonorrhea. But it didn't.[25] The Caldwells figured there had to be another factor working to slow HIV in some places while accelerating it in the "AIDS Belt."

Being anthropologists, the Caldwells looked for answers in cultural practices. Did societies that practiced ritual scarification—slashes or cuts on arms, legs, or faces—have more HIV, perhaps from sharing cutting tools? What about societies that were matrilineal? Did the greater degree of economic and sexual freedom for women in these societies make HIV move faster? Or slower? Could societies in which men tended to marry at a later age subtly encourage prostitution, and hence infection? Or what about cultures that prohibited sex for many months or years after the birth of babies? Perhaps that led to more extramarital sexual relationships and, ultimately, more HIV? Any one of the Caldwells' possible explanations had a rough kind of plausibility. But when they compared maps of these cultural practices to the regions where AIDS was the worst, none fit. Then they turned to the map of circumcision patterns and HIV rates that had been produced by John Bongaarts and Priscilla Reining.[26] To an amazing degree, the noncircumcising regions overlapped with the AIDS Belt.

With this realization the Caldwells had essentially caught up with Reining and others who had been documenting this connection for years. But the Caldwells then made one more leap in their analysis, concluding that circumcision was a key factor, but one that did not operate in isolation. Many societies worldwide, after all, did not circumcise but had little HIV. That included most of Europe and Asia. The Caldwells came to believe that what was happening in parts of Africa resulted from a mixture of factors—low rates of circumcision combined with high rates of multiple sexual relationships—that was uniquely suited to sparking and sustaining a major AIDS epidemic. In March 1996, the Caldwells wrote in the popular magazine *Scientific American*:

> Because the virus progresses through different segments of the population at markedly distinct rates, the best chances for combating AIDS everywhere lie in targeting education and prevention programs at high-risk groups—homosexuals, prostitutes and their clients, intravenous drug users, and promiscuous men and women. In sub-Saharan Africa, circumcision could be offered as a reinforcement of other protective measures. Universal male circumcision, which is likely to prove unacceptable, would nevertheless probably reduce the level of infection in the AIDS belt to the much lower numbers seen in the rest of sub-Saharan Africa.[27]

Through the Caldwells' academic language it was now possible for the first time to see a coherent description of what distinguished AIDS in the hardest-hit parts of Africa from nearly everywhere else. It took a certain kind of sexual behavior plus a preponderance of uncircumcised men for AIDS to go from a bad problem to an absolute catastrophe. The dynamics would prove different in places where intravenous drug use or sex between men drove the epidemic forward. But in places where AIDS was fundamentally a disease contracted and spread through heterosexual relationships, both relatively high rates of multiple sex partners and low rates of circumcision needed to exist for a major, sustained epidemic to take hold. Take away either element of the tinderbox and the spread of HIV would slow. Take away both, and the flame of the epidemic would falter and eventually fizzle out.

But the conversation in global public health in that era wasn't about sexual culture or circumcision. It was about expensive new drugs that had finally given sick people in wealthy countries hope, and the looming battle over whether this medicine would ever reach the poorer corners of the planet where HIV was rampant. Not surprisingly, this fueled the instinct among many in global public health to view the fight against AIDS as fundamentally one about resources. And getting resources from a reluctant world—one that rarely seemed to economize when it came to buying tanks or investing in new airports—meant playing politics in a more aggressive way.

Among the energetic new generation of health officials drawn to UNAIDS in the mid-1990s was Elizabeth Pisani, a journalist-turned-epidemiologist who helped write the agency's annual global reports that became benchmarks for describing the spread of the epidemic. Emblazoned with red ribbons and filled with glossy images of people suffering from—or, in a few cases, triumphing over—AIDS, the global reports were regularly quoted by policy makers, researchers, and journalists. And for several years, Pisani—who was smart and ribald and had the sense of consuming mission that fueled the early days at UNAIDS—often had her fingers on the keyboard. In her 2008 book, *The Wisdom of Whores*, Pisani wrote that UN officials had worked to convince wealthy nations that AIDS epidemics would spill over their borders, putting everyone in peril. Pisani called this "beating it up," an expression in British journalism to describe the act of hyping stories: "The first thing we needed to do was get more money for HIV prevention. That meant convincing rich countries that they should worry about AIDS in poor countries."[28]

This was made easier because the crude tracking tools available at the time made it hard to know the extent of HIV in many countries, or of its future spread. Shifting the date of first infection on an epidemiological model, for example, could make tremendous differences in predictions about its path. And after years in which the WHO's technical staff had offered cautious estimates of HIV rates, the new statistical gatekeepers in Geneva began revising numbers upward. Piot has dismissed Pisani's account as a "caricature" of the process of generating HIV estimates at UNAIDS.[29] And many others in the AIDS community believed, given the uncertainty of the epidemic's trajectory, that it was better to run the risk of overestimating than underestimating—and of failing to rouse a complacent world. Yet Pisani's account suggests that UNAIDS knowingly shaped data for maximum political impact. One common tactic, she wrote, was to describe epidemics not in terms of the total number of people infected but of changes in the infection rate. That made places with small but growing epidemics appear in particular danger, even when compared to those living among larger but more stable ones. Another was to imply that a single study of infection rates among a narrow group—for example, women with other sexually transmitted diseases in a particularly high-risk community in India—represented the likely future for all women in that vast, diverse nation.[30] Pisani called such plays "sleight of hand."

She and her colleagues believed—perhaps correctly—that the funding would stay modest as long as gay men, drug addicts, or men cheating on their wives were the public face of the epidemic. And so the stories of suffering women and children, the classic "innocent victims," became staples of UNAIDS reports—and so too in much of the journalism and policy work inspired by them. The other UN agencies that were supposed to be working together under the umbrella of UNAIDS also had powerful institutional incentives in how the epidemic was portrayed. More alarm, and more politically appealing victims, meant more programs, more staff, more money. (Pisani called AIDS a "honeypot" for UN agencies.[31]) If AIDS was not just a public health problem but also a substantial development issue, the World Bank and the United Nations Development Program had a claim to rival the WHO's original one. If its list of victims included children, so did UNICEF. Soon UNAIDS became like a snowball tumbling downhill. The original six partner agencies grew to ten as those overseeing drug abuse, refugees, hunger, and labor relations joined in with their own programs targeting AIDS. So at the same time that some researchers were beginning to sharpen the scientific understanding of the essential nature of HIV—how it spread, who got it, and how it could be most effectively prevented—UNAIDS and its partners were blurring the picture again. Pisani offered her revelations not as mea culpas but as explanations. "We weren't making anything up. But once we got the numbers, we were certainly presenting them in their worst light. We did it consciously. I think all of us at that time thought that the beat-ups were more than justified; they were necessary."

Viewed from the perspective of the time, it's easy to sympathize with Pisani's logic. But some outside researchers came to believe that officials in

Geneva had their thumbs on the scales. And in the same era—as evidence was quietly growing about the importance of male circumcision and multiple sexual partnerships—UNAIDS and other global health authorities continued to focus overwhelmingly on condom promotion in their prevention strategies. The lessons of Uganda went largely unheeded.[32] Circumcision had all but vanished from the radar.

THINGS JUST FELL APART

Christmas Day 1997 broke warm and holy on Homa Mountain as it sat high above the gray vastness of Lake Victoria. Not far from here was the birthplace of Barack Obama, Sr., the Kenyan father of the unlikeliest of American politicians, a man who eleven years later would send his father's homeland into celebration by becoming president of the United States. But on this day there was little hint of those happier days to come. Instead there were signs of a great tragedy gathering force here in Luoland, the western Kenyan home of the Luo ethnic group. It was at once plain to see—in the thinning faces of its victims, the empty eyes of the widows and children left behind—and yet as hidden as secret love. Those who spoke openly of the new sickness blamed prostitutes, witches, outsiders of all sorts. Few suspected that soon it would spread far beyond the margins of a proud society, that it would undermine even the best of families, dragging their private lives out of the darkness of their mud huts and into the bright Kenyan sunlight.

It was communities such as this that would pay the highest price for the world's failure to slow the spread of HIV. The AIDS epidemic almost certainly would have found this place just over the border from Uganda. But swifter, smarter action might have made a difference for the family of Audi

Waga, whose ten sons gathered this Christmas Day under an *orunda* tree, in Kakdhimu village, on the south face of Homa Mountain. Their huts shared some fifty acres of rust-colored hillside, and though Audi himself was long dead, four of his five wives were nearby to prepare the day's feast. A goat had been slaughtered for stew and steaks. The tilapia brought from the lake fried in sizzling vats of oil. The women also toiled over boiling pots of *ugali*, the Kenyan cornmeal mush that they would form into a mound so solid it could be sliced with knives. As the meal gradually came together over many hours, the boys and the younger of the girls—those who weren't already apprenticing for marriage by helping with the cooking—played hide-and-seek or soccer, the idle pastimes of childhood.

With Audi gone the patriarch was now the eldest son from his first marriage, the stern-faced Joseph Okendi. He was a school headmaster, an elder of the village, and a founder of its Seventh-Day Adventist Church. As a Christian, Joseph Okendi had only one wife. As a Seventh-Day Adventist, he was plain in his dress, reticent in his manner, and apparently restrained in his appetites, shunning alcohol even on festive occasions such as this. He and his brothers were more than mere neighbors and friends. Together they formed the backbone of a family that moved almost as one, supporting each other and all of their relations. In this world, each member had a secure place within a clear—if sharply circumscribed—future. And on this day, as on every Christmas Day as long as anyone here could remember, they talked and joked. Information was shared, family dramas were resolved.

But there was one problem for which there was no obvious solution. One of the brothers, Dickens, looked as if he was wasting away. He had a persistent cough and strange rashes. Village gossip had fixed on a likely cause. Dickens, who had two wives, reportedly had also had an extramarital affair. Rumors were that this woman died of the strange new disease. And now Dickens seemed to be on that same awful road. Mystery and suspicion still accompanied any mention of AIDS. But one thing already was clear to those in Kakdhimu village: AIDS was the product of sex and, therefore, almost certainly of sin. Many regarded it as God's judgment, and many had reason to fear that they too might soon be judged.

As the wives, grandmothers, and aunts began assembling the feast, some of the next generation of boys gathered nearby to dine with their fathers and uncles. One boy, at age nineteen, was already larger than the others. He was Collins Omondi Okendi, the eldest son of Joseph Okendi and hence likely to become the patriarch of the family's next generation. The position would give him responsibility not only for the affairs of his brothers but for all of their wives and children as well. Even his uncles and aunts, and his own mother and grandmother, would have to answer to the will of Collins whenever Joseph Okendi died.

Collins, who had a round face, high cheekbones, and dimples that appeared whenever he flashed his warm smile, didn't entirely believe the rumors about Dickens having AIDS.[1] But when Collins looked at his uncle, he had to admit that the man looked like death in an outsized suit. Collins was not the only one worried. Yet as the feasting continued for hours, as the sun slipped behind the hills and lanterns flamed to life, it was possible to imagine that nothing ever really would change here. With darkness taking hold, the women began clearing the table of fish bones and dirty plates, and the men walked off in groups, happily escorting one another to their huts under the Kenyan sky. Through the terrible years that followed, Collins would recall this as his family's last perfect moment.

The Luo who settled the eastern and southern shores of Lake Victoria were Nilotic, descended not from the Bantu speakers who migrated from West Africa but from another group that moved downward from the Nile River Basin of South Sudan. Where the Bantu were settled farmers, the Luo were fishermen and herdsmen. They measured wealth in cattle owned, wives married, and sons produced. They did not circumcise their boys, though there were exceptions, including Collins. His father arranged for the procedure because Collins often swam and bathed with children of other ethnic groups, and otherwise might have felt out of place. The general lack of the practice helped define Luos as ethnically distinct from the mainly Bantu-speaking groups that dominated Kenya. The Luo also had their own painful coming-of-age ritual: pulling six lower front teeth from boys and girls on the cusp of adulthood.[2]

For several centuries the Luo lived in villages such as Kakdhimu, following traditional ways still faintly visible in the Audi family's Christmas feast in 1997. The Luo were polygamous and patriarchal, meaning families were organized around a male elder whose authority was passed down through his line of sons. New wives moved from their homes to join their husbands on the family homestead. The children those marriages produced belonged to their father, not their mother. And a woman who left, or who was cast out for violating some social norm, departed alone. The rules of the system were plain to see in the geography of a Luo homestead. The main hut, typically inhabited by a man and his first wife, had a door facing downhill, toward the lake. The other wives had their own huts, in which the husband might stay for days or weeks at a time, and their doors also faced downhill. Each son as he approached adulthood built his own bachelor hut, called a *simba*, and its door faced uphill, in a sign of permanent respect toward his father. Young women were gradually married off in arrangements made not by themselves but by their parents, who received a bride price in the form of cattle or goats as their daughters moved away to be with the families of their new husbands.

There were rules governing sexual behavior, which had both practical functions—producing children and expanding the family's wealth—and ritual ones.[3] So essential was the production of babies that when a man failed to impregnate his wife over the course of years, relatives quietly would arrange for another man, often his own brother, to discreetly do the job. The village encouraged men who successfully sired children and acquired a degree of wealth to take more wives and father more children. Most major moments of Luo life also had related sexual rituals. New plantings and harvests had to be accompanied by intercourse, and the patriarch and his first wife were supposed to perform this act first, before the junior brothers could follow with their own wives. Weddings and funerals carried sexual duties as well, and a new widow needed to be "cleansed" of foul spirits through sex with a designated man, or she faced permanent shunning by her late husband's family as well as her own. These ritual sex acts were complete only if the man's semen entered the womb of the woman—a traditional belief that, as the AIDS epidemic took hold, made condoms more difficult for many Luo to accept.

"Some Luos say 'How do you eat bananas without peeling them?'" explained Elizaphan Ager Kirowo, a retired history teacher and member of the Luo Council of Elders. "When you are wearing a condom, you are doing nothing."

Sexual experimentation was allowed before marriage. A boy could go so far as to rub his penis between the thighs of his girlfriend; ejaculation was permitted so long as there was no penetration of her vagina.[4] That singular act was saved for the wedding day, and for her eventual husband. Only between a man and his wives was semen supposed to be exchanged. Elders had once explained such things to young Luos at evening classes given to adolescent boys and girls. The boys gathered together at a grandfather's hut, called the *duol*, and girls gathered with grandmothers at a separate hut called the *siwindhe*. Village elders overseeing these classes explained the rules and responsibilities of adulthood, and made it clear that all Luos had a duty to procreate, to expand their families and the village. Virginity was expected of both bride and groom, although not, of course, as a man took his second or third wife.

Wedding rituals themselves were elaborate affairs lasting weeks or months of intermittent ceremony, feasting and gifts. After a man negotiated a bride price with a young woman's family, he and his brothers led her—by force if necessary—to his family's home. A certain degree of resistance from the new bride was expected before she yielded her virginity in what was essentially a public act of deflowering, witnessed by a small, designated group of close relations. The blood that flowed was then smeared on a broken piece of crockery and given to the bride's sisters. They were then supposed to sing triumphantly all the way back to their home village, carrying this visible evidence of the bride's virginity, secure in the knowledge that the family's reputation had been upheld.[5]

That moment amounted to the great divide in the life of female Luos in precolonial Kenya. Before losing her virginity to her husband, a Luo woman was the responsibility of her father or her senior brother. Afterward she was the province of her husband. And if a woman's husband died, the family would arrange for her to have a new one, often a brother or cousin of the man who had passed away, who would inherit her and her children. There was no room in this society for single women or divorcees or widows who took care of themselves. Women had no more standing than children, and less stature within the village than their own grown sons.

This system lacked many of the basic freedoms and rights that most Kenyan women today regard as essential. Rivalries often developed among co-wives as well. The Luo word for "co-wife" is *nyieka*; it also means "jealous."[6] But the system, despite its inherent injustices, offered an abundance of companionship within an ever-expanding universe of sisters, sisters-in-law, aunts, and co-wives. Together they shared the daily burdens of raising children, tending crops, and cooking for their families. The system also offered Luo men and women security in a world without police, schools, or a formal welfare system. Only the most unimaginable of disasters could destroy a sprawling Luo family.

Under colonialism, Kenya belonged to the British, just as Congo belonged to the Belgians and Cameroon to the Germans.[7] These colonies were about more than prestige. They were supposed to deliver wealth to businesses and government coffers back home. That meant somehow harnessing a profoundly foreign land—vast swaths of Kenyan countryside dotted with tiny villages connected by rivers or footpaths—into a functioning economic system. Transportation was the first imperative. And the British decided that the solution was to build a railway all the way from the Indian Ocean to Lake Victoria, 450 miles away. When it reached its western terminus in Kisumu in 1901, the rail line brought a pulsing artery of the world's most powerful empire straight into Luoland. The final train station, where goods and people transferred from railcars to waiting steamships, came within about twenty miles by boat from Homa Mountain. This gradually pulled Luoland, previously remote from European imperial power, much more tightly into the economic and political orbit of the colonial capital of Nairobi. On that rail-road came government administrators, businessmen, and missionaries. Just as the European powers sliced up Africa into zones of political influence, var-ious Christian denominations had zones of spiritual influence on the conti-nent. The Seventh-Day Adventist Church developed missions in the part of Luoland that included Homa Mountain.

In 1906, a Canadian, a German, and a Malawian schooled in London established the church's first outpost just a few miles from Kakdhimu village. On 320 acres of land bought for $244, they built a stone mission house, a carpentry shop, and a school.[8] In a few years they had hundreds of attendees, who gathered on Saturdays under Seventh-Day Adventist doctrine decreeing this as the proper day of worship. The missionaries also began converting the Luo language into written form for the first time, and they brought the re-gion its first printing press as well, all for better spreading word of Christian salvation to this corner of Africa. The Luo the missionaries first encountered had no Western medicines. They had no schoolhouses. They had no Bibles, nor any written document to guide their religious or ritual lives. So the Ad-ventists began providing education, health care, and the first stirrings of modern economic development. These—combined with the highly devel-oped sense of hospitality that is a cultural bedrock in Luoland and much of Africa—helped early missionaries become welcome fixtures of Luo commu-nities.

Canadian Arthur Carscallen, who was among the trio of original Adven-tists in the region, reported in 1910:

> Less than four years have passed since we found this place lying in heathen darkness, the people never having had a word spoken to them about the gospel. Not one word of their own language had been reduced to writing. Now several have expressed a desire to become Christians, and have given up the old ways and cus-toms; and many can read and write in their own language. The future seems bright with promise.[9]

The missionaries baptized their first group of sixteen Luos the following year; a year after that, in 1912, there were twenty-four. World War I interrupted the mission's growth, and its facilities were looted, but the Adventists were back in strength in the 1920s. Among the first elements of Luo culture to disappear were the evening classes that had prepared them for marriage and adulthood. As the missionaries gradually reshaped Luo family life, the priorities of European capitalism reshaped Luo economic life. British colonial authorities imposed hut taxes that had to be paid in hard currency. Cash wasn't a major feature of precolonial Kenya's rural villages. But suddenly it was essential. Luo now had to grow cash crops such as cotton in fields that once grew corn. Or Luo sons had to venture far from home to work in the new cities or on colonial plantations, where tea bushes and sugar cane needed tending by thousands of hands. These crops offered little in the way of nutritional value for hungry Africans, but they offered plenty of commercial value on international markets. Under this new economic regime the soil was Kenyan, the sweat was Kenyan, but the profits flowed to bank accounts in faraway London.

These new economic forces gradually drained life away from Luo homesteads. Some young men who traveled for jobs returned only for holidays, or not at all. Some young women found they too could earn cash and a rough kind of freedom selling their bodies to men who spent months at a time away from home. The determination of missionaries to end polygamy fueled this, as it had in Congo and elsewhere in Africa. When a polygamous husband renounced his second or third wives in obedience to Christian doctrine, the prospects for these new ex-wives to remarry were severely limited. So was their ability to care for their children. As more and more Luo found work far from their home villages—three out of four men had worked somewhere else by the 1970s—Luo daughters increasingly began migrating along the same colonial pathways followed by Luo sons.[10]

There were other signs that monogamy was not working out as the missionaries had preached it. A man born of a polygamous family often did not lose his appetite for multiple women, even if he only had one official wife. The mobility allowed by the new roads and railways created many opportunities for extramarital sex. In a faint cultural echo of the cows that a husband had to pay for his new bride, intercourse became a new kind of currency as the ancient rules of the homestead dissolved. Some women offered sex in exchange for transport, rent, or access to the best catch that fishermen pulled from Lake Victoria. The bars and boardinghouses of market towns that grew along the new transit routes, meanwhile, offered plentiful options for men no longer inclined to take on the expense and responsibilities of a new wife. The Luo called these extramarital affairs *miel loka*. The translation was: dancing across the river.

Though Collins Okendi idealized the Christmas feast of 1997 as the end of a glorious traditional era in his family, the old Luo rules already were loosening when his grandfather founded his homestead on the side of Homa Mountain. Though Audi would have five wives—in the words of Collins, "My grandfather died a pagan"—the next generation was adopting Western ways and hewing closer to the moral instructions of the Adventists. Most of Audi's sons joined the church. Several had only one wife, though not, as it turned out, only one sex partner. These shifting sexual roles made the children and grandchildren of Audi Waga much more vulnerable to HIV than his own grandfather, or their wives, would have been.

The virus moved through Luo communities with unusual speed for another reason as well. Without Bantu initiation rituals Luo men had intact foreskins that HIV could infect relatively easily. When the epidemic peaked in Kenya, the adult HIV rate reached roughly 10 percent. Among the Luo, the only major ethnic group not to circumcise, it was about 20 percent, meaning one in five adults had HIV at any given time.[11] In 1997 alone a government district administrator reported issuing an unprecedented three hundred burial permits in a Luo community of just two thousand people.[12] Village funerals, once an imperative for all relations and neighbors to attend, began coming so quickly in Luo villages that families divided themselves up into several teams that took turns attending services. Only in that way could they make certain that each funeral included a respectable number of mourners.

Joseph Okendi used to warn his children about the new disease that was moving through Luoland, and already had killed some distant cousins. But it remained a largely abstract concern, perhaps nothing more than a scare tactic to keep restless teenagers in line. "You guys are going to die of AIDS," Joseph Okendi warned in his church-elder voice. "Be close to God."

Yet when Joseph Okendi's own brother, Dickens, became gravely ill, nobody confronted the issue openly. Collins saw his uncle Dickens two days before his death in January 1998, just a few weeks after that family Christmas feast. During the visit Dickens was so weak that he couldn't eat, couldn't walk, couldn't use the bathroom without assistance. Yet he never admitted the obvious—that he had AIDS. When Dickens died few accepted the likelihood that he had gotten the disease through the long-rumored dalliance outside his marriage. Instead, blame within Kakdhimu village focused on the supposed failings of his wife. Much of the family, her natural source of support at the moment she became a widow and the sole parent of four children, blamed her for the disease that killed her husband.

Such easy assumptions became harder to defend as AIDS worked its way through the family, and especially as the respected patriarch himself, Joseph Okendi, developed strange bumps on his chest. They were raised whitish sores, with red dots in the middle. Doctors called them herpes zoster, a classic opportunistic infection that could develop as AIDS wore down the immune systems of its victims. Joseph Okendi would cover as many as he could with a shirt, but Collins caught a glimpse of the sores one day and knew trouble was ahead. The pain was so intense during outbreaks that Joseph Okendi—headmaster, village elder, churchman—would sometimes break down and weep like a child. "Things just fell apart," Collins later recalled.

Collins fled to Nairobi for work, for school, for distance from his disintegrating family. But there was no way to avoid AIDS. His roommate in Nairobi, a beloved cousin named Ben, got sick while Collins was living there. Rumors spread of other illnesses within the family. And though neither of his parents would tell him the full truth, Collins knew that his father's condition was deteriorating too. When another uncle, Washington, died in a car accident in February 2001, Collins returned home. He was broke, as he often was, so he turned to one of his professors at the University of Nairobi for help. The man gave him bus fare, several old shirts, and fifty dollars in cash, and he made Collins promise he would use the money to buy new shoes for himself. But when Collins arrived back in Luoland, it was clear the money was needed elsewhere.

At the funeral Joseph Okendi had a ghostly look that quickly was becoming familiar to Collins. As he mingled among his relatives, one interrupted to break the terrible news. Collins's father had collapsed on his way to the toilet. There, just steps from the door, Collins found Joseph Okendi crumpled on the floor in his best suit, covered in his own vomit and diarrhea. As Collins helped his father to the bathroom he vowed to draw on every available resource to somehow get him medical care. If Collins couldn't save his father, he at least would make sure he died with some vestige of dignity.

The fifty dollars in his pocket was enough for a taxi to a hospital in Kisii, a large, nearby town where Joseph Okendi had spent years working as a teacher, but Collins wasn't sure he had enough left over for the medical bills. So he parked at the home of a relative and settled his father inside. Then Collins went back outside in search of financial help. This was, after all, the family's former home and a place where Joseph Okendi long had been an esteemed member of the community. As Collins moved through the town he ran into a group of three teachers, all fellow Luos and all former colleagues of Joseph Okendi. Collins asked for help and also, in hopes of bolstering his father's spirits, a visit. But rumor of AIDS had preceded them. The teachers politely declined the invitation as they handed Collins six hundred shillings—about eight dollars—and walked away.

"Be strong," one told Collins as they departed. "He will recover."

Collins was gradually accepting that his father would never recover, but he was determined to at least make Joseph Okendi look presentable. So in the washroom at the house Collins slowly pulled off his father's clothes and washed his frail, emaciated body. Collins began thinking of himself as a young man without a father, as the son of a man dead from a feared, taboo disease. And unlike when Dickens had died, Collins permitted himself no illusions about what brought the virus into the family. He once had spotted his father with another woman in a market town about twenty miles from Kakdhimu village. For all his stern warnings to his children, Joseph Okendi had failed to protect himself from HIV. He had danced across the river.

The final moments between father and son came within the walls of Kisii Hospital, as plastic tubes dripped fluid into Joseph Okendi's wasted veins. Though often remote, his tone turned intimate as he spoke to Collins. "You really have to be responsible," Joseph Okendi warned his son. "Your mother is my wife. That's my responsibility. But your brothers are your brothers. . . . Don't disappoint your brothers."

With that Collins understood that Joseph Okendi was handing over control of the family. Collins, the new patriarch, was just twenty-two years old. Joseph Okendi also made clear that it was time for Collins to return to Nairobi, and to his studies, which alone offered the possibility of a career that would allow him to support a family that sprawled far beyond just his brothers and sister. Soon cousins, as well as aunts and even his grandfather's surviving wives, would turn to Collins for help. "This place does not belong to you," Joseph Okendi said, sending Collins away from Luoland and back to Nairobi. "Just go."

Joseph Okendi then reached into his wallet and pulled out three hundred shillings, about four dollars, and gave it to his son. Collins tried to refuse the money but gradually accepted that he had to obey this final act of paternal love.

Two months later, Joseph Okendi died. When the news reached Collins in Nairobi he again rushed home for a funeral, and arrived to even more bad news. His mother, Eunice Okendi, once tall, plump, and formidable, looked shrunken in her dress. She had a persistent cough. And she was not the only one.

In his final years Joseph Okendi had built one of Kakdhimu village's largest homes, with solid mud walls, glass windows, and a front door made of wrought iron. It sat high on Homa Mountain, and on most days the sun twinkled off Lake Victoria in the distance. The home's floors were dirt, and there was no plumbing or electricity. But by the village's meager standards it was a palace, about three times the size of most of the huts there. This was to be the hard-earned retirement home for Joseph and Eunice Okendi in the twilight of lives well lived, after decades of service to school, to church, to family. But the death of Joseph Okendi ended that dream. Eunice remained in the much smaller, thatch-roofed hut next door. Without her husband's income, she quickly fell into poverty. The one-time headmaster's wife survived by cutting sisal plants and meticulously weaving their coarse fibers into ropes, an arduous task even for young hands. Two days of toil might produce ten ropes, which she could sell at thirty shillings each. If she sold them all, that might bring the equivalent of two dollars a day to feed herself and the two children still living at home.

As Christmas approached in 2001, the first after the death of Joseph Okendi, Collins stayed in Nairobi. For the fifth straight year, there was no family reunion. It took the generosity of a visiting cousin to give Eunice and her children enough food to eat on Christmas day. Collins was struggling as well, and though he now had a job at the university library and sent home what money he could, it was never enough. The pace of deaths in the family, meanwhile, was quickening. A five-year-old cousin, the first AIDS victim of a new generation, passed away. Then Collins's mother died, in August 2002. An aunt passed in 2003, followed by another of Joseph Okendi's brothers. Collins could feel himself hardening to the deaths. He and other surviving relatives began calling the funerals that now came several times a year "our family get-togethers." And still, the deaths kept coming, almost faster than Collins could keep count. Out of the ten sons and five daughters of Audi Waga, seven died of AIDS between the 1997 Christmas feast and 2010. So did thirteen of their spouses and three of their children, for a combined AIDS death toll over two generations of Collins Okendi's family of twenty-three. Collins knew of ten more relatives who had the disease but were surviving, for now, on antiretroviral drugs that had recently become available.

With each passing Christmas, as the sun rose on Homa Mountain, the family homestead was closer to ruin. Many of the mud homes simply melted back into the soil, washed away by the heavy rains that rolled in from the lake. Stones not much bigger than footballs marked the modest burial mounds, though these too, under Luo tradition, were supposed to disappear into the soil after a few years. Collins still visited occasionally, doling out cash and smiles to those who remained, but he rarely stayed for more than a few hours at a time. As much as he loved Kakdhimu village, and what was left of his family, it was not quite his home anymore. He was no longer sure he had one.

"This is a ruined family," Collins said as his massive shoulders slumped. "Gone."

BOOK III

THE HUMBLING

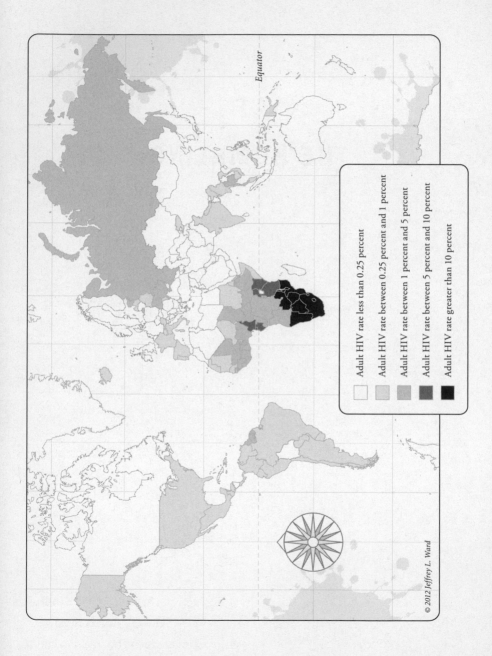

Equator

Adult HIV rate less than 0.25 percent

Adult HIV rate between 0.25 percent and 1 percent

Adult HIV rate between 1 percent and 5 percent

Adult HIV rate between 5 percent and 10 percent

Adult HIV rate greater than 10 percent

© 2012 Jeffrey L. Ward

X FACTOR

D aniel Halperin was an intellectually restless anthropologist at the University of California, Berkeley, when his mother made a fateful trip to the dentist's office. Halperin had led a peripatetic career, but by his midthirties he had turned his attention to exploring the epidemiological and cultural underpinnings of HIV. He had been reading ravenously on the subject for years but remained stumped by one of the most vexing questions of the epidemic: Why was it so severe in some places but not in others? Halperin was certain that variations in sexual cultures were a factor but, by themselves, they didn't seem to fully explain the curious patterns of infection seen around the world, and especially within Africa itself. He felt sure there was something else at work. Halperin started calling it "the X factor." And he buried himself in the massive stacks of Berkeley's Public Health Library, one of the world's finest collections of writing on human disease, in his quest for answers.

Then, in 1996, his mother came across the March issue of *Scientific American* during that visit to the dentist. It featured an intriguing article about AIDS in Africa. Particularly striking were a series of maps that compared rates of HIV with concentrations of male circumcision and other factors that might be related to the spread of the virus. She jotted down the name of the article, then suggested that her son look it up.

He soon did, and the revelations offered by anthopologists John and Pat Caldwell altered the trajectory of Halperin's professional life. He had long been a student of human communities, chatting up anyone willing to share their views about how their societies worked, what they valued and why. While there was something ultimately diffuse about much of Halperin's earlier career, the Caldwells helped change that by offering up, in the pages of the *Scientific American*, what he instantly felt sure was the elusive X factor underlying the spread of HIV in Africa. The biology made intuitive sense to him, and the map of the AIDS Belt and patterns of male circumcision seemed to fit together too neatly for them to be unrelated. He had now found a cause worthy of his energies.

Halperin had grown up in San Francisco, an energetic kid who from an early age displayed a knack for uncommonly intense inquiry. His friends were amazed, and occasionally irritated, at his ability to argue passionately on one side of a debate and then, a few days later, on the opposite one.

He started college at sixteen but was in no hurry to finish it. Halperin spent more than a year in his late teens wandering solo through South America and would return often to Latin America over the years. Between long periods of travel, work, and research in developing countries, he managed to publish several academic articles and earn three degrees at the University of California, Berkeley, including a doctorate in medical anthropology. Along the way Halperin took extended detours away from anything resembling a conventional career path. He studied jazz saxophone full-time for several years. For another stretch he was trained in dance-movement therapy at Hahnemann Medical College in Philadelphia and worked with deaf children and other patients at various psychiatric facilities. For another Halperin studied African-derived spirit possession rituals in Brazil and wrote a seven-hundred-page doctoral dissertation on them. The title, "Dancing at the Edge of Chaos," faintly mirrored his own life, in which periods of passionate devotion to a subject alternated with others in which he seemed to lose focus. To

pay his bills during these years of intellectual and artistic wandering, he occasionally worked as a cab driver.

This choice of jobs, by pure chance, kept Halperin in regular contact
with San Francisco's large, vibrant gay community at a time when it was one
of the centers of the American AIDS epidemic in the 1980s. In his taxi he
drove men, some visibly ill, out to clubs or bathhouses, and he would wait
outside—with the meter running—while his customers engaged in the kind
of easy sex that helped spread HIV. Halperin also took note as this fast-lane
scene slowed in the face of mounting deaths. As bathhouses fell out of fashion and gay men began seeking steadier relationships, he listened to their
stories, filing them away in his mind for a purpose he couldn't yet imagine.

One man left a particularly strong impression. Halperin was volunteering
at an AIDS hospice in Oakland, offering massages to victims of the disease
in the final months of their lives, when he developed a sense of kinship with
one whose body felt so fragile that Halperin could barely touch it. The man
was highly educated, Jewish, about his age. They could have been cousins.
Halperin tried to ease his pain, but the man felt brittle beneath his fingers,
as if his bones might snap. Not knowing what else to do, he eventually
switched to another technique that a spiritual healer in Brazil once showed
him but that Halperin had seldom tried before. Slowly he moved his hands
across the man's body about an inch from the skin but not touching it. This
seemed to offer some relief. But soon he sensed a kind of warmth emanating
from the man's sunken chest, just below his heart. There alone, Halperin laid
his hand lightly on the man's skin. A week later, at the next session, he heard
the bad news: "I have a tumor," the man said with a mixture of amazement
and resignation. "It's exactly where you put your hands."

Soon after, he was dead.

Friends would sometimes question Halperin on his choice of volunteer
activities. Fear of HIV was running especially high in the San Francisco Bay
Area. Casual flings had regained the taboo status they had before the Sexual
Revolution. Authorities were urging vigilance in using condoms for every
sexual act across a broad range of relationships. In that climate of rising
anxiety, massaging men with AIDS did not necessarily seem the safest move.

"Do you use gloves?" some friends asked.

"Not usually," Halperin would reply.

"Aren't you worried you'll get AIDS?"

He wasn't. In fact, Halperin suspected that the risks of HIV infection for a straight man such as himself were overblown. This was an era when some activists, concerned that victims of the epidemic would remain politically marginalized, were pursuing a concerted campaign of "degaying" AIDS, according to author Gabriel Rotello. The slogan of this effort was AIDS DOES NOT DISCRIMINATE. And it was remarkably effective given that few heterosexuals, aside from hemophiliacs and drug users, were getting HIV. The main exception was in the African American community, where higher rates of injecting drug use and incarceration spread the virus faster than within other ethnic groups in the United States; unprotected anal sex in prison was much more common than in the outside world. Stigmas against homosexuality in the black community also resulted in higher levels of bisexual activity, as African American men were more likely to hide their attractions to other men. And studies have increasingly suggested that higher rates of concurrency are also a cause of higher HIV rates among African American heterosexuals.[1] But overall, much of middle-class America faced little risk, even though waves of straight college students and others unlikely to contract HIV began overwhelming testing centers, terrified that having had oral sex without protection might lead to dying from AIDS.[2] In *Last Night in Paradise*, author and culture critic Katie Roiphe described how even public school teachers joined in the effort to instill fear in their students that "unsafe sex" could easily lead to illness and death. She quoted a sex-ed teacher of ninth graders at one suburban New Jersey school as saying, "I know AIDS isn't spreading into the mainstream heterosexual community at the rates they originally predicted. But I don't want the kids to know that. I don't want them to relax too much. I'd rather they were scared."[3]

Halperin's instinct was to push back, not against the idea that AIDS was serious—he knew it was—but against the idea that it was equally serious everywhere, and among all communities. The article by the Caldwells sharpened his sense that HIV struck where it did for particular, knowable reasons.

Rather than retreat from the issue, secure in the knowledge that he faced a better chance of being hit by lightning than of getting AIDS, he decided to learn more about the people and places that faced much higher risk.

This question drew Halperin to South Africa, that inspiring nation that appeared—from a distance, anyway—to have defeated racial oppression, revived a stagnant economy, and installed a vibrant new democracy just a couple of years after the fall of apartheid in 1994. The explosion of HIV infection in East Africa's Great Lakes Region had by this time begun to slow, though the biggest waves of death still lay ahead.[4] The epidemic, meanwhile, was moving forcefully south, into mining towns, dusty slums, and sprawling rural areas where traditions were eroding faster than modernization was improving lives.[5] The virus had been following the roads and railways built over a century earlier at the behest of Europeans and those of European descent, but now it was beginning to rage mainly among black South Africans at the moment they finally were beginning to achieve basic freedoms. But for the nation's first postapartheid president, the iconic Nelson Mandela, there was no Museveni moment. Though a gifted leader, Mandela was of a generation of African men not accustomed to discussing sexual matters in public, across traditional boundaries of age and gender. And so Mandela said nothing, leaving the matter to his health minister and a medical community that was considered the most advanced on the continent but nonetheless found itself utterly unprepared for the onslaught of HIV.

That is what Halperin encountered when he flew there for the first time, in 1996, with a suitcase full of donated ARVs. The medicine was an offering to the small, energetic group of treatment activists who had formed there. But he mainly wanted to talk with South Africa's leading AIDS thinkers about new approaches to fighting HIV's ruinous spread there, including the possibility that male circumcision might play a useful role in slowing it down. But the scientists Halperin met there did not view the procedure as anything like the relatively cheap, easy prevention tool that he was beginning to think it could become. South Africa is a polyglot nation, with no single dominant group. The nearly 10 percent of the population that is white is split between speakers of English and speakers of Afrikaans, a variation on the seventeenth-century Dutch spoken by the country's first European settlers. The nonwhite majority includes Indians (Hindus and Muslims) as well as so-called coloreds, a mixed-race mélange that amounts to its own distinct ethnic group. And even black South Africa consisted of several distinct groups divided by language, custom, and, in some cases, histories of mutual distrust. When the new South Africa was born under Mandela the government adopted eleven official languages and ordered the nation's public broadcaster, SABC, to deliver news in every one.

The two biggest South African ethnic groups were the Zulus and the Xhosas. And despite generations of extensive intermarriage and routine contact in business, politics, sports, and every other imaginable aspect of life, a sense of rivalry prevailed. To make matters worse, the ruling African National Congress was seen as being largely Xhosa—Mandela was a Xhosa, as were so many other party leaders that they acquired the nickname Xhosa Nostra—and their most powerful competitors for the loyalty of South Africa's black voters was the Zulu-led Inkatha Freedom Party. The spread of HIV played into this stew in complex ways. AIDS hit first and most disastrously among the Zulus, a fiercely proud group that felt marginalized under ANC rule and hadn't circumcised their boys for nearly two centuries. The Xhosas, traditionally a circumcising group, had more political clout and an HIV epidemic that was spreading more slowly.[6]

Halperin's notion that circumcision might be made widely available, regardless of ethnicity, seemed to strike at the heart of these sensitivities. Most of the South African experts he met, like those elsewhere, expressed doubt that if more men became circumcised, this could slow the spread of HIV. The South Africans argued that other factors must explain the especially ruinous HIV rates among the Zulus, which in some areas would soon see one third of their working-age adults infected. "You're first of all crazy," one prominent South African doctor told Halperin. "And even if it were scientifically true, the Zulus would never agree to it." Others just rolled their eyes.

But he got a different reaction when he raised the subject directly with Zulu men and women. As he traveled the country on this and subsequent visits, he often picked up hitchhikers for what ended up being mobile interviews. He also chatted up customers in bars or simply wandered into townships—often with a companion he had met somewhere else—in search of conversation. This was Halperin's anthropologist side at work, and it yielded insights hard to get any other way. Most of the men and women he met were puzzled by the suggestion that circumcision might protect against HIV, but nearly all said they favored the procedure anyway, in part because foreskins could be hard to keep clean.[7] Halperin came to believe that even though Zulus and some other groups had stopped practicing circumcision generations earlier, many would welcome the procedure if it were made widely available and affordable. When Halperin—often introduced as "Doctor" because of his PhD in anthropology—later conducted focus groups to discuss circumcision with Zulus and other African men, often one or two from each group would approach him afterward in hopes that he could conduct the procedure personally.[8] Some even offered money on the spot.

As Halperin continued his research, he became even more convinced that male circumcision was key to understanding the path of AIDS through Africa—and, potentially, to reversing it. Not only did the patterns of circumcision appear to explain the remarkable variations of HIV rates in African countries, they also offered explanations for the puzzling variations within some countries. In Mozambique, for example, the epidemic was raging in parts of the country where circumcision was rare while it remained much more subdued in the areas where the practice was still a cultural norm.[9] In addition, the association between male circumcision and HIV seemed to be similarly strong in some parts of Asia. Thailand and the Philippines, for example, both had major commercial sex industries. But the Philippines, where male circumcision was nearly universal, never developed an HIV epidemic that approached the one in Thailand, where few men were circumcised. The epidemic in India, where circumcision was uncommon, was much worse than that in its ethnically similar neighbors Bangladesh and Pakistan, which were overwhelmingly Muslim and, as a result, routinely circumcised boys.[10]

Circumcision patterns also seemed to offer insights into why a heterosexual epidemic had taken off in the Caribbean nation of Haiti. Though Michael Worobey's genetic research made it clear that HIV jumped from Congo to Haiti in the 1960s, it wasn't clear why a major epidemic took hold there. Haiti, which Halperin visited regularly over his years of HIV work, had paid a high price, not only in sickness and death but in the collapse of its once lucrative tourism industry. The issue of what role Haiti played in the early movements of the epidemic remains, even today, profoundly sensitive.[11] But Halperin eventually became curious about an even earlier bit of history. Most Haitians were descendants of enslaved men and women captured hundreds of years earlier in parts of West Africa where Bantu initiation rites—including male circumcision—were routine. Had circumcision continued through the generations, it would have helped Haiti resist HIV's spread. But few Haitian men now are circumcised.[12] The epidemic would eventually have spread to the Americas anyway; air travel and the extent of HIV in Africa made that virtually certain. But without the slave trade's impact on African initiation traditions, the virus might have arrived later, and moved much more slowly, in Haiti itself.

Such insights stoked Halperin's enthusiasm about the potential for new approaches to preventing the spread of HIV. The combination of multiple sex partners and low circumcision rates seemed like an obvious area for more attention. About this same time he came across the writings of Edward Green, the anthropologist who had first noticed the shifting sexual patterns in Uganda and had been working to garner attention for that subject ever since. He and Green soon began communicating regularly by phone and e-mail, in conversations that fueled Halperin's enthusiasm for prevention strategies beyond condom promotion and other common approaches. He debuted this emerging philosophy in a 1998 paper titled, "Rethinking AIDS Prevention: Towards a Second Stage in HIV Risk Reduction." In the article, published in *Global AIDSLink Newsletter*, Halperin noted that promotion campaigns had struggled to get condom usage rates to levels that would stop the virus from spreading. Then he asked, "In addition to condom promotion, then, what *other* information could be disseminated for more effective HIV prevention?"

He suggested that new strategies should focus more on treating other sexually transmitted infections, encouraging people to reduce their numbers of sex partners, and offering circumcision to men who want it.[13] "Male circumcision, in effect, has probably prevented more HIV infections worldwide than all other factors combined. In virtually every country facing a severe heterosexually spread AIDS epidemic, most men are uncircumcised."

Global AIDSLink Newsletter was hardly *The Lancet* in terms of prestige or impact, but Halperin had laid out an argument that he would refine and expand over the next dozen years. In this he shifted among roles as a research scientist, a synthesizer of ideas first produced by others, and as an agitator for more aggressive action on approaches he thought were overlooked. Much of this work was done with colleagues who had useful—and somewhat different—credentials. Among the most valuable of these relationships was with Robert Bailey of the University of Illinois, Chicago. Though also an anthropologist and epidemiologist, Bailey then had a more traditional résumé and more experience in Africa. He also had a deep-voiced, authoritative manner that played well when paired with Halperin's more urgent, passionate style. But both could trace their interest in male circumcision to the Caldwells' article in *Scientific American*. And both made a point of asking Africans themselves, especially those from groups that didn't routinely circumcise, what they thought of the procedure.

In most surveys, including some seminal ones conducted by Bailey and his colleagues around Lake Victoria, Africans generally viewed circumcision favorably.[14] But the issue had largely fallen off the radar of the global public health community in the years after the Pumwani research in 1989. As early as 1994 Canadian researcher Stephen Moses had reviewed thirty epidemiological studies that noted the circumcision status of its subjects as well as rates of HIV infection.[15] All but four found a connection between heightened infection risk and lack of circumcision. But when he submitted the findings to the WHO in 1994—at the agency's request—the matter deadlocked in a high-level committee that ended up doing nothing.[16] Criticisms of the science tended to focus on doubts over whether circumcision by itself slowed HIV or if lower infection rates and high circumcision rates both resulted from the same cause, such as ethnicity or religion.[17] Male circumcision, meanwhile, was not common in the home countries of many officials in a global health infrastructure still dominated by Westerners. In continental Europe the practice was confined mainly to Jews, Muslims, and African immigrants. The debate gradually settled into a loose regional divide. European scientists were nearly uniformly opposed to any alteration to male genitalia if the evidence of its protective effect was less than bulletproof.[18] But North American scientists were split, with a large group of skeptics and a smaller number who thought the idea had enough promise to merit more study. As these debates played out, interest in circumcision was growing among some African men, studies at the time showed.[19] Expanding services would have had other benefits: Circumcised men have lower rates of some sexually transmitted diseases and penile cancer, and their female sex partners are less likely to get cervical cancer.[20]

Getting circumcision taken seriously again would require generating more attention. Together Halperin and Bailey wrote an article: "Male Circumcision and HIV: 10 Years and Counting," which appeared in *The Lancet* in 1999, a full decade after the original Pumwani research had appeared in that same journal. It reiterated the Kenyan finding that uncircumcised men were eight times more likely to get HIV and cited the same dozens of studies that Stephen Moses had in his report to the WHO, and some newer ones as well. Halperin created a chart showing that all of the nations with the worst AIDS epidemics had, along with risky sexual behavior, fewer than 20 percent of their men circumcised, and that nearly all of the nations with circumcision rates over 80 percent had relatively low HIV rates.[21] But what made this piece different from previous articles was its unmistakable tone of advocacy. It was a call to action, and a call to account for past inaction. "The hour has passed for the international health community to recognize the compelling evidence that shows a significant association between lack of male circumcision and

HIV Rates and Male Circumcision in Africa

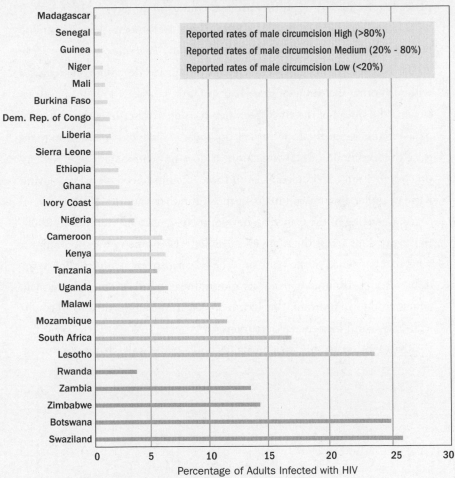

Reported rates of male circumcision High (>80%)
Reported rates of male circumcision Medium (20% - 80%)
Reported rates of male circumcision Low (<20%)

Percentage of Adults Infected with HIV

HIV rates and reported male circumcision status for all sub-Saharan African nations in which population-based surveys have been conducted. These comparisons obscure regional variations within countries such as Kenya, Mozambique, and Tanzania, where HIV is usually much higher in areas where circumcision is not routinely practiced. In such countries as Lesotho and South Africa, the traditional initiation rituals often do not involve removal of the entire foreskin (chapter 18, endnotes 6 and 9). And in Rwanda far fewer men report multiple sex partners than in most other sub-Saharan African countries (chapter 10, endnote 11). For more information, see www.circlist.com/archive/library/ HalperinTimberg.html.

HIV infection." Offering routine, voluntary, and safe circumcisions in the places hardest hit by AIDS, they argued, could save millions of lives.[22]

But a report released by UNAIDS in June 2000 made clear the intense resistance this argument faced. That year's annual *Report on the Global HIV/AIDS Epidemic* celebrated vaccines, HIV testing, and development strategies without noting that none of these had so far demonstrated much success in slowing the spread of the virus. Yet when it came to describing the impact of male circumcision, the tone turned dubious: "While circumcision may reduce the likelihood of HIV infection, it does not eliminate it. . . . Relying on circumcision for protection is, in these circumstances, a bit like playing Russian roulette with two bullets in the gun rather than three."[23]

At a meeting of the Canadian development agency in Montreal in 2000, Bailey saw Piot across the room and decided it was time to question him on what had happened in the years since he was part of the University of Manitoba research team as it made its groundbreaking discoveries about male circumcision in Pumwani. But Piot made clear he had no time for the subject. "Oh God," he sighed. "Circumcision."

Then Piot turned his back on Bailey and silently walked away.[24]

THE INTERESTS OF THE ANC

n January 1997, representatives of an unlikely group of medical researchers came before President Mandela and his cabinet in Pretoria, the capital of South Africa. The core of the group consisted of a lab assistant, her businessman husband, and two heart surgeons. Their experience in infectious disease and the complex world of medical research was minimal. But even so they were coming to announce that they had solved a riddle that long had eluded better known, better funded researchers in the United States and Europe. These South Africans said they had a drug that would cure AIDS, not just hold off its symptoms the way ARVs did. Plus, it was cheap, easy to produce, and discovered in Africa. After the initial pitch from the researchers, apparently healthy people spoke in emotional terms of how they had recovered from the disease after just a few months of treatment, as arm patches infused the medicine into their veins. When the presentation ended the cabinet members rose from their chairs in spontaneous applause.

The name of the drug was Virodene. High-level talk began almost immediately of direct government investment in developing and producing it on a scale that, the researchers claimed, might vanquish AIDS everywhere while bringing prestige and profits to South Africa. Particularly smitten was President Mandela's second in command, Deputy President Thabo Mbeki, who had arranged for the presentation by the Virodene researchers. Mbeki's signature issue during this era was his belief that South Africa was in the vanguard of what he called an "African Renaissance" that would recast impressions of the beleaguered continent. Virodene, if it worked, fit this narrative perfectly—as an African solution to an African problem. "There was this sense that this drug would be the thing that offset the perception . . . of Africans as substandard and less than capable," Quarraisha Karim, the head of South Africa's national AIDS program, later told *The Washington Post*.[1] "All eyes were upon [the ANC], and the expectations were very high, and they were really trying to find their feet, but they didn't want to exercise caution. . . . This was driven by this need to show the world: 'Yes, Africans can do this. We can do this.' Virodene became our redemption."

But not everybody saw it that way. South Africa's independent drug regulatory agency, the Medicines Control Council—generally called the MCC, and roughly equivalent to the Food and Drug Administration in the United States—summoned the researchers two days after the cabinet meeting. The MCC demanded to know Virodene's active ingredient and whether they had permission to test it on human subjects. The regulators soon announced that the drug was made from "a highly toxic industrial solvent which may cause irreversible fatal liver damage" and banned further human testing.[2] And so began a battle that would define and distort South Africa's efforts to grapple with AIDS during a decisive period, as rates of infection and deaths soared to levels seen in few other places in the world.

Between that cabinet meeting in 1997 and 2002, when the government finally relented in its drive to develop Virodene, ARVs went from prohibitively expensive to relatively available in several African countries. A few other nations made meaningful strides in facing the underlying forces driving HIV's spread. And some began positioning themselves to use the massive surge of international AIDS funding to make general improvements in their struggling public health systems. Meanwhile, South Africa—the richest and most admired nation on the continent, still aglow from apartheid's relatively peaceful end—managed none of these things. Scientists writing in the *Journal of Acquired Immune Deficiency Syndromes* would later estimate that years of delays in embracing ARVs led to the untimely deaths of 330,000 South Africans, as well as avoidable infections among 35,000 babies born to mothers with HIV.[3]

The key actor in this sad tale was Mbeki, the crown prince of the ruling African National Congress. The British-trained economist, often caricatured as a stuffy, pipe-smoking intellectual, was smart, savvy, funny, erudite, a thinker of tremendous intensity and self-assurance. In an era when Mandela's charisma made him the hero of the antiapartheid movement, Mbeki was its leading intellectual. When it came time to entice more dollars out of twitchy Western donors, Mbeki went on the road. When it came time to schmooze the foreign correspondents whose dispatches kept South Africa's plight in the world's gaze, Mbeki poured the cognac and turned on the charm. And when it came time to lay the groundwork for a workable peace with a determined foe, it was Mbeki who had some of the first contacts with emissaries from the apartheid government. Mbeki also worked to keep the militant wing of the ANC in check through years of painful, halting talks. Mbeki believed a peaceful transfer of power was possible for South Africa's black majority. Long after most ANC members grew restless, he doggedly stuck to his conviction.[4]

Mbeki became the day-to-day manager of the South African government in 1997, as Mandela, though still president, became something akin to a head of state. Virodene was an Mbeki priority from the beginning of this period, and he remained heavily involved in the drug's development as he rose to the top leadership job in the ANC and, eventually, the presidency of South Africa itself in 1999. There may never be an adequate explanation of Mbeki's many missteps in fighting the epidemic. But several factors clearly were at work. He grew supremely irritated at portrayals of Africans as inherently hypersexual and diseased. He distrusted that drug companies, which often had tested their medicine on Africans while pricing the resulting products beyond their reach, had the best interests of his citizens at heart. And he found many of the simplistic explanations of the epidemic and how it spread unconvincing. Mbeki was a man who came of age in 1960s Britain, where the ANC had sent their future leader to the University of Sussex, and he spent most of his life in exile in various world capitals. Throughout, Mbeki lived among considerable sexual experimentation yet would have seen few, if any, of his European friends getting HIV even as millions of Africans were falling prey. He later told *The Washington Post* in a 2003 interview, after years of living back in South Africa, that he didn't have a single friend or relation that he knew to have died from AIDS—provoking howls of outrage in a nation with more sick people than any other.[5]

A meticulous analyst of South African political history, James Myburgh, has traced Mbeki's mishandling of AIDS to his moments of first contact with the Virodene researchers in the months leading up to that memorable appearance before the cabinet.[6] His fixation on this experimental drug at a time of rising global enthusiasm for ARVs amounted to Mbeki's first tentative steps away from the international scientific consensus on AIDS. It also appears to have planted a seed of uncertainty in Mbeki's mind—one perhaps not entirely unlike Halperin's instinct that there had to be an X factor to explain the patterns of the epidemic's spread—about the portrayals of AIDS offered by most experts. But while Halperin turned to the Berkeley Public Health Library for answers, and eventually found a convincing one by way of the Caldwells' article in *Scientific American*, Mbeki went another way.

In the same years that Mbeki was growing skeptical of antiretroviral drugs, a generation of South African doctors and activists were embracing them with nearly religious fervor. Francois Venter, a lanky, energetic medical student who was training at Johannesburg Hospital in 1997, had his transformative experience with a hemophiliac man who had gotten HIV from infected blood products. The man was laying wasted and nearly lifeless on his hospital bed when his physicians managed to procure some ARVs. The doctors delivered them to the patient through a feeding tube—the man was too weak to swallow pills—and though nothing happened at first, he began to regain strength after a few days. Before long he could sit up in bed, eat, and talk. A few weeks later the man rose unassisted from what everyone had assumed was his deathbed and checked himself out of the hospital. "It was phenomenal," Venter recalled. "It was nothing short of a miracle."

Postapartheid South Africa had all of the elements of the tinderbox. The transport system was robust, the economy was strong, and concurrent relationships were common.[7] Male circumcision, meanwhile, had been in decline, as the nation continued to Westernize. By 2002, more than five million South Africans would have HIV, meaning that roughly one in eight infected people worldwide lived in a single nation with a population of just forty-five million.[8] As these infections progressed, weary, bone-thin, coughing, feverish men and women began lining up at clinics and hospitals across the country seeking relief. Doctors usually could do little but bear witness to the suffering. A dose of antibiotics might control infections ravaging the bodies of their patients. Acetaminophen might ease their pain. But day after day the doctors watched helplessly as their patients died. Shops sprang up along roadsides offering cheap caskets. Cemeteries ran out of plots, forcing communities to open new ones. And survivors found they spent much of their free time attending funerals for friends and relations. Pages of South African newspapers gradually filled with pictures of smiling men and women, adorning what appeared to be announcements of college graduation, marriage, or some other milestone of young adult life. But they were death notices.

Western medicine—and that is the kind that dominated in South Africa, the most Westernized nation on the continent—is focused mainly on curing illnesses. The experience of AIDS only sharpened this tendency. Many South African doctors began desperately scrounging around for any ARVs they could get. They signed up for studies bankrolled by American and European universities. They developed pipelines to the emerging generic industries in India and elsewhere. And some, including Venter, began smuggling costly ARVs into South Africa. Activists routinely traveled abroad with empty suitcases, collected as many pills as they could, then snuck them home past customs agents. One flight attendant for a major European airline came in a couple of times a month with a bag bulging with medicine. Venter and the other doctors at Johannesburg Hospital gradually built up enough capacity to keep several hundred patients alive. Some of those contributed their own money for their pills, which often were the cheapest, most toxic on the market. But most had little cash to offer.

"I'd say to patients, 'How much can you afford?'" Venter recalled. Based on the answer, he would reply: "This is what I can give you. It's not very good, but it'll buy you a couple months."

To understand the Virodene debacle it helps to recall that South Africa was in the midst of an awkward transition from all-white rule to multiracial democracy. Under apartheid nearly every important position within the South African government was held by whites, and given that whites had a massively disproportionate share of the nation's advanced degrees and technical expertise, there was no quick, easy way to change the racial makeup of senior government offices.[9] The ANC under Mbeki's leadership often acted like an occupying power that barely tolerated the agencies that it nominally ruled but, in practice, depended on for a range of crucial bureaucratic functions. Among those agencies still relying on the skills of white technocrats was the MCC. Their resistance to Virodene infuriated Mbeki. He summoned the MCC's leaders to private meetings, and the ANC issued statements suggesting that those who opposed Virodene wanted to see more black people die of AIDS. South Africa's intelligence service, meanwhile, paid at least one unexpected visit to an MCC bureaucrat.[10] Yet the agency stood its ground. Repeatedly, and in the face of the full weight of the ruling party and government, the MCC refused to allow Virodene to be tested on South Africans. Serious talk soon began of revoking the independence of the agency so that the political appointees of the health ministry could potentially overrule its decisions. Ultimately the health minister removed its recalcitrant staff and reconstituted it under new leadership. But the new crop of MCC officials, reviewing the same information, refused as well.

Mbeki did not back down either. "I and many others will not rest until the efficacy or otherwise of Virodene is established scientifically," Mbeki wrote in an op-ed piece appearing in several newspapers in 1998. "If nothing else, all those infected by HIV/AIDS need to know as a matter of urgency. The cruel games of those who do not care should not be allowed to set the national agenda."

But gradually the development of Virodene shifted out of the public realm and into a private one that still answered to Mbeki. A group of ANC activists began raising money to support the drug's development and arrange for its testing in other African countries, where official resistance was more easily overcome. The leader of the fundraising effort was Max Maisela, a former head of the national post office with close ties to both Mbeki and the health minister. Maisela, in an interview years later, acknowledged that he organized $3.5 million from investors, including former members of the ANC's armed wing.[11] The orders to embark on the venture, he said, came from Mbeki and were transmitted through the party's treasurer general.[12] Maisela described how he was deeply moved by the growing numbers of AIDS deaths. He helped where he could with financial support but also was eager to make a bigger impact against AIDS. In retrospect, he said, he wished he had studied Virodene and its research team more carefully before becoming the ANC's lead agent on the project. "That's basically how I operated, as a cadre of the ANC," recalled Maisela. "You don't do your own thing. You act in the interests of the ANC."

As Mbeki and the party deepened their involvement in Virodene, its researchers enjoyed an unusually close relationship with top government and party officials. On October 21, 1999, the researchers sent two batches of documents directly to Mbeki at Pretoria's Union Building, the equivalent of the White House for South Africa. One set of documents announced that they had successfully procured a U.S. patent for Virodene. The second was a compilation of literature questioning the safety of the antiretroviral drug AZT, a natural rival in the AIDS treatment market should Virodene ever win regulatory approval. AZT was one of the first ARVs developed and was more toxic than newer ones. It also was among the cheapest ARVs and had been the centerpiece of a South African pilot program to keep pregnant women from giving HIV to their babies. The government had canceled the program in 1998, citing its cost, and the decision had sparked enduring controversy that had not cooled by the time the Virodene researchers sent the documents to Mbeki.

Reports of the dangers of AZT appear to have shocked him deeply.[13] In a speech one week later, Mbeki said:

We are confronted with the scourge of HIV-AIDS against which we must leave no stone unturned to save ourselves from the catastrophe which this disease poses. Concerned to respond appropriately to this threat, many in our country have called on the government to make the drug AZT available in our public health system. Two matters in this regard have been brought to our attention. One of these is that there are legal cases pending in this country, the United Kingdom, and the United States against AZT on the basis that this drug is harmful to health.

There also exists a large volume of scientific literature alleging that, among other things, the toxicity of this drug is such that it is in fact a danger to health. These are matters of great concern to the government, as it would be irresponsible for us not to heed the dire warnings which medical researchers have been making. I have therefore asked the Minister of Health, as a matter of urgency, to go into all these matters so that, to the extent that is possible, we ourselves, including our country's medical authorities, are certain of where the truth lies.[14]

Pushing Mbeki along were the Virodene researchers, the deepening political and financial investments of ANC activists, and the hope that the drug might eventually end the epidemic. "Every step of the way, he had a hand in it," said a former top government official under Mbeki, speaking on condition of anonymity. "The possibility of a breakthrough by South African scientists was something that made people involved in it very, very passionate."[15]

As the development of Virodene progressed in secret the South African government expressed public concern about the cost of ARVs. The sheer numbers of infected people, combined with the high prices demanded by their Western manufacturers, threatened to swamp the nation's budget and hamstring its ambitious social agenda. The health ministry estimated in 1999 that providing the medicine to every South African with HIV would cost ten times the total national health budget, to battle a single, incurable disease.

The ANC had promised massive infrastructure investments—a million new homes, vast new programs to deliver electricity, paved roads, and clean water—with the goal of undoing some of apartheid's legacy. When it came to AIDS, government officials often said that preventing new infections was more fiscally sensible than putting untold numbers of people on expensive treatment regimens for the rest of their lives. But Virodene, by offering the hope of a cheap, easy cure, had the effect of undermining that focus on prevention, as did the increasingly urgent political demands from those already infected.

On December 1, 1998, the eleventh annual World AIDS Day, activists in South Africa joined their counterparts from around the world in seeking to raise awareness about the epidemic. Speaking out was a hallowed ritual within the movement, but denial ran especially deep among the epidemics in Africa. Many who suspected they had HIV refused to be tested, to seek what amounted to confirmation of a death sentence at a time when few had any hope of getting relief from ARVs. Those who knew they carried the infection rarely told even their families or closest friends out of fear of being shunned, chased from home, left to die desperate and alone.

One of the first people to publicly challenge the stigma in South Africa was Gugu Dlamini, a thirty-six-year-old field-worker for a national association of people with HIV. She lived in the KwaMashu township outside of Durban. On World AIDS Day, Dlamini, a single mother of a thirteen-year-old daughter, spoke out on Zulu-language radio and television stations to warn about the disease.[16] The backlash was immediate. Many of her neighbors accused Dlamini of tarring the reputation of their community by acknowledging that she had HIV. Weeks after Dlamini's announcement a man punched and slapped her. That night a mob formed outside her house and attacked. She was kicked, hit by sticks, pummeled with stones. Dlamini died the next day, having paid the ultimate price for her act of bravery. "She was a nice, bright woman, and now her child is an orphan because of AIDS," Mercy Makhalemele, a fellow AIDS activist, told *The New York Times*. "But not because she died of it. Because she was trying to exercise her constitutional right to freedom of speech."

The death of Gugu Dlamini sparked concern across the world. Mbeki issued a sympathetic statement, as did Peter Piot at UNAIDS. Editorial writers from several nations voiced their outrage. But the most important action came within South Africa, where an assertive new activist group had formed shortly before Dlamini's death. The Treatment Action Campaign drew some of its inspiration from international AIDS activists but had the distinct flavor of South Africa, where marches and other forms of direct action were key elements of both union activism and the struggle against apartheid. Soon the group printed the first batch of T-shirts that would become the most visible signature of their new struggle. It said, in bold capital letters, HIV POSITIVE. And for those who wore them—a group that included both people who were infected and some who were not—it was a statement more of attitude than of virological status. With that the Treatment Action Campaign was on its way to becoming the most powerful AIDS activist group in Africa, one that was a moral beacon in its call for treating those infected with HIV. The group's founder and leader, a gay man infected with HIV, Zackie Achmat, for several years refused to take ARVs until they were made available to all South Africans with HIV. This courageous stance brought Achmat to the edge of death before he relented. The group was especially effective in forcing the government to finally provide ARVs to pregnant women and their babies, helping prevent infection among tens of thousands of newborns.

The Treatment Action Campaign was a crucial foil against Mbeki's worst decisions, and many people are alive today because of their fight to make ARVs available more quickly. But something also went wrong in a way that few noticed at the time. The fight over AIDS in South Africa became dominated by a debate over whether and how to treat those who already had HIV. Mbeki and some within the ANC were fixated on Virodene; doctors and activists wanted ARVs. Attention shifted, meanwhile, away from prevention efforts. The Treatment Action Campaign lobbied for better access to condoms and clinical services, such as HIV testing. The government continued to promote these approaches, as well as initiatives aimed at encouraging young people to make wiser decisions. But broader—and more complex—

questions about sexual culture went largely unaddressed, as did the role of male circumcision.[17]

Among those who became an early member of the Treatment Action Campaign was Francois Venter. He proudly donned the HIV POSITIVE T-shirt, marched in rallies, and lobbied the government for universal access to ARVs even as he helped keep alive the Johannesburg Hospital's informal airlift of the drugs. Venter also was among a group of South African AIDS doctors who cared for prominent members of society—high government officials, leading ANC players and their families, and the Treatment Action Campaign's own activists—when they needed the drugs to stave off death. Yet Venter gradually began to sense the problem. Even as ARVs were beginning to trickle into the country, the flow of new infections was a relentless gush, with no sign of easing off. "Nobody," he said, "marches in the streets to stay negative."

[20]

POVERTY TRAP

Mbeki's public attack on AZT triggered furious rebuke, both within South Africa and from around the world. The British drug manufacturer Glaxo Wellcome was among the first to rise in defense of AZT, but soon doctors, editorial writers, rival politicians, human rights activists, religious leaders, and the MCC all challenged Mbeki's assertion that the dangers of AZT outweighed the benefits. Drug regulators from around the world had long ago approved it despite its known levels of toxicity, on the grounds that it was among the few drugs capable of controlling HIV. It performed poorly against the virus when used by itself but became a routine part of the combination therapies used against AIDS beginning in the mid-1990s. Short courses of AZT by itself, meanwhile, could prevent infection in someone raped by a man carrying the virus. And providing several doses to a woman in labor as well as to her newborn baby can significantly reduce risk of transmission to the infant. The United Nations, whose officials choose their words carefully even when confronting warmongers and genocidal maniacs, rapped Mbeki in remarkably blunt language. Joseph Perriëns, who oversaw care and support services for UNAIDS, told reporters that AZT's side effects, such as nausea and anemia, were similar to those of many other drugs. He suggested to the Associated Press that Mbeki was "not doing his people a service."[1] He was even harsher in comments to *The New York Times:* "There is toxicity, but this is not a sweet, this is a drug. To combat a fatal disease, it is perfectly acceptable to use drugs slightly more toxic than an aspirin."[2]

Mbeki responded not by backing down but by digging in. He read deeply into the writings of a group of prominent denialists who disputed most of the scientific consensus about AIDS, including its causes and remedies. This group included several prominent Americans, such as the University of California–Berkeley molecular biologist Peter Duesberg and Nobel Prize–winning biochemist Kary Mullis, and their work was easily available on Web sites that Mbeki found while surfing the Internet.[3] These scientists argued that the common understanding of AIDS was wrong, that HIV did not cause the syndrome of wasting, fevers, diarrhea, and other maladies generally associated with AIDS. Instead, they said, these symptoms resulted from other diseases long common to poor Africans but rarely treated adequately. And because HIV wasn't the cause, antiretroviral drugs were a hoax foisted on unsuspecting patients at massive cost by pharmaceutical companies, regardless of dangerous side effects.

Soon Mbeki had established formal contact with another prominent denialist, American biochemist David Rasnick. It's easy to see the appeal of this group's theories, especially to a man who regarded much of the international conversation about AIDS—its peculiar nexus of sex, drugs, and sin—as an affront to his deeply felt sense of African nationalism. Mbeki biographer Mark Gevisser wrote:

> He had come to believe that the AIDS epidemic was the latest racist weapon in the arsenal of the Afropessimists, and was being exploited by "Big Pharma" (the pharmaceutical industry), which was using AIDS activists as its stooges as it dumped expensive products on unsuspecting Africans, while the real causes of AIDS were Africa's ongoing poverty and underdevelopment. . . . Mbeki came to see the AIDS discourse as a slight on African masculinity and the latest manifestation of a centuries-old discourse that

pathologised Africans as near-savage in their libidinal excess, and
thus irredeemable vectors of disease.[4]

Mbeki's approach exasperated mainstream AIDS activists, researchers,
and policy makers. He sent a list of questions to Rasnick and, in early 2000,
convened what Mbeki called an expert advisory panel, nearly half of which
consisted of denialists. Some mainstream scientists who had agreed to work
on the panel withdrew in protest, and an increasingly isolated Mbeki grew
even angrier. He wrote to President Clinton that he and the AIDS denialists
were enduring a "campaign of intellectual intimidation and terrorism" that
he compared to the days when heretics were burned at the stake.[5]

Mbeki also developed an alternative narrative for AIDS that, despite the
controversy generated by most of his ideas, would gradually accrete into
the conventional wisdom about the epidemic. He argued that HIV may be
one cause of the illnesses suffered by those diagnosed with AIDS, but more
broadly the symptoms attributed to the disease were manifestations of other
maladies common to Africa because of its poverty and poor medical care.
This was a core tenet of the denialist line. He once complained to Gevisser
that the WHO said AIDS was responsible for only 16 percent of illness
on the continent, and yet the conversation overwhelmingly was about just
this one disease. "What about the other 84 percent? Why are we not talking
about that? And these are killer diseases."

Mbeki's campaign to change the world's thinking on AIDS culminated
at the Thirteenth International AIDS Conference in July 2000. The conference was, for the first time ever, held on the continent most affected by the
disease, in Durban. For months AIDS activists threatened to boycott the
event, and five thousand scientists (including Halperin) eventually signed a
declaration aimed at Mbeki declaring that evidence for the connection between HIV and AIDS was "clear-cut, exhaustive and unambiguous, meeting
the highest standard of science."[6] His office replied that the declaration was
worthy only of "the dustbin." But a subtler response came from Mbeki him-

self when he spoke on the conference's opening night to an audience of thousands.

The speech, which soon found a permanent home on dissident Web sites, was vintage Mbeki: gracious and thoughtful, as if he was simply adding another brick to the world's ever-growing edifice of knowledge. Mbeki quoted extensively from a 1995 WHO report noting the biggest source of illness and death on the planet was not a particular disease but extreme poverty. Mbeki was not only an economist; he also was a Marxist for several years earlier in his career.[7] And by reframing the debate in terms of development inequities, Mbeki finally brought the discussion of AIDS to familiar terrain. The rising tide of sickness in South Africa, he said, did not result from "a single virus" but from the legacy of abuse visited upon the continent, leaving it too poor and chaotic to deliver the basic care essential to health and longevity. "The world's biggest killer and the greatest cause of ill health and suffering across the globe, including South Africa, is extreme poverty."[8]

There was no applause as Mbeki delivered the speech. Instead, hundreds of delegates walked out. The reviews from those who remained behind ranged from disappointed to outraged. But, looking back, Mbeki had a point. Most of the diseases in Africa did spread most easily in the poorest places and could be controlled with basic medical care. He listed a series of maladies—cholera, river blindness, guinea worm, and many others—that thrived mainly because of Africa's poverty.

The irony was that only one disease named in the speech defied Mbeki's analysis: AIDS. Substantial resources were essential to deliver the medicines that controlled the symptoms of those already infected. But when it came to understanding the broader path of the epidemic across the planet, HIV did not require poverty. It certainly spread widely in some impoverished regions, but a degree of economic vitality helped the virus move more easily. Here was a disease that hit hardest in the urban dynamos of San Francisco, Bangkok, Moscow, New York, and Johannesburg. Within Africa it had tended to spread fastest among the more affluent and educated.[9] HIV wasn't rampant in parched northern Nigeria or in the dusty Darfuri scrubland of western

Sudan. It never spread very widely in the impoverished seaside cities of Free-town or Mogadishu. Instead it was devastating South Africa, the continent's most developed nation.

Yet alone among the points that Mbeki sought to make in his crusade on AIDS policy, the one that ended up taking hold was his argument connecting poverty to HIV. The Clinton administration echoed it. Poverty increasingly became a key element of UN reports about the epidemic. Activists embraced it and soon began arguing that debt relief for poor nations would slow HIV's spread. And in this circuitous way Mbeki legitimized one of the core arguments of the denialists, helping to make it part of the increasingly flawed conventional wisdom about AIDS.

Two months after the AIDS conference left Durban, South Africa's secret program to test Virodene got underway in the Indian Ocean city of Dar es Salaam, Tanzania. The investor group, with the support of the ANC, had managed an initial test of the safety of their experimental drug among twenty men at a London laboratory. The test found that while Virodene was generally well tolerated in small doses, one subject developed signs of liver trouble, perhaps because of its combination with the common, over-the-counter pain-killer acetaminophen.[10] For the next phase of testing, which was to focus on whether Virodene successfully cured the symptoms of AIDS, the investors started administering the experimental drug at a military hospital and a private clinic, Chadibwa, said to be owned by the Tanzanian army's chief of staff. AIDS activist groups helped recruit hundreds of volunteers, who received Virodene through patches on their arms. Documents show that the health ministry in Tanzania knew about the trials, but the researchers did not initially get approval from the National Institute for Medical Research, the regulatory body responsible for overseeing medical trials there. In addition, consent forms signed by some participants described Virodene as experimental but did not explicitly say that it might cause serious side effects and had been banned from human use in South Africa.

The trials had a heavily South African flavor. A South African research company handled the data. Spies dispatched by South Africa's intelligence service kept tabs on the tests, which ran from September 2000 through March 2001. Mbeki's health minister, Manto Tshabalala-Msimang, who once had lived in Tanzania during her years as an ANC exile, paid a personal visit that January and toured one of the trial sites.[11] As patients began receiving Virodene, an initial sense of hope prevailed, said several of the former subjects in interviews years later. They stayed in clean, modern wards and ate more and better food than most were accustomed to. They also received painkillers for the dizziness and headaches that often accompanied doses of the medicine.[12] "When you have AIDS and you know there's no cure, anybody tells you anything you tend to believe it," said Yahaya Ramadhan, a lean, soft-spoken former seaman who said he participated in the trials.

But gradually some of the participants began reporting problems, including abdominal swelling, pain, and other symptoms that could have indicated liver abnormalities. Friends and relatives of several others said the Virodene trials were followed in a matter of weeks or months by progression into AIDS. Connecting cause and effect in a medical trial is perilous, and in this case perhaps impossible. A representative for the researchers denied that any medical problems resulted from the use of Virodene but said it was not surprising when participants, all of whom had HIV, deteriorated and eventually died after the trials ended.[13]

A, B, AND C

The push to scale-up AIDS treatment had the effect of sapping attention and intellectual energy from prevention, but the failings in that area did not result from a shortage of money. The overall funding for HIV programs increased steadily across Africa as the AIDS community succeeded in framing the battle as one fundamentally about resources. Between 1996, when UNAIDS began operations, and 2001 the estimated global spending on AIDS grew from $292 million to $1.6 billion, a fivefold increase in as many years. And the surge had just begun.[1] That same year the United Nations held a special session on the epidemic and pledged even more increases in the years ahead. The market responded with a proliferation of new groups and programs created to put this money to use.

Watching the flow of new money up close was Suzanne Leclerc-Madlala, an American anthropologist who had met a Zulu man in college and moved home with him to a South African township during the height of the anti-apartheid struggle of the mid-1980s. She watched with horror as HIV started taking hold in the decade that followed, right as apartheid was finally crumbling. She eventually moved to Durban and took a position at the University of KwaZulu-Natal, keeping a keen eye on the social and sexual dynamics that helped the virus spread so swiftly through both townships and cities.

Leclerc-Madlala also noticed that as the scramble for AIDS funds took hold, few programs seemed aimed at the root causes of the epidemic. "Oh my word," she later recalled, "what they've done with AIDS money over the years is amazing."[2]

Leclerc-Madlala joined the board of directors of South Africa's largest umbrella group for AIDS organizations at a time when they were proliferating with tremendous speed. Cadres of well-paid consultants, many from North America and Europe, flew in to help develop mission statements, work plans, and elaborate charts of corporate structures. This was mainly for the consumption of Western donor groups, including governments, international agencies such as UNAIDS, and private philanthropies. The key to success, Leclerc-Madlala said, was deploying the right buzz words in proposals. "Peer education" was a winner. So was "condom promotion," "voluntary HIV testing," or anything having to do with youth, who were an especially popular target for interventions, even though research had made clear that most new infections occurred in adults.[3] New nonprofit groups began appearing out of nowhere, claiming to focus on causes that donors favored, such as helping AIDS orphans. "Would funders see through this? Or was it just a matter of time?" Leclerc-Madlala recalled asking herself. "But they never did."

Especially effective at winning the AIDS money flowing into South Africa was a program called loveLife, supported by the U.S.-based philanthropies, the Kaiser Family Foundation and the Bill and Melinda Gates Foundation. They donated millions of dollars to support loveLife's chain of youth centers, mass media campaigns, and publicity events, which included a sailboat journey from Cape Town to Rio de Janeiro.[4] The goal was to inspire optimism about the future, on the premise that it would make youth more likely to protect themselves from HIV. The program also prized frank talk about sex, but their most prominent initiative was a series of billboards that left most South Africans confused. A giant lime-green one featured the words SCORE and RED CARD, with corresponding checkoff boxes beside them.[5] Other colorful loveLife billboards featured similarly cryptic word pairs: YOUR BODY/ANYBODY, CLIMAX/ANTICLIMAX, or DROP DEAD GORGEOUS/THE DROP.

Another showed a naked white woman with a zipper for a mouth.[6] The text for this one—as with many loveLife billboards in a nation where those most at risk for HIV preferred Zulu, Xhosa, or another African language—was in English. It simply said GENDER. Many South Africans also worried that loveLife glamorized sex at a time when caution was more appropriate. One loveLife pamphlet featured this piece of text: "Yes, Yes, Yes, sex is on our minds and in the air. Sex is going to be part of the rest of our lives. Thank your body, thank your hormones—this can be such fun!"[7]

When Halperin first encountered loveLife on a visit to South Africa, he was aghast that the billboards and other promotional materials could be so stunningly off point. His conversations with hundreds of young South Africans, conducted both in focus groups and less formal settings, convinced him that the the loveLife messages also puzzled their target audience. Halperin chatted with a group of young men in Mogwase, a town a few hours from Johannesburg, and found only one who made the connection with HIV. Halperin recounted that anecdote in a critical opinion piece about the program that he wrote for *The Washington Post* in 2001.[8] The article also touched

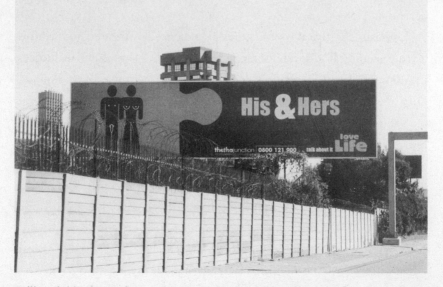

Billboards like this one began appearing in South Africa in the late 1990s.

on a theme becoming increasingly important to Halperin: Even as money poured into fighting AIDS, the things he thought worked best were generally inexpensive. Circumcising a man cost forty dollars or less per procedure. Campaigns on sexual behavior were most effective when they drew upon the grassroots energies of community and religious leaders. In fact, Halperin began to worry that the unprecedented sums flowing into African AIDS programs from richer parts of the world were already distorting the response. The emphasis was on deliverable commodities, imported Western strategies, and consultants' reports. The AIDS war was becoming increasingly bureaucratized.

During years of regular visits to southern Africa, including a semester spent teaching a graduate course on AIDS at the University of the Witwatersrand in Johannesburg, Halperin ventured into the urban slums where most black Africans lived and talked to them about the epidemic. When he asked young people what they thought about AIDS, one common reply was: "It's a good business to be in nowadays."

During one of these stays in South Africa, Halperin learned of an unusual opportunity: USAID was looking for an HIV prevention adviser, preferably one with a background in that collection of concepts called "behavior change." He had been exploring the subject for several years, and he had credentials from stints as an academic researcher at the University of California campuses in San Francisco and Berkeley. He applied to USAID and soon landed the job. In September 2001, Halperin moved into a cubicle in Washington, D.C.'s federal core, as one of the U.S. government's most unlikely bureaucrats.

President George W. Bush had recently come to power, fueled in part by the evangelical zeal of fundamentalist Christian supporters, and he espoused a clear-cut moral worldview that prized bold action, especially if it involved breaking from the policies of his predecessor, Bill Clinton. That produced an administration open to rethinking traditional approaches on issues such as fighting AIDS, but also one in which many rank-and-file bureaucrats were suspicious of newcomers who might be agents of the new regime. Halperin was a Jewish, Berkeley-trained anthropologist who normally would have been more at home in a Democratic administration. But he also had developed ideas on HIV that soon startled some of his new coworkers. Halperin's focus on male circumcision put him well outside of the mainstream, but perhaps even more unconventional were his ideas about sexual culture, Uganda's HIV decline, and the limits of condom promotion in Africa.

He first began raising some of his ideas in a brown-bag lunch meeting with a few of his USAID colleagues just weeks after he was hired. It was intended to be a relaxed introduction to Halperin and his research, but when he began suggesting that Ugandans had beaten back HIV not mainly by using condoms but through broader changes in their sexual culture, the room turned tense. Halperin watched uncomfortably as the faces of some of his new colleagues grew pink, then scarlet. As they began pushing back against his arguments, Halperin struggled to hold his ground before sensing that he had won at least one convert. Jim Shelton, who had just published a widely cited journal article about the need to raise levels of condom distribution, urged the others to hear Halperin out.[9] "Please, please be quiet," Shelton said. "Let him talk."

Halperin had no ambition to build a career within the ranks of a government agency. He simply wanted to bring in some new ideas that he believed might make a difference against HIV, and to this job he brought an approach that more resembled that of an energetic college professor than a bureaucrat. He began to arrange a series of events to generate attention for these ideas, beginning with a February 2002 session simply called "What Happened in Uganda?" He booked a nearby conference room and lined up an unusual panel of speakers, including anthropologist Edward Green, a well-known epidemic modeler named John Stover, and a Ugandan physician, Vinand Nantulya, who had been among Yoweri Museveni's scientific advisers during the Zero Grazing years. Then Halperin initiated an epic e-mail campaign. Several sleepless nights later more than eighty people from Washington's HIV-policy elite crowded into the conference room. The attendance was so far beyond expectations that many had to watch from an overflow room with a video link.

Green opened with his account of Uganda's campaign—the use of fear, the invoking of traditional morality, the cooperation of church leaders and cultural figures. Stover followed with an analysis of how both Uganda and Kenya saw condom usage rates rise through the 1990s. Yet only Ugandans also reduced their numbers of sex partners, which helped explain the major decline in HIV transmission.[10] The kicker was Nantulya, who described the terror that accompanied the arrival of AIDS in his own village in western Uganda. His community first blamed outsiders, witchcraft, and sin but eventually accepted that they could make themselves safer through more careful sexual choices. That meant young people could delay the start of their sexual lives, older ones could stick to a single partner at a time, and those unwilling or unable to use either of the other two approaches could use condoms. Nantulya's final slide showed a pie chart with three equal-size sections: One was marked A for "abstinence," one was marked B for "be faithful," and one was marked C for "condoms." "You see," Nantulya concluded. "There's something for everybody. A, B, and C. Not A, B, or C."[11]

The three messages had coexisted uneasily in AIDS prevention campaigns in Africa and some other parts of the world for years. The letters advocated rather different approaches to sexual life. The teenage virgin, the faithful wife, and the philandering husband all could find a place within the range of options for preventing infection. This flexibility had advantages, especially when attempting to craft messages across cultures. A Tanzanian priest in the late 1980s created a popular parable based on the biblical floods, except that instead of one boat there were three—one labeled abstinence," one labeled "fidelity," and one labeled "rubber lifeboat," for condoms.[12]

But the simplicity of ABC masked divisions over how best to fight HIV. Religious leaders of many faiths worried that aggressive condom promotion campaigns eased moral sanctions on sex outside of marriage while subtly encouraging prostitution and other casual encounters. A Kenyan cardinal and a Muslim imam burned condoms and AIDS awareness booklets in Nairobi in 1995.[13] A poster at a condom-burning demonstration in Nairobi the following year struck at the heart of the fragile compromise that was ABC: SEX EDUCATION IS NOT THE CURE FOR AIDS; ABSTINENCE AND FIDELITY ARE. In the United States, Christian conservatives after the Republican takeover of Congress in 1994 had pushed for abstinence programs in American high school and cut the budgets of international family planning initiatives. This had enraged those who saw declining birth rates as key to improving educational opportunities and development in poor nations, where women often had few reliable options for avoiding unwanted pregnancies.[14] ABC offered a way to lead AIDS policy beyond this polarization. But its political appeal also masked its weaknesses as a prevention strategy. Each letter could, in theory, prevent HIV, but the slogan had a way of oversimplifying human behavior. The parable of the three lifeboats during the biblical flood sometimes was drawn with wooden planks connecting them together. That signaled an acceptance that an enlightened person might shift among the strategies, based on changing conditions. But sexuality was more complex for many people, and almost certainly less rational. Some might have one steady relationship that was avowedly monogamous while also maintaining casual ones on the side. Long stretches of abstinence might be followed by longer stretches of having two or three partners concurrently. And even devoted spouses might stray during the months of separation common in regions such as southern Africa, where miners and other laborers often worked hundreds of miles from home. ABC offered potential answers in each situation, but not everyone managed to shift from boat to boat without sometimes tripping over the side.

A USAID report on the Uganda presentations, featuring Nantulya's Power-Point slide, would eventually help galvanize interest in ABC within the Bush administration and beyond. But few within Halperin's agency shared his enthusiasm for the subject, or for male circumcision.[15] One of his supervisors closed the door to his office and admonished, "Don't talk about circumcision anymore. Don't talk about ABC." Another superior, much higher up in the food chain, was even more blunt; he called Halperin up to his office and screamed: "I don't want to hear another fucking thing about ABC. We do condoms!"[16] But a third USAID colleague, Jeff Spieler, gave advice that Halperin took more seriously: "In my twenty years here I've learned it's better to ask for forgiveness than to ask for permission."

Along the way Halperin also developed an unlikely ally in a former missionary doctor and public health expert who had worked for Christian aid groups in Africa, Anne Peterson. She was a moderate Republican whom Bush had appointed as director of global health at USAID and, though many rungs higher than Halperin in the bureaucratic hierarchy, Peterson expressed interest in his alternative narrative about what was driving the epidemic and how best to fight it. Halperin was uneasy about his growing dealings with some conservatives but, like Peterson, had come to appreciate the political power of marrying their fixation on abstinence with the liberal fixation on condoms. So Halperin and some colleagues launched a study into the scientific underpinnings of the ABC strategy, comparing trends in sexual behavior and HIV rates in three countries where the epidemic had subsided with those in three other countries where it had continued unabated.[17]

Peterson gave him the institutional backing to keep working on such ideas, an effort that included organizing several other research conferences.[18] ABC also began catching on in the press. The left-leaning *New Republic* magazine ran an article in 2002 describing the Zero Grazing campaign under the provocative headline SEX CHANGE: UGANDA V. CONDOMS.[19] The piece quoted Green, the anthropologist, speaking skeptically about the role of condom promotion. "I'm a flaming liberal, don't go to church, never voted for a Republican in my life. But if you say the things I've said . . . the religious people love you and the people in public health get suspicious."[20] The *National Review* followed up with its own article, which was infused with the increasingly partisan atmospherics of the HIV prevention debate. The issue featured a striking black cover that had the name of the conservative magazine in red, the color that long had symbolized concern for AIDS. The image showed an African woman holding what appeared to be her young son, who peered directly into the camera. The headline said, WHY LIBERALISM CAN'T BEAT AIDS.

ON THE JERICHO ROAD

The United States was deep into a war in Afghanistan and was heading toward a second one in Iraq when President Bush addressed a joint session of Congress for his State of the Union address on January 28, 2003. With thousands of U.S. troops deploying to a Middle Eastern desert, the speech was dominated by the president's justification for the impending invasion. But before shifting into a list of reasons for attacking Saddam Hussein, Bush showed the softer side of his political persona—what his handlers famously branded "compassionate conservatism"—with an announcement that the United States would create a massive new program to fight AIDS in poor countries. The sweep of Bush's ambition surprised all but a small circle of advisers. The price tag alone, $15 billion over five years, was beyond anything tried before in international global health. But the goals Bush set were, if anything, more of a stretch. He vowed to prevent seven million new infections and treat two million people who already were sick, mostly in Africa.[1] "Ladies and gentlemen," Bush told Congress and the world, "seldom has history offered a greater opportunity to do so much for so many. We have confronted, and will continue to confront, HIV/AIDS in our own country. And to meet a severe and urgent crisis abroad, tonight I propose the Emergency Plan for AIDS Relief, a work of mercy beyond all current international efforts to help the people of Africa."

The plan, which came to be known by its acronym PEPFAR, for the President's Emergency Plan for AIDS Relief, grew from a combination of factors. Perhaps none was more important than the deepening concern about AIDS by prominent Christian conservatives such as Senator Jesse Helms (R-N.C.) and evangelical leaders Rick Warren and Franklin Graham, son of the famed televangelist Billy Graham. Many Christian leaders during the 1980s had shunned AIDS victims, viewed at the time mainly as gay men and drug addicts. But that attitude had shifted by the dawn of the new century, a time when people with HIV more often were portrayed as African babies or wives of men who brought the virus home through extramarital affairs. On a strictly political level, the massive humanitarian gesture of PEPFAR also offered an alternative narrative for those hoping to keep the Bush administration from being entirely defined by war. By the end of his eight years in office, even many of his critics were calling PEPFAR one of the president's finest achievements. But some closer to the program would be more reticent in their praise, judging it to have shared many of the flaws that Bush also brought to his war in Iraq—a quest for quick results over long-term planning, an eagerness to declare victory on a battlefield still contested, and a failure to grasp how profound cultural differences could undermine the effectiveness of initiatives conceived in Washington.[2]

The program would bear the stamp of Bush's thinking about fighting HIV. USAID relied to some degree on the contributions of epidemiologists, as well as on anthropologists and other social scientists. But the White House's own group of advisers was skewed heavily toward physicians, mostly from the CDC and National Institutes of Health. The plan they produced relied heavily on biomedical tools, such as antiretroviral drugs and HIV testing. Nobody would have asked a social scientist to prescribe antiretroviral drugs to a patient, but frequently physicians were called upon to craft prevention strategies that would have benefited from deeper understandings of how pathogens moved through communities. Jim Shelton, one of Halperin's colleagues at

USAID and himself a physician, once compared this to asking people skilled in hand-to-hand combat to organize an entire war.

Bush made clear his preference for the doctor's worldview in a remark to Joseph O'Neill, a physician, moments before the president announced his appointment as the White House AIDS director. "We have to start treating this like a public health issue," the president told O'Neill, "like a disease."[3] The anecdote reflected the administration's rejection of some of the social agendas that had infused the debate over the epidemic at institutions such as UNAIDS and the World Bank. But it also meant that there was little room at the table for those who had devoted careers to developing complex epidemiological analyses rooted in cultural subtleties. Bush called for a more straightforward approach as he launched what amounted to the most ambitious house call in the history of the world.

PEPFAR grew from the shifting politics of Washington and, perhaps inevitably, was shaped by the careful balancing of constituencies there. Even when it came to designating which nations would be "focus countries" worthy of the heaviest doses of funding, geopolitical forces came into play. The Bush administration did not want PEPFAR to be seen as aiming exclusively at Africa, so the tiny nation of Guyana, in South America, and Vietnam, in Southeast Asia, were added to the list, despite having far smaller epidemics than most African countries. And even within the continent, there were choices driven by factors other than objective evaluations of where the need was greatest. Rwanda, which had become a favorite of the aid industry in the aftermath of its devastating genocide in 1994, became one of the fifteen focus countries, while neighboring Burundi, with a similar epidemic, did not. Key U.S. allies Nigeria and Ethiopia, both of which had HIV rates in the single digits, were deemed focus countries, thereby worthy of hundreds of millions of dollars in PEPFAR money a year. Meanwhile, Swaziland, Lesotho, and Malawi—among the hardest hit in the world, with infection rates many times higher than some of the focus countries—did not make the cut and, as a result, received much less support. Zimbabwe, a relatively populous nation with a horrific epidemic and some of the medical infrastructure necessary to deliver the life-extending drugs, did not become a PEPFAR focus country because the government under dictator Robert Mugabe was hostile to the United States as well as unusually corrupt.[4]

Other political considerations involved powerful domestic constituencies. Under Bush's plan, the allocation for prevention was 20 percent of the total, or $3 billion, for his goal of preventing seven million infections. That was equivalent to $429 for each of the HIV infections he had vowed to avert. The ABC formulation, meanwhile, had caught the attention of conservatives who saw an opportunity to expand AIDS programs beyond their traditional focus on condoms while also promoting an issue they cared about: limiting sex to marriage. This opened the door to expanding the participation of religious groups, many of whom already were on the front lines of the fight against AIDS in Africa but often had no interest in promoting condoms. The debate quickly became polarized, with conservatives rallying for A and liberals for C as each disparaged the other side's preferred strategy. The intensity of the debate sucked the oxygen away from almost everything else, including the part of the slogan that Halperin and a small but growing number of colleagues believed offered the best chance to curb HIV—the B for "be faithful," a shorthand for the broad concept of partner reduction. They started calling it the "neglected middle child" of ABC.[5]

Conservatives pushed for what became known as the "abstinence until marriage" earmark, which set aside one third of the entire AIDS prevention budget for that strategy. There was little science to back this other than the uncertain claim that teens who delayed the start of their sexual lives had a role in helping control Uganda's epidemic. But while some delay had been documented by Stoneburner, Green, Carael, and others, it wasn't clear how important it ultimately was. If a Ugandan woman started having sex at age seventeen rather than sixteen, that merely pushed her risk for infection into the future. But her lifetime risk depended far more on her overall number of sexual partners and the extensiveness of the sexual networks she joined.[6] When the story of Uganda moved to the center of the debate, scientist David Serwadda traveled from Kampala to Washington to lobby Congress, and to try to correct the record on what had happened in his country. But he quickly came to believe that supporters of abstinence programs didn't care. Uganda's example seemed merely politically convenient. What was far more urgent, Serwadda concluded, was satisfying the demands of Christian conservatives. In one Capitol Hill meeting he attempted to convince a Republican senator that abstinence had not been essential to fighting AIDS in Uganda, but the senator made clear that it would be now. "Look here, young man," he told Serwadda, who was then in his forties and beginning to sprout gray hair, "the only way you are going to have money in Africa for AIDS is if the bill passes this way."[7]

Back in his hotel room that night Serwadda ordered a glass of whiskey and took out his frustrations on the keys of his laptop computer. A few hours later he had drafted an opinion piece for *The Washington Post* arguing that the brewing Bush program already was veering off track because of its insistence on abstinence programs. "As a physician who has been involved in Uganda's response to AIDS for 20 years, I fear that one small part of what led to Uganda's success—promoting sexual abstinence—is being overemphasized in policy debates," Serwadda wrote. "While abstinence has played an important role in Uganda, it has not been a magic bullet."[8]

But in the end, the senator was right: Abstinence funding was essential to the political compromise that made PEPFAR possible, and it didn't ultimately matter what role it actually had played in Uganda.[9] Soon after Serwadda's visit, the White House agreed to support the abstinence earmark, amounting to $1 billion for encouraging young people to forgo premarital sex. This was nearly triple the entire U.S. government's global AIDS budget only three years earlier. Yet again in the fight against the epidemic, science had butted up against politics. And, yet again, politics had won.

Some of Bush's critics portrayed PEPFAR as a diversionary tactic in a presidency devoted to war. But Bush and his aides always described PEPFAR as a moral imperative. At the height of the push to get the program approved by Congress he gave a Rose Garden speech in which he invoked the biblical story of the Good Samaritan: "When we see this kind of preventable suffering, when we see a plague leaving graves and orphans across a continent, we must act. When we see the wounded traveler on the road to Jericho, we will not—America will not—pass to the other side of the road."

Embedded in that metaphor was an idea of African suffering that had infused the West's approach to the continent since the days of David Livingstone. The wounded traveler on the Jericho road had been assaulted and robbed by others. The Good Samaritan was acting to ease injuries for which he had no direct responsibility. But the story of AIDS was more complex. President Bush, of course, bore no personal responsibility for sparking and spreading the epidemic, but he was the leader of a Western world that had never fully accepted its essential role in helping create a deeply troubled modern Africa. Like a physician examining a sick patient, Bush saw the disease and reached for available remedies. And his determination to invest billions of dollars in those remedies merits praise. Yet PEPFAR—along with the Gates foundation and other major international donors—subtly entrenched a story line that obscured the fundamental narrative of the epidemic, casting the West as inherently wise, generous, and benevolent. It just happened upon a wounded traveler and offered help.[10] The suffering Africans were simply victims, helpless to help themselves.

PEPFAR mythologized Museveni's campaign in Uganda but otherwise hewed to the essential Western tendency to disregard what Africans themselves knew about their epidemics and societies. Even African officials in line to benefit from PEPFAR sometimes bristled at what they saw as arrogance in its delivery, and at the relentless need for bureaucrats to produce the numbers to fulfill Bush's pledge.[11] Mozambican officials, for example, were astonished to learn that the United States had designated it a PEPFAR focus country without consulting with them first. They also fought back against the American insistence on brand-name drugs at a time when the country had already decided to treat its citizens with generics that were cheaper and easier to take. The subject became even touchier when Bush appointed a former drug company executive, Randall Tobias, as the head of PEPFAR, fueling suspicions that the administration was more interested in protecting

Big Pharma than African lives.[12] The health minister in Zambia complained that plans for fighting AIDS in the nation had "all come from Washington."[13]

As Bush's plan got rolling, condoms would arrive by the hundreds of millions, more than ever before. Donor-funded abstinence programs proliferated in nations that in most cases had never had them before. Hospices and hospitals got major new investments, as did government contractors and American universities, who shipped well-salaried doctors, researchers, and consultants to Africa by the hundreds. ARVs, meanwhile, eventually began reaching millions of people who otherwise might never have gotten them. But as African governments fell in line, and did their best to keep up with the myriad rules and reports PEPFAR demanded, homegrown initiative faltered in many parts of the continent. There would be no more Francos or Philly Lutaayas sternly warning their countrymen against the growing plague. Even as the story of Uganda was celebrated, the conditions that had encouraged its political, religious, and cultural leaders to teach their nation how to save itself were all but gone. Sam Okware, one of Museveni's earliest advisers, and who had watched the build up of a donor-driven AIDS industry in Uganda remarked sadly, "The whole thing is too big now, too heavy. . . . It has adapted too much to international guidelines instead of sticking to our own methods, which were very controversial at first but which worked."[14]

GORDON AND THANDI

Several months after Bush's State of the Union address, I made my first trip to South Africa as an aspiring foreign correspondent for *The Washington Post*. I had worked several newspaper jobs, including stints covering Virginia politics and Washington, D.C.'s mayor, and I was eager for new challenges. When I inquired about a job coming open—as Johannesburg bureau chief, with a purview including the southern third of the continent and much of West Africa as well—the foreign editor gently suggested South Africa might make a good vacation spot for somebody angling to get on staff. So I took ten days off and flew to Johannesburg, armed with little more than a list of potential contacts and a credit card. A colleague of mine, the *Post*'s biotechnology writer, Justin Gillis, had begun schooling me on AIDS, taking me through the basics of opportunistic infections, protease inhibitors, and how to make HIV testing more routine. Gillis had the brain of a scientist, capable of sorting through reams of data in search of the immutable building blocks of truth. And what I imbibed over a series of tutorials was the AIDS epidemic as biomedical challenge. What we never talked about was sex, or culture, or the complex realities that I would soon find in Africa.

After landing in Johannesburg I rented a car, bought a cell phone, and began making calls. I soon had places to stay around the country but no real idea how to find the face of AIDS. A few parts of southern Africa had higher infection rates, but at the center of everything was Johannesburg, the cosmopolitan metropolis known simply as Joburg. A night in a bed-and-breakfast in nearby Soweto, where reports made clear the epidemic was even worse, offered few insights. Everybody was willing to talk about HIV. I could see the puzzling loveLife billboards. But if there was a war on, it was all but invisible. Where were the walking skeletons? The shrunken babies? The wailing mothers that I had seen on every news report and fundraising appeal I'd ever seen about AIDS?

To find that, it turned out, took some work. Most South Africans who had HIV didn't know it, and those who knew it often kept the secret to themselves, or told only a small circle of friends and relatives. When visible symptoms of disease set in, many with AIDS returned to family homesteads in far-off villages in the countryside. Eventually, as the wasting, the diarrhea, and the coughs became too much, victims ended up in hospitals, hospices, or the back rooms of a parent's home, waiting for death. As I began to understand I finally saw how AIDS had continually tricked those living among it. The disease spent years moving invisibly through societies as little more than a rumor, a vague nightmare you could wish away every dawn. By the time most people could see the epidemic in the faces of those they loved, critical years had been lost. For many it was already too late.

I soon got my first unvarnished glimpse of the epidemic, in the town of Tugela Ferry, in the heart of the rocky, rugged homeland of the Zulus. As I drove down a steep mountainside into town, I passed goats and cattle wandering on the road, along many pedestrians for whom a car such as my little Nissan was an impossible luxury. The ferry for which the town was named had stopped running long before, but there was a bridge now, along with several low-slung concrete blocks of shops and a large hospital founded generations earlier by Church of Scotland missionaries but now run by the provincial government.

After pulling into the lot and parking my car, I wandered into the rudimentary lobby and began asking around for a doctor whose name I had been given by a journalist friend. That doctor was away, but another, Dr. Theo van der Merwe, a gentle, brown-haired missionary, greeted me instead and offered a tour. The wards were a series of large, open rooms jammed with metal beds that afforded no privacy to the dying. As we walked through, I began to have the sickening feeling that strikes almost everyone when first encountering the devastation of AIDS. The doctor said that half of the hospital's three hundred beds were taken up by people in late stages of the disease. But my own eyes suggested the ratio was even worse. Most patients seemed bony and slack, with pallid skin and an unmistakable air of exhaustion. Van der Merwe put me under strict orders not to mention the words AIDS or HIV. I was to use instead a neutral term, "the virus," as we walked by row after row of deteriorating bodies. I spoke no Zulu, so a conversation with a patient was unlikely, but in the eyes of these men and women I saw a dullness, a hopelessness that made me want to look away.

After our rounds through the wards we stopped in a room that had a series of small doors along the wall. It was the morgue. The facility was designed to hold forty-eight bodies in cold storage for the few days before relatives collected their dead, and the capacity once had been sufficient. But AIDS had overwhelmed the hospital. Now several dead babies had to share a single drawer and, during particularly bad stretches, even the bodies of adults got doubled and tripled up. Van der Merwe told me that a handful of the hospital's patients got ARVs through private insurance plans. A university study might soon allow the doctors to put more adults on an effective combination of ARVs. But I knew they weren't likely to arrive in time to help the men and women I had seen lying silent and miserable in the wards.

I didn't know it at the time, but a political shift was underway that would soon change the situation. Drug companies, at the demand of activists and the urging of UNAIDS and other international groups, were lowering their prices, and generic drugs were becoming more widely available. Now that Bush had made his $15 billion commitment, South Africa's cabinet was dropping its objections to ARVs. There was another, quieter factor as well in this decision. The final results from the Virodene trials in Tanzania had arrived in early 2002, and they were disappointing. Patients in the trial, by some measures, had seen their health improve, but the drug demonstrated no impact on HIV. Virodene's investors continued to make the case that the medicine could work, just through a different mechanism than expected. But South Africa's intelligence service, which was monitoring the trial, had picked up hints of troubling side effects, and President Mbeki's point man on the project, Maisela, had gotten so fed up with the Virodene investors that he had walked away from it in disgust. A few weeks after the results arrived, South Africa's cabinet agreed to launch a new program using ARVs to curb transmission between infected mothers and their babies. The activists at the Treatment Action Campaign and beyond had won. Rumor soon spread that an even bigger initiative—to deliver antiretroviral drugs to all AIDS patients who needed them—was in the works. Tellingly, Mbeki had fallen into a sullen silence on the subject.

I returned to South Africa the following year, in 2004, as the *Post*'s Johannesburg bureau chief, but my enthusiasm for attempting to understand AIDS soon ran into barriers familiar to almost all correspondents posted on the continent. Most American readers, it seemed, would happily donate some pennies to the cause via their taxes or church collection plates. But they displayed much less interest in yet another tale of suffering Africans dying of some dreadful disease. Meanwhile, the easy availability of ARVs in wealthy countries had made the epidemic seem like a diminishing threat in the places where most of the *Post*'s readers lived. There remained, of course, significant spread of HIV among gay men. African American communities, in Washington and elsewhere, also were relatively hard hit. But many people in the United States generally had begun viewing AIDS as a chronic disease similar to diabetes—unpleasant but essentially controllable with the right medicine. One veteran colleague said to me, with a tone verging on the unkind, "Sooner or later you're going to realize that there's a lot more going on than AIDS."

She wasn't wrong. South Africa was heading into its third presidential election since the end of apartheid. Politics consumed the airwaves and casual conversation nationwide. But unlike campaigns I had covered in the United States, the questions weren't about who was going to win but by how much.

The ANC had a lock on the electorate, and though much of that stemmed from the party's historic role in bringing multiracial democracy to the nation, I found myself wondering why there wasn't a political price to be paid for Mbeki's disastrous handling of AIDS. If it was a political issue here, it was hard to tell beyond the ranting of small opposition groups. In Johannesburg's plush northern suburbs, where the rising African middle class thronged to shopping malls that had a polished-stone glossiness to rival their counterparts in suburban Washington, I would do imaginary head counts ("One, two, three, four, five—you have HIV," I'd say to myself). But there was no sense of looming disaster as shoppers stocked up on Diesel jeans and Nike track shoes. Even South Africa's famously confrontational AIDS activists had decided to keep a low profile during election season. I wanted to know: Why?

That question brought me to Gordon Mthembu and his fiancée, Thandi Mzukwa, the leaders of the Sinethemba Support Group. They lived, like many black South Africans, in a township created by apartheid's architects to keep cheap labor near cities but not so close that white people might have to encounter blacks as something other than maids or gardeners. In the post-apartheid era, this ugly dynamic largely held, although the townships also had plenty of unemployed men and women, as well as concentrations of HIV startling even by the standards of southern Africa. The Sinethemba Support Group worked in the dusty township of Katlehong, a few miles southeast of Johannesburg. And its leaders, although affectionate and apparently much in love, disagreed about whether to vote to reelect Mbeki and the ANC.

Mzukwa, twenty-nine, was slim and sweet-faced, with a tendency to end her sentences with the Zulu word "*ne?*" which rhymes with the Canadian word "eh?" and has a similar linguistic function. In the previous year she had lost her niece and her best friend to AIDS—needlessly, she knew, because ARVs were available just a five-hour drive away in Botswana. To punish Mbeki, Mzukwa was preparing to vote for the main opposition party. "If he did provide ARVs in time, my friend would be still alive," she said angrily. [1]

Mthembu, forty, whose voice had the leathery tones of a heavy smoker, was just as frustrated but unwilling to abandon the ANC. During the struggle against apartheid, he had led a gang of youths in burning down buildings belonging to the government and its sympathizers. They even torched the mayor's house. Mthembu spent six years in prison for these acts of insurrection. So when he mentioned that he'd like to punch Mbeki, I believed him. "For people living with HIV," Mthembu told me, "we were condemned by the president." Yet with that same leathery voice he made clear that he was not about to abandon a party at the core of his self-identity. The other twenty-four members of the Sinethemba Support Group felt the same way.

Mthembu and Mzukwa also wanted me to interview a member of the group who recently had gotten sick enough to be admitted to a nearby hospital. We met there the next day, and while I was happy to see the couple again, I was not prepared for what greeted me next—a pungent, overripe odor that consumed us as soon as we stepped into the hospital. There was a potent antiseptic, probably ammonia, lurking in there, but it was losing out to the cadaverous stench of sickness. Soon I would know this aroma of an overwhelmed African hospital all too well. But on this warm April day it was unnerving. My stomach clenched as we climbed the battered stairway and entered a large open ward painted a drab institutional green. Amid a long row of steel hospital beds was the man we had come to see, Dennis Makgobojane. The virus had not settled for eating away his immune system, his muscles, his skin. It seemed to be consuming his soul. Thin and dazed, with a bag of his urine hanging from a hook, Makgobojane's dignity was draining away along with his life. His dark, watery eyes wandered emptily, unable to focus, unable to fix on anything in a world fading to black.

He somehow propped himself up for my visit, that legendary African hospitality asserting itself as a reflex, and attempted to answer my questions. And I did my best to ask them. In the bubble of the reporter-subject relationship, I distracted myself as best I could from the larger truths: I was a well-salaried white American whose life, with a little luck, stretched out ahead of me for decades; he was a poor African dying from a disease both preventable and, for those with the means, largely controllable. I managed to get the spelling of his name (M-A-K-G-O-B-O-J-A-N-E), his age (forty-three), his life details (unemployed, with two daughters). And then I asked him about the election a week away. Makgobojane mumbled a few sentences about how he wished the government would do more about AIDS, but he planned to vote for the ANC. I dutifully scribbled down the quotes, then looked for an excuse to leave as quickly as possible. The conversation appeared to have exhausted him. So I thanked Makgobojane after just a few minutes and began to back away. One of his friends from the support group stayed behind as the rest of us climbed back down the stairs into the fresh air, free from the stench, free from the hospital's troubling odors of death and injustice.

Outside I thanked the support group members for their help and climbed back into my tidy white Nissan, with its oddly reassuring rental-car smell. Soon I was driving back to the comfort of Johannesburg's Hyatt hotel, where I could immerse myself in the familiar world of words and deadlines amid a soothing decor that I had come to think of as Africa-by-Disney. As I settled into the hotel's plush central lobby, I began typing away on my Dell laptop while eating a light dinner. The words that flowed from my fingers were hardly poetry, but they worked well enough for my first dispatch from Africa. I filed a couple of hours later over the hotel's Wi-Fi link and tried to relax. After an exchange of some computer messages it became clear that my editor back in Washington was satisfied. And as midnight approached, I was beginning to ease into a second glass of red wine when my cell phone rang. It was one of the members of the AIDS support group: Makgobojane was dead.

I awkwardly expressed my condolences. Then, confused and a bit distraught, I called back my editor. I had seen a few bodies during my years as a journalist, but the passing of a man just a few hours after I had pestered him with political questions jolted me. Now I was not at all pleased with my story. It needed more. Anger? Empathy? Urgency? In the version that was heading into tomorrow's newspaper, I hadn't even quoted him, couldn't figure out how to use a dying man's final words. My sensible editor, understanding far better than I did that death alone does not impress newspaper readers, convinced me to keep the story essentially intact. The original version had ended with a pair of quotes from frustrated members of the support group who already had lost two of their friends to the epidemic. My editor and I added a new last sentence reading, "A few hours later, a third member of their support group, Dennis Makgobojane, 43, died of AIDS."

That night, as the wine continued to soften my defenses, I entered a dark cave of doubt and self-recrimination. Journalism had never felt so futile.

I was invited to Makgobojane's funeral on the Saturday after his death. It was a quiet day; my wife and children were still back in Washington preparing for the move to Johannesburg. So I drove to Katlehong and, with my note-book and pen put away, began to absorb the township in a new way. For starters, Katlehong, though indeed dusty, had a working-class vibe I had missed on my first visits. This was no shantytown of tin shacks and sewage in the streets. The homes were made of sturdy concrete blocks or bricks. Many of the roads were paved. Kids were kicking balls around. There was even an ice cream truck ding-a-lingling its way through the township, with a posse of ecstatic kids trailing behind. The small house that Mthembu and Mzukwa shared with ten relatives—a number I had slipped into the story for its potential shock value—seemed cramped but livable as I assessed it as a guest rather than as a reporter. The weather was warm and sunny most days, so people spent hours outside, chatting, visiting neighbors, hanging out at the informal bars called "shebeens," which were mostly just the front yard of somebody with a big refrigerator full of beer.

For the service, we gathered at Makgobojane's family home. A tent in the front yard cast shade on rows of folding chairs facing a casket and a tall, thin minister in black robes. He also was a member of the support group, mean-ing he too had HIV, and though he spoke in Zulu, I could easily grasp the sentiments as he talked about their late friend. Standing beside the minis-ter were Mzukwa and other women from the support group, wearing red T-shirts that boldly proclaimed HIV POSITIVE, in what amounted to the signature uniform of AIDS activism in South Africa. The women sang what at first sounded like South African versions of church hymns, with call-and-response melodies enlivened by Zulu rhythms. But they were AIDS protest songs about defeating stigma and living positively with the disease. The soul-ful music continued at the cemetery, where the remains of Makgobojane were lowered into the ground. We took turns tossing clumps of dirt onto the casket while friends and relatives said a few final prayers. Then it was time for the party.

Back at his house, there were mounds of food, along with a steady supply of beer and liquor. It was my first encounter with a South African feast. Many more were to come, because most holidays, funerals, weddings, graduation parties, and church cookouts there eventually took on the familiar, festive contours of what I first experienced that day. At these events, more than at any other time during my years on the continent, I felt wrapped in Africa's embrace. Wide-hipped grannies decked out in colorful floral dresses and all but immovable in their favorite chairs would take turns holding court and passing around the youngest children. Older kids would play in the yard or in the street just beyond the gate. And almost everyone between twenty and sixty would drink their way to chatty, then tipsy, and often beyond. Late afternoon would bring thumping, bumping dance music emanating from giant speakers rented for the occasion.

At these parties I also began to sense the powerful connections among even distant relatives here. The family trees were hard for me to grasp, or maybe they were just irrelevant. Every contemporary was called a cousin, a sister, or a brother, and everyone in middle age a mother, father, aunt, or uncle. Every older woman or man was a grandmother or grandfather, and afforded a measure of respect beyond what I typically saw back home in the United States. In South African townships, everyone seemed knitted together by a web of kinship and obligation and, more often than not, affection. Those with decent jobs at a shoe shop or in an office helped support four, six, or ten relations. And my family and I—despite being outsiders with the same pale skin as this community's longtime oppressors—were always treated like friends. Even the poorest South Africans would invite us into their homes, pour some tea, and dig out their last tin of sugar cookies. At parties, the women toiling in the kitchens would emerge to chat, and make sure we all had some food. If we settled in even a little bit, one of the ladies would dash off to get us heaping chunks of meat, stewed greens, the boiled cornmeal concoction called *pap*, and tangy salads made of shiny red beets.

During our years in South Africa, my wife and I were forever leaving these events too early, loading our sleepy kids into their car seats just as the evening was getting festive. But at this funeral, with my family far away, I relaxed and chatted with Mthembu, Mzukwa, and the rest of the support group. Through my beery haze it gradually dawned on me that they were nothing like the dour, passive AIDS victims often portrayed in fundraising appeals or on political programs. Still in their HIV POSITIVE T-shirts, most had a cigarette in one hand and a drink in the other. Many spoke solemnly about the need to use condoms every time they had sex, and they all seemed concerned that they might reinfect each other with new strains of the virus if they weren't careful. But it seemed to me that other messages about protecting themselves had not sunk in very deeply. Just hours after burying a friend the imperative of celebrating life overpowered the fear of death. Or maybe something I couldn't grasp was going on. In any case, it didn't fit the preconceptions I had loaded into my brain before leaving the United States.

Such dissonant moments often are the first sign of a good story looming in the distance. So I made plans to keep in touch with the Sinethemba Support Group. Perhaps I could even write about the wedding of Mthembu and Mzukwa when it came, to illuminate the complexities I was beginning to sense. How would they plan a future together? Manage their health? Start a family? The idea seemed to offer a way past the "AIDS fatigue" of readers by offering up accessible characters whose lives they could understand, maybe even share for the few minutes it took to read a newspaper story. But my imagination had gotten ahead of me. The next time I saw the two of them was several months later, at Mzukwa's thirtieth birthday party. This time my wife and kids were along for the festivities in Katlehong, which were as raucous as ever. But the wedding had been canceled. In the intervening months Mthembu had gotten another woman pregnant, which Mzukwa took as a profound betrayal. In the age of AIDS, such arrangements carried the additional problem of increased risk of infection, or in their case, possible reinfection.[2]

That same day I found a minute to speak privately with Mthembu, this man whose fierceness and conviction once had inspired me. So who was this other woman? And what about the condoms everybody swore they were using every single time they had sex?

"Ah, Craig," he replied in a sheepish, smoke-cured voice. Then he began to tell me the truth about condoms. Some people used them some of the time. Some people never used them. Almost nobody used them all of the time.

A MARSHALL PLAN FOR BOTSWANA

A s the Bush administration embarked on its historic mission to fight AIDS, a glimpse into the future—reassuring in some ways, worrying in others—was available by visiting one remarkable nation: Botswana. Few African countries emerged from colonialism with so little damage. For a century, Europeans had scoured southern Africa for gold and copper while building cities whose laws and structures subjugated the people who had traditionally lived there. But most of Botswana, by contrast, was considered worthless desert. The British declared it a protectorate, under the name Bechuanaland, and built a colonial city, Francistown, in the far northeastern corner. When independence arrived in 1966 the nation was poor but peaceful, with a degree of ethnic homogeneity that helped it to avoid the power struggles that consumed so many newly liberated African nations. Diamonds were found soon after across much of Botswana's midsection, and a well-led government used the resulting profits—Botswana would become the world's largest producer of gem-quality diamonds—to develop the nation. It built modern highways and water systems. A gleaming new capital rose on the southeastern edge of the Kalahari Desert. Subsidized education became available to citizens, up through doctorates at universities abroad. In this shining era before AIDS took hold Botswana's currency, the pula, became known as the "African pound" because of its sturdiness.

The allure of governmental competence attracted international aid groups like few other African nations. Gaborone, the capital, was an easy, pleasant place to live and work, with its gleaming new malls and excellent road network, and the urban bustle of Johannesburg was just a few hours' drive away. But during Botswana's fast-forward modernization, some problems grew. Marriage had fallen out of fashion, as it had in many southern African nations—and a culture of casual sex had taken hold.[1] HIV spread, and international donors responded with slick condom promotion campaigns that inadvertently reinforced this element of the culture. On Halperin's first visit to Botswana, in 1999, he observed an event organized by Population Services International, a Washington-based nonprofit group that specialized in the social marketing of condoms. At the event, billed as a "teenage bikini beauty contest," girls strode provocatively across the stage clothed in little more than Brazilian-style swimwear consisting of strings and tiny patches of fabric. A packed crowd full of beer-drinking men, meanwhile, yelled out lewd comments and swayed to throbbing American rap music with lyrics such as "I want to fuck you, ho!"

Halperin was no prude. He had been a child of San Francisco's free love era in the early 1970s and had hung out interviewing prostitutes during his South American travels. But he was amazed that this performance, featuring girls who in some cases appeared to still be in middle school, qualified as AIDS-prevention programming in a country with one of the world's highest HIV rates. The only reference to the epidemic were the occasional admonishments by the event's emcee to "always use a Lovers Plus rubber!"—a plug for the condom brand sponsored by Population Services International and underwritten by American tax dollars.[2]

When it came to treating HIV, Botswana—with 1.6 million people scattered across an arid, landlocked mass nearly the size of Texas—did much better. Having already built a relatively sophisticated public health system by 2001, Botswana had a foundation on which to create Africa's first major program to distribute ARVs. The program, called Masa (a Setswana word meaning "new dawn"), combined the nation's diamond wealth with resources and expertise offered by high-powered Western donors, mostly the Gates foundation and the drugmaker Merck. Bill Gates had become interested in health issues in poorer countries after reading one of the World Bank's annual development reports; he realized that for what amounted to pennies on the dollar he could parlay his software fortune into programs that could save millions of African lives.[3] "This is about life and death," Gates later told *The Washington Post*. "This is about the degree to which we've improved human existence, and how much more we have to go. Years from now people will look back at this era and ask: Was humanity improving?" Gates had the good fortune of sharing this passion with his wife, Melinda, who was similarly moved by the poverty she encountered during visits to Africa.

These insights were not unique but the resources that the Gateses brought to bear were. The assets of the Gates foundation swelled into the tens of billions of dollars, making it not only the largest charitable foundation in the world but the largest in the history of the world. The single biggest chunk was dedicated to global health initiatives. A massive initial investment went into deploying existing vaccines in poor countries that couldn't afford them. Gates also invested heavily in malaria, tuberculosis, river blindness, guinea worms. But many of the highest profile initiatives were for AIDS. And Gates brought a worldview to this problem that was not merely Western; it was hyper-Western. Gates and the people he hired, including a heavy contingent of research scientists and physicians, were methodical, data-driven, results-oriented, and they displayed abiding faith in the answers produced by labs. They bet on vaccine research, vaginal gels, and antiretroviral drugs along with HIV testing and condoms, both male and female. If there was a great idea lurking in some laboratory or factory that might save Africans from AIDS, the Gates foundation was going to spare no expense in finding and developing it.[4]

This approach made the Gateses natural partners when American pharmaceutical giant Merck went looking for help in brightening its public image—AIDS activists had long portrayed Big Pharma as greed incarnate—and to test whether its medicines could be deployed effectively in the nations with the highest HIV rates.[5] The resulting initiative, the African Comprehensive HIV and AIDS Partnerships, or ACHAP for short, forcefully challenged lingering doubts about the usefulness of ARVs on the continent. Gates and Merck each pledged $50 million to the effort, which also had support from the Harvard AIDS Institute and donations from other drugmakers. Soon tiny Botswana found itself the proving ground for the Western model for fighting the epidemic in Africa. Here there would not be any doubts that the national government was providing the political will many AIDS experts believed was essential to the battle. Botswana's president was the Oxford-educated Festus Mogae, one of the continent's most respected leaders, and he was as vocal, forthright, and open to outside help as Mbeki in neighboring South Africa was surly and distrustful. As excitement built about the Gates/Merck project, supporters began calling it the Marshall Plan for Botswana.[6]

Yet hints of trouble already were perceptible in the initial decisions ACHAP made. Resources in the first budget it released were understandably skewed toward treating existing infections. Leading the drug-treatment effort was the suave, Harvard-trained physician Ernest Darkoh, who later told South Africa's *Sunday Times* newspaper, "You can talk prevention all you want. The realities of infection were so high, Botswana needed treatment."[7] The relatively limited resources for prevention, meanwhile, went overwhelmingly to promoting and distributing condoms. Men there reported using them frequently for casual sex but, as elsewhere, not with regular partners. Little emphasis was put on addressing sexual behavior more broadly, and on the dangers of maintaining multiple relationships.[8] The formula pioneered years earlier by the Ugandans—which relied on fear and the invoking of traditional social and moral codes to slow the spread of infection—was turned on its head. Some prevention messages in Botswana went so far as to celebrate a culture of sexual conquest; one Population Services International billboard showed a soccer ball wrapped in condoms and the words EVEN THE

BEST BALLERS TAKE A SAFE DUNK WITH IT.[9] Edgier still was another billboard
that featured a boxing glove and a condom, with the line IT CAN TAKE THE
FIERCEST PUNCHES![10]

Professor Serara Selelo-Mogwe of the University of Botswana was a community elder in the African sense—not just old but wise and widely respected. She had founded the university's school of nursing in the 1970s, and
for decades had been a national authority on public health.[11] When AIDS
first became a consuming national problem she began to speak out against
donor-funded prevention strategies that she believed were out of sync with
African cultures. Now, several years later, Selelo-Mogwe was increasingly
frail but still convinced, as Uganda's leaders had been, that AIDS was the
manifestation of undesirable social changes, part of a shift from communal
African values toward more individualistic, pleasure-seeking Western ones.
Her voice filled with disgust as she described crossing the campus each Monday morning, stepping carefully past the beer bottles and used condoms that
were the detritus of yet another raucous weekend.[12]

She believed that the people of Botswana needed to spend more time
talking to one another about the traditional elements of their culture—
respect for family, elders, each other—and less time listening to all of the
Western experts who were descending on their country. Selelo-Mogwe said
the influx of outsiders revived a colonial dynamic in which the most revered
knowledge and power came from outside and Africans themselves often fell
into subservient, passive roles. "Some white person comes with utter garbage,
something that will not work in our culture, and it's hallelujah! Praise the
Lord!" Selelo-Mogwe said, as sarcasm filled her frail, quivering voice. "We
don't question anything that comes from the Western world."

Instead, Selelo-Mogwe favored the revival of more traditional notions
of morality, along with the resurrection of the coming-of-age schools that
in precolonial times were a central feature of adolescent life in Botswana.
Christian missionaries had taken the lead in stamping them out—and with
them, the routine circumcisions once received by adolescent boys there—as
heathen. The final blow came in the 1980s, at a time when the HIV epidemic

was making its first inroads into Botswana. Some community elders were trying to revive the initiation schools, but outside experts at the time were expressing concern about circumcision, which in Africa had often involved cutting the foreskins of several boys in a row with a single unwashed blade.[13] And so one of the final vestiges of a fading tradition disappeared. "If we had tried to build on our own culture," said Selelo-Mogwe years later, "it might have made a difference."

The seriousness of that wrong turn became even clearer in 2005, when a team of researchers offered the most convincing proof yet that male circumcision could be an important tool in preventing AIDS. Momentum for more definitive evidence had been growing for several years, especially since the announcement of powerful new findings in 2000. These included a rigorous review of all existing evidence that estimated circumcision could reduce men's risk of HIV infection by half or more.[14] Soon after, government-backed medical research agencies in Canada, the United States, and France decided to devote millions of dollars to underwriting sophisticated studies to measure the potential impact of male circumcision on reducing vulnerability to HIV.

Randomized controlled trials are routinely used to test the safety and effectiveness of new drugs, and their protocols are so strict—with every possible measure taken to isolate a single cause and its particular effects—that they are considered the gold standard of evidence for health interventions. Three such trials got underway almost simultaneously early in the decade, but the first one to report results was led by French researcher Bertran Auvert, who conducted his trial in the Orange Farm township south of Johannesburg. The setting was not nearly so bucolic as the name suggests. Like many of the places where black South Africans lived, it was a chaotic sprawl of shanties mixed in with some sturdier homes, a few shops, a few paved roads, and a railroad station. The township was lightly governed and prone to occasional outbursts of violence. If prosperity and social equality were on their way in this postapartheid world, it was too far over the horizon for most there to see. Orange Farm also had an absolutely ruinous AIDS epidemic, with something on the order of one quarter of all working age adults infected with

HIV. Most of the men who lived there were Zulus, meaning few were circumcised.[15]

Western anxieties about running a major trial on circumcision long had focused on concerns that Africans would view it as an unwelcome cultural intrusion on the most private part of a man's anatomy. Yet Auvert had no trouble recruiting participants.[16] The bigger worry among the researchers was an ethical dilemma common to medical trials: If you think you have a tool to keep people from getting sick, how can you withhold it from half of the study's volunteers? It was a troubling issue, but years of ideological and bureaucratic resistance to circumcision had made the larger stakes even clearer. Without randomized controlled trials offering definitive proof, international health authorities would never accept the procedure's power to slow the spread of HIV. The choice essentially was to accept the ethical quandary and proceed or give up on a tool that could eventually prevent millions of infections.[17]

Auvert's team recruited 3,274 uncircumcised men age eighteen to twenty-four and provided them all with HIV tests, standard counseling about safer sex, and access to free condoms. Then half were chosen at random to be circumcised. This group became the experimental arm of the trial, while the other men, foreskins still intact, became the control arm against which circumcision's effects could be measured. Both groups agreed to follow-up visits, including additional HIV tests and interviews about sexual behavior over the next two years. But after only fourteen months the men who were not circumcised already had developed more than twice as many infections as the ones who had been circumcised.[18] The split was impressive but also troubling. With such clear results, the data and safety monitoring board established to protect the subjects of the study concluded that there was no ethical way to continue the experiment. The board instructed the researchers to immediately shut down the trial and offer circumcision to all of the volunteers.

After years in which ethical concerns had help keep HIV prevention programs from offering circumcision, the dynamic had suddenly reversed itself: It became unethical not to offer it. The decision cast years of official inaction on the subject in a new light. When Auvert and his colleagues analyzed the data they found that, by the most conservative analysis, circumcision appeared to make the men 60 percent less likely to contract HIV. A more comprehensive analysis that also counted the experiences of men who were not supposed to get circumcised but did anyway estimated the protective effect at a remarkable 76 percent.[19] This wasn't for a single encounter. It was for the entire study period, involving dozens and in some cases hundreds of sex acts.

Auvert presented the findings at an international AIDS meeting in Rio de Janiero in July 2005. Although some skepticism remained—UNAIDS issued a cautious statement saying that the agency "notes with considerable interest" the results but stopped well short of endorsing the procedure—that began to evaporate when the results of the two other circumcision trials were announced the following year.[20] In Uganda, where Johns Hopkins University researchers had recruited nearly five thousand men, the group that had been circumcised had half as many HIV infections as those who were not. Robert Bailey, Halperin's coauthor on their "10 Years and Counting" article in *The Lancet,* was the lead researcher in Kenya, where the trial included nearly twenty-eight hundred men. In this trial, the split between the two groups was even wider.

In none of the trials were there any significant differences in any other factor, such as age, income, numbers of sexual partners, or visits to prostitutes. As with Orange Farm trial, the monitoring boards for the Uganda and Kenya trials concluded that the initial results were too powerful to allow them to continue. The National Institutes for Health, which was providing most of the funding for both trials, shut them down and offered all the men free circumcisions. Two decades of scientific debate over whether circumcision slowed the spread of HIV were now finally resolved. Never again would researchers be allowed to recruit men for a trial testing whether circumcision offered protection from HIV.[21] It was established scientific fact. And the protection offered by male circumcision was roughly what Halperin and Bailey had suggested in their *Lancet* article. "Six years, probably a million new infections, maybe $30 million later, you get the same answer we had staring us in the face in 2000," Bailey said later.[22]

• • •

In Botswana, the nation's groundbreaking treatment program struggled through its first couple of years. Many of the patients who came to clinics initially were terribly ill, just weeks away from death and in need of dramatic medical interventions. And those who had HIV but were not yet visibly sick often stayed away, unable or unwilling to confront the stigma. The project had planned on putting nineteen thousand patients on treatment in its first year but managed to attract only about three thousand by the end of 2002—a small percentage of the estimated one hundred thousand people who needed the drugs immediately.[23] Gradually, though, Botswana's treatment program gained strength, as did aggressive HIV testing initiatives that helped people discover they needed medical help.

By July 2005 the number of people being treated in Botswana was pushing past forty thousand, and the national mood had begun to lighten. During a visit to Gaborone I interviewed Harriet Isaacs, fifty-nine, a lean but energetic civil servant who had been on antiretroviral drugs for three years and was thoroughly acclimated to the regime. With retirement just a few months away she was optimistic about a future of playing with her grandchildren and, perhaps, maybe even with her great-grandchildren. "I'm confident that I can go up to one hundred [years old] now," she told me, as her eyes twinkled from behind her glasses.

The fact that Isaacs readily gave her name also was a good sign. Only a couple of years earlier journalists often had to resort to obscuring identities when writing about AIDS, because people with the disease would not otherwise describe their troubles publicly. But with medicine increasingly available, and the certainty of imminent death lifting, shame and stigma began to ease as well. AIDS became a more routine part of life. The pace of deaths, meanwhile, began to slow. Reliable numbers were hard to pin down, because physicians often sought to protect the reputations of their patients by listing a different cause of death—meningitis or tuberculosis—when someone died of AIDS. Yet people in Botswana could feel the shift when they gradually got their weekends back. Before ARVs arrived it wasn't unusual to spend two or three weekends a month traveling to funerals. As Masa built up its roster of patients, death began to resume its appropriate place as only an occasional part of life.

But Botswana's success had a troubling side as well. As AIDS became a manageable disease, new people were being infected nearly as fast as ever. Eventually there was a drop in infection rates among young people but not like the sharp declines across most age groups seen in Uganda and some other nations that didn't have Botswana's resources. Substantial reversals of the epidemic in Kenya, Zambia, and Rwanda had been tracked by researchers, and soon would be in Ethiopia and Malawi as well. There seemed to be similar reasons in most of these places: Amid the terror of mass death, sexual mores changed. [24] Those who kept the most partners tended to get sick and die, along with many of their spouses. Those left behind tended to be more careful. That was a terrible way for a society to learn how to protect itself from AIDS, but what was happening in Botswana was also far from ideal. Amid all of the HIV testing, condom promotion, and miraculous drugs, the epidemic was becoming normalized but not beaten into retreat.[25]

WHAT SHALL WE DO?

A s ARVs became almost universally available in Botswana, the comparisons with South Africa became increasingly explicit. While South African AIDS deaths were blamed on Mbeki's reluctance to embrace ARVs, deaths delayed in Botswana were attributed to the good deeds of the Gateses, Merck, and President Mogae. The reality, of course, was subtler. Botswana was certainly a better nation in which to have AIDS; the medicine was free and plentiful, delivered in modern clinics, and available to nearly everyone. But the ultimate goal of a life free from HIV remained elusive in both nations, and Botswana's infection rate stayed even higher than South Africa's. Meanwhile, a third nation in the region, Zimbabwe, was on a different path.

As far back as 1987, a time when few Zimbabweans believed that AIDS even existed, musician Oliver Mtukudzi wrote a song called "Stay with One Woman" as part of a WHO competition to publicize the disease. Mtukudzi was a tall, elegant showman known for his all-night performances and a deep, mournful voice that wove social themes into energetic guitar melodies. The government sometimes employed similar messages. A 1992 poster warned YOUR NEXT SEXUAL PARTNER COULD BE THAT VERY SPECIAL PERSON. THE ONE THAT GIVES YOU AIDS.[1] President Mugabe himself, soon to become notorious for political repression and economic catastrophe, talked about AIDS more often than most African leaders. Even First Lady Grace Mugabe, a former government secretary who started dating the president while his first wife was dying of kidney disease, instructed Zimbabweans that "appropriate behavior based on good old morality and self-respect is the surest and cheapest long-term solution to the problem of AIDS."[2]

Although the Zimbabwean response tilted toward messages about sexual behavior, condoms also had a substantial role. International donors shipped them to the country by the tens of millions, and men in Zimbabwe, like those in Botswana, began reporting condom use with prostitutes and other casual partners at rates that were among the highest in the world. Some promotion campaigns were remarkably explicit for a nation in which talking about sex remained taboo in many settings. In one initiative funded by international donors, troupes of lightly clad women traveled the country performing seductive dances, complete with gyrating hips and thrusting pelvises, that more closely resembled actual copulation than anything airing on MTV. Sometimes the dancers would arrive in remote villages with strobe lights and disco machines, said David Wilson, the University of Zimbabwe social scientist who managed the project and later became head of AIDS programs at the World Bank. The faces in the crowd, where some were aghast and many others clearly titillated, made Wilson begin to rethink the entire approach. When he questioned other Zimbabweans involved, they felt equally uneasy, but they also felt powerless to craft alternative messages. "We are so embarrassed," they told Wilson, "but this is what they told us to do." He later recalled, "We didn't really think it was on the table to tell people how to live their sex lives. Condoms were such a simple solution."[3]

As the epidemic continued to grow and AIDS deaths mounted, Zimbabweans began to turn on each other. In 1994 a member of parliament demanded that all pregnant women with HIV be killed to prevent the disease's transmission.[4] One of the nation's most important liberation heroes, Vice President Joshua Nkomo, lost his son to AIDS two years later and angrily declared at the burial that the disease was "harvested by whites to obliterate blacks."[5] Those few who knew they had HIV, meanwhile, kept the secret to themselves for fear of the shunning that often happened when communities began to suspect that somebody with the dreaded disease lived among them. The absence of antiretroviral drugs sharpened the terror that swept Zimbabwe, as death rates began mounting in the late 1990s. Mtukudzi, who lost his younger brother and four band members to the disease, wrote a new AIDS-themed song, "Todii," that was released in 1999. The lyrics were far from the Old Testament directives of Franco's "Attention na SIDA" and were more cryptic than Philly Lutaaya's "Alone and Frightened." Instead, "Todii" described the torment of watching loved ones die of AIDS but offered no explicit solutions. The title, a Shona word that means "what shall we do?" was also the chorus, repeated in both Shona and English. Mtukudzi later recalled in an interview, "The aim of it was to make people talk."[6]

They did. Many Zimbabweans approached him to ask, "Are you really sure this disease is real?" Mtukudzi would reply, "Yes. I'm really sure. My brother died of it." A video of "Todii" released the following year showed a family in which the man of the house was coughing and struggling to eat as he gradually died of AIDS. A chilling series of images showed mourners tossing handfuls of dirt onto a casket moments after it had been lowered into the earth. Several particularly unnerving shots came from within the grave itself as the dirt rained down and Mtukudzi sang, "What shall we do?" It's worth noting the subject of that plaintive question: "We." Mtukudzi was a Zimbabwean talking to fellow Zimbabweans, and that gave the message moral power that outsiders could never invoke.

Amid this heightening awareness of AIDS, Zimbabwe's politics and economy took a drastic turn. Serious inflation began in 1998. By 2000, a powerful opposition movement was challenging Mugabe's hold on power, and self-proclaimed veterans of the 1970s liberation war began invading commercial farms owned by the white families that still controlled most of the nation's best land. While Mugabe resorted to repression and brutality to maintain control, aid groups fled, inflation spiked, and food grew scarce. Soon Zimbabwe had few friends among the Western nations that were underwriting the fight against AIDS. At the very moment when international political will was building to provide ARVs to Africans, Zimbabwe largely took itself out of the game. As the trickle of medicine flowing into Botswana and other African nations gradually grew into a gush, it would largely bypass Zimbabwe.

UNAIDS's 2005 annual epidemic update report bore the agency's signature mix of alarm sprinkled with a few uplifting stories and a declaration that only massive new funding could prevent the epidemic from getting even worse. A top WHO official told journalists in a conference call, "It's increasing everywhere."[7] Yet hidden in the report was news that the HIV rate in Zimbabwe had mysteriously dropped. Researchers who later investigated found that a familiar mix of terror and death had caused a fundamental change in the country's sexual culture. The key years turned out to be from about 1999 to 2003, while the earnings of Zimbabweans were beginning to plummet. Condom use with casual partners and prostitutes during that time remained high, and teens on average were waiting a few months longer to start having sex. But the most important shift, as in Uganda, was a broad retreat from having multiple sex partners. Between 1999 and 2005 the number of married men reporting sex outside of their marriages declined by 30 percent. In one especially rigorous study in the eastern Manicaland region, the numbers of both men and women reporting concurrent sex partners dropped by 40 percent. An adult HIV rate that had peaked at an estimated 29 percent in the late 1990s gradually fell nearly in half, to 16 percent.[8] High rates of death were clearly a factor, but, as in Uganda, there also were other forces at play in what had become the largest HIV decline ever recorded in southern Africa.

Halperin traveled to Zimbabwe to study what had happened and found a broad social awakening to the reality of AIDS and a message reasonably well unified across a range of government, religious, and cultural leaders. The potentially complicating factor, however, was Zimbabwe's economic and political self-destruction. The moral authority Mugabe enjoyed during the early years of his presidency had disappeared by then. The most educated Zimbabweans, meanwhile, were fleeing by the hundreds of thousands. Those left behind grew poorer by the month. And the public health system had frayed almost completely.

During a reporting trip to Zimbabwe in 2006, I met a woman, Faris Kungara, forty-one, who had gone to a government hospital with searing headaches. The doctor told Kungara, a round-faced mother of three who survived by selling vegetables on the street, that she likely had cryptococcal meningitis, a common opportunistic infection for people with AIDS. He ordered a painful spinal tap to confirm the diagnosis, but instead of sending the test off to a lab and commencing treatment, the doctor handed Kungara a small, red-topped vial of her own spinal fluid and a bill for 6.1 million Zimbabwean dollars—at that point, about $60 in U.S. currency. The doctor explained that only after Kungara paid the bill would he complete the test and prescribe treatment. When Kungara said she didn't have the money the doctor told her, "You go and find it," and sent her away. When I interviewed Kungara months after this episode she still had the vial of her spinal fluid, the unpaid bill, and the headaches that left her all but unable to earn money or care for her family.

Horrible though her story was—and Zimbabwe was full of such tales, and even worse ones as the nation developed hyperinflation that rivaled Germany's in the 1930s—Halperin and his research colleagues found a silver lining when it came to the spread of HIV. Men increasingly lacked the cash to hang out in bars and pick up women. Poverty wasn't spreading infection, as many experts long contended. Something closer to the reverse was true. As financial security collapsed, ministers reported more couples in their church pews, and extramarital girlfriends became too expensive for most men to manage. "That's by and large now the preserve of the wealthy," said Pastor Elliot Mandaza of New Life Covenant Church in Harare, the capital. "It's hard enough to look after number one." Even prostitutes, at less than one U.S. dollar a visit, were becoming an unaffordable luxury. Rising death rates also inspired profound shifts in behavior.[9] Even bars tried to curb the casual hookups once commonly available there, firing crews of "cleaning women" who had picked up extra money by selling sex. Frank Muhamba, who owned a bar called Ghetto Blues in a bedroom community south of Harare, said spreading HIV had become bad for business. "Before we could go to a bar, and we'd find ten women wanting us." Now, he added, "we will go home without talking to any of those girls. . . . They will kill us."[10]

The precise relationship between Zimbabwe's mounting financial disaster and falling HIV rates was complicated.[11] The main declines in new HIV infections due to behavior change appeared to come relatively early in the crisis.[12] Yet the nation's story offered another example of how broad shifts in sexual culture could curb HIV. And it challenged the idea that major infusions of outside money and expertise were necessary to reverse the epidemic. Zimbabweans did it, as Ugandans had before them, with relatively little in the way of external help. It's possible, though unpleasant to contemplate, that the shortage of ARVs had the perverse effect of helping prevent infections. The terrible surge of deaths across Zimbabwe, combined with the lack of realistic hope that ARVs would soon arrive on a large scale, chastened a troubled nation.[13]

The comparison with Uganda was not the only important one. Zimbabwe has a border with Botswana that runs for more than four hundred miles, and though the nations were dominated by different ethnic groups, they could easily have ended up as a single nation but for the quirks of the region's colonial history.[14] Yet on HIV, Botswana had turned in one direction while Zimbabwe ended up heading in nearly the opposite one. Those circumstances yielded notably different trajectories in their AIDS epidemics. This begged a troubling question: Was it possible to reverse the HIV epidemic while also helping millions of people already caught in its tide?

Halperin had gone to work for the U.S. government to help untangle such questions. But the politics of the job were wearing on him. On one occasion he was part of a USAID team that went to Brazil to assess its epidemic and recommend ways that the agency could help. It was near the beginning of the U.S. war in Iraq, a time when many nations had brittle relations with the Bush administration. Brazil was deservedly famous for having achieved widespread access to ARVs, in part by producing its own generics, and it did not need assistance regarding treatment issues. The head of Brazil's national

AIDS program meanwhile had no interest in the ABC strategy. Instead he asked for an assistance package centered on providing a billion condoms for use by Brazilian youth. The USAID team pointed out gently that such an initiative, in the absence of any effort promoting other approaches, such as abstinence, would be unlikely to win approval in the upper reaches of the Bush administration. But more important, Halperin and his colleagues argued, prevention efforts should target the most important means of transmission. In Brazil that was anal sex.[15]

The subject was taboo but the act popular, between men and men and also between men and women.[16] Halperin and some HIV experts in the country soon developed a plan built around a proposed new product—a redesigned, tougher condom that would be built and marketed specifically for anal sex. When Brazil rejected the $40 million package offered by the U.S. government they portrayed it publicly as a moral stand against PEPFAR's controversial requirement that recipient groups all sign pledges opposing prostitution. But behind closed doors, the conversation was decidedly more coarse. One national official said angrily to the USAID delegation, "Fuck your ABC abstinence crap. We don't need your dollars."

The ABC formulation that Halperin had helped develop within the Bush administration had created such a polarizing backlash that he spent much of 2004 helping to craft a consensus statement that backed all three elements of the prevention strategy but put the emphasis back where he thought it belonged: on the B, for "be faithful" and partner reduction. Halperin and a group of colleagues went through furious rounds of outreach and negotiation and managed to get 149 leading scientists, policy makers, activists, and religious leaders from sixteen countries to sign it for an article on seeking "common ground on HIV prevention" that appeared in *The Lancet* shortly before World AIDS Day, December 1. The list included one superstar, South African archbishop Desmond Tutu, and almost included two other notables who might have given the document more political impact. Both Peter Piot, at UNAIDS, and Randall Tobias, head of the PEPFAR program, came close to signing the consensus statement, along with the heads of USAID, the CDC, and the National Institutes of Health. But at a critical hour the Bush White

House, unhappy that the document also supported condoms, forbade anyone representing the U.S. government from putting their name on it. And
when most U.S. officials pulled out, so did Piot. Halperin sidestepped the
prohibition by citing his nominal relationship with the faculty of the University of California, San Francisco, even though he had been at USAID for
more than three years. Another USAID coauthor was listed simply as representing "Washington, D.C."[17]

Such experiences heightened Halperin's desire to be based closer to the
front lines of the African epidemics. The declaration in *The Lancet* had provoked controversy within the administration, and meanwhile, his longtime
defender at USAID, Global Health Director Anne Peterson, had drawn the
ire of some conservative Christian leaders, who eventually pressured her into
leaving.[18] Rumors started spreading soon after that Halperin might be the
next to go. He moved first, by getting himself posted to southern Africa as
the agency's HIV prevention adviser for the region. He successfully argued
for a home base in Mbabane, the capital of Swaziland, a tiny mountain
kingdom that had the highest estimated HIV rate in the world and was even
farther away from Washington politics. The Swazis had abandoned male
circumcision in the 1800s, and the society remained in flux between the
traditional polygamy practiced by their king, Mswati III, who had thirteen
wives and counting, and a more modern variant that often involved several
ongoing relationships but fewer actual spouses. Halperin figured it was a
place where the need was massive and where relatively focused investments
of time and resources might yield substantial results.

During a preliminary visit in August of 2004, he stood outside his room
at Mbabane's Mountain Inn and gazed down the Ezulwini Valley, a rocky
green expanse so beautiful it made him want to stay forever. If he had any
regrets about leaving the United States and putting his family through the
drama of an international move, they disappeared during a subsequent phone
call with the USAID home office. Peterson's replacement was Kent Hill, an
evangelical Christian who some in the administration may have expected to
be in sync with its conservative supporters. Halperin and most of his USAID
colleagues generally found Hill, who had a steady, agreeable manner, to be

nonideological, with views grounded in evidence. But shortly after Halperin and his family arrived in Swaziland, he was walking down a dirt lane one moonlit night when he received a call from Hill in Washington—eight thousand miles, seven time zones, and an entire moral universe away. It turned out Hill was looking for a new head of the office of HIV/AIDS, an increasingly prominent post within USAID. Halperin was friendly with one of the potential candidates, a veteran foreign service official who had strong HIV credentials as well as a record of positive experiences working with Christian groups in Africa. He was happy to endorse her, but he stopped in his tracks when Hill pushed for a bit of personal information that Halperin regarded as off-limits. "Does she go to church every week?" Hill asked. "I need people who cannot just talk the talk but walk the walk."[19]

While Halperin would not have been very surprised to hear this from some others within the administration, he was stunned that it came from Hill. For once, Halperin was speechless.

RAYMOND THE GREAT

n December 2005 I drove through the vast dry flatness of the South African highveld, swigging Vanilla Cokes and chomping on biltong, the spicy dried meat that amounted to the national snack food there. Past cattle herds and corn fields, past shantytowns and smokestacks, I pushed my car to speeds I wouldn't have dared on the heavily patrolled highways back home in the United States. And after about three and a half hours of numbing sameness, of towns I couldn't tell apart without a map, I rounded a steep, pine-covered slope and found myself entering Swaziland.

The dramatic mountain highway, with its palate of lush greens and rocky grays, soothed my weary eyes. But more important, Swaziland offered a way into the intriguing circumcision news bubbling out of the scientific community. The Swazis were embracing the emerging evidence more openly than any other nation in southern Africa. The impetus came, I would soon learn, from Halperin. The heads of PEPFAR were still treating the subject warily as a potential tool against HIV, but he knew how crucial it could be in a society where few men were circumcised. So he had became a roving advocate for expanding access to the procedure. During Halperin's years in Washington, he had used USAID funding to help start circumcision pilot projects in Zambia, South Africa, and Haiti. But in Swaziland he was more directly involved, crisscrossing the country with PowerPoint presentations full of charts, stats, and maps. Along the way he developed relationships with traditional leaders, health officials, and physicians, including one who had a weekly radio show during which the host soon was espousing the benefits of male circumcision. Halperin also managed to pry loose almost $150,000 in USAID money to underwrite a pilot project offering circumcision to Swazi men. At the first free clinic, which had been advertised through a small ad inside one of the nation's two newspapers, so many men lined up that the

Communications materials for Swaziland's pilot male circumcision
project in 2005–2006 also included messages about sexual behavior.

assembled doctors couldn't attend to them all. News reports called the resulting unrest Swaziland's "circumcision riot."[1]

When I pulled into Mbabane I could see it had the familiar look of small,
hilly cities I had known on the east coast of the United States, where down-
towns of older brick buildings gradually were being supplanted by newer
suburban-style developments. Halperin's office was at the headquarters for
the National Emergency Response Council on HIV and AIDS, which despite the impressive name was housed in a two-story concrete building that
doubled as the main gas station on the road in from South Africa. It was
there, amid the diesel fumes and highway roar, that Halperin and I first met.

We shook hands and indulged in some chitchat before deciding to walk
to a nearby restaurant for lunch. As the crow flies, the place was no more
than a few hundred feet away, but for most humans the journey involved
walking out onto the sidewalk, up the main drag, and back through the
parking lot of a shopping complex. Halperin, however, had determined that
he could shave a few minutes off the journey to the restaurant by going
overland, through the gas station parking lot, along the weedy, muddy back-
side of a building, over a small footbridge. And so we went, careening off
the path of conventional wisdom in our first moments of face-to-face ac-
quaintance.

He opted for a soft sell on our first lunch together. I told him that I had
growing doubts about how the AIDS epidemic typically was portrayed. And
I made clear that the circumcision news, while somewhat baffling, felt like a
fresh angle that deserved more exploration. Halperin filled me in on the
details of the Orange Farm study and explained how removing a man's foreskin eliminated the cells most vulnerable to HIV while also making the skin
tougher and more resistant to infection. "It gets like your finger," he said
helpfully, if somewhat crudely, aiming his index finger in my direction. He
also talked about the centrality of sexual culture to understanding an AIDS
epidemic, and in particular the networks established in a society with high
rates of concurrent relationships. A key to this was something Halperin called
the "acute infection phase," in which HIV surged through a newly infected

person for a few weeks or months before easing, as the immune system gradually responded. It was during this time—a span of increased contagiousness when there are few signs of illness and a standard HIV test usually cannot even detect the infection—when much of the epidemic's transmission happened.[2] That helped allow the virus to race through the nation's extensive sexual networks.

When I asked about Swaziland's notoriously high HIV rate, which UNAIDS had put at 38.8 percent of adults, Halperin readily agreed that the epidemic threatened the future of a nation of just a million people. But he also offered a word of caution. The UNAIDS reports, he said, relied mainly on statistics generated by clinics tending to pregnant woman, and on average they were younger, more urban, and more sexually active than the population as a whole. Each of these factors made these women more likely to have HIV than the average adult in a country. So while Swaziland's adult infection rate was surely horrific, and probably was still the worst in the world, it almost certainly wasn't brushing up against 40 percent. In any case, Swaziland was soon due to be the subject of one of a new generation of surveys, funded by USAID, that would help settle the matter. Such a survey—called Demographic and Health Survey, or DHS for short—used census-style sampling techniques to collect data on issues such as access to family planning information and health services. But in nations with major AIDS epidemics, these surveyors in recent years had begun also including HIV testing to better track the path of infection.

The results of these studies suggested that the UNAIDS picture of the epidemic was off in some important respects. The numbers in most individual nations were considerably lower when measured using these more rigorous sampling techniques.[3] And the overall weight of the data, from these surveys and other independent studies, suggested that HIV wasn't on the rise worldwide, or on the verge of breaking loose across new frontiers of Asia or Latin America. It had peaked and was gradually retreating, as all epidemics eventually do, nearly everywhere across the globe.[4] That didn't mean AIDS wasn't a serious problem. It was a catastrophe in southern Africa and some other places. Even by Halperin's conservative guess, one in four of Swaziland's working age adults had HIV. And though the number of new infections probably had begun falling, it was falling much more slowly than in some other African countries that had more effective prevention strategies. But he also maintained that the spread of HIV was more confined than generally understood, with the hardest hit nations almost exclusively located in Africa's AIDS Belt. Having just written a story for the *Post* about the most recent UNAIDS report, and having uncritically quoted a UN official saying "It's increasing everywhere," I began to feel like a very large fool.

Halperin followed up with a series of e-mails about HIV and what was wrong with the way the world was fighting it. He urged me to pay particular attention to a few places that had done better. Top on that list was Uganda. And soon in my e-mail inbox, a series of documents arrived that included a *Science* article, from April 2004, on the miracle that had been Zero Grazing.

I didn't yet know the remarkable backstory of the article, of how authors Rand Stoneburner and Daniel Low-Beer had struggled for years to win acceptance for the findings they first began to glimpse one night in Kampala in 1995. Nor did I know how UNAIDS decided not to hire Stoneburner but did offer a job to Carael, whose interpretation of Uganda's HIV decline—with its emphasis on condoms over partner reduction—essentially became the officially sanctioned history within the new agency and the conventional wisdom among experts everywhere.[5] Halperin, it turned out, had a bit part in undoing that when, during his time at USAID in Washington, he officially requested the Ugandan behavioral data for one of the studies he oversaw.[6] Stoneburner, then doing some consulting work for that USAID study, then was able to finally complete the research he had started so many years before.

The result turned out to be worth the wait. With the full data set at last, Stoneburner and Low-Beer reexamined the results of those crucial questionnaires, along with other information. This allowed them to produce the definitive account of why Uganda's epidemic retreated. The window for this accomplishment had been narrow, from the late 1980s to the early 1990s.

Rates of new infection leveled off after that, meaning the decisive changes
had to have happened during Museveni's Zero Grazing campaign and as
Lutaaya was dying. The time frame made it clear that the favored narratives of the Left and Right—which still were transfixed by their battle pitting
condoms against abstinence programs—were both fundamentally flawed.
Condom promotion programs began on a serious scale in Uganda in the
mid-1990s, as progress against the epidemic's spread was, in retrospect, just
beginning to slow. Formal abstinence programs began even later, during a
long period of stagnation in the fight against HIV. What did fit the time
frame were the broad shifts away from multiple sex partners that, thanks to
the full data sets from 1989 and 1995, Stoneburner and Low-Beer finally
were able to document with authority. For these changes, they credited Uganda's use of "rational fear," "credible communication," and "indigenous responses at the community level" arguing that they caused crucial changes in
sexual behaviors.[7] Along with the massive mortality at the time, these led to
the nation's historic HIV decline.

The *Science* article did not challenge Michel Caraël by name, but UNAIDS
did not emerge unscathed. Stoneburner and Low-Beer noted that the agency
long had reported that condoms and teen abstinence were keys to Uganda's
success. The authors added drily, "An important and perhaps overlooked
measure of behavior change in Uganda between 1989 and 1995 was a 60%
reduction in persons reporting casual sexual partnerships in the past year,
evident in urban and rural populations." And they gently noted, "The behavioral changes in the data have been evident since 1996." Stoneburner and
Low-Beer concluded that Uganda's campaign was equivalent to a vaccine that
was 80 percent effective. In a separate analysis, they estimated that if a Zero
Grazing–style strategy had been followed in South Africa, which had a larger
population and even higher HIV rates than Uganda, it could have prevented
more than three million infections.[8]

Halperin's cautions about the UNAIDS estimates surprised me, but soon he
e-mailed me several of the DHS reports he had mentioned. Produced by a
company called ORC Macro, based in suburban Washington, the reports had

informed some of the agency's most successful initiatives, such as helping to provide family planning services for poor women across the world. The DHS survey techniques were not well suited for measuring sensitive sexual matters, because research teams generally posed their questions in an individual's home, often with other family members or neighbors within earshot. But at gauging the extent of HIV in an area, the DHS reports were a step beyond the standard methods used before.[9] The blood tests were voluntary, but in most countries a large majority of survey participants agreed to them. This allowed what often was the first detailed mapping of infection patterns across all parts of a society, and it allowed analysis of infection rates by a number of measures such as age, education, reported sexual behavior, and, eventually, male circumcision status as well.

The surveys found that people who reported more lifetime sexual partners were much more likely to have HIV. So were city dwellers, educated people, and relatively affluent ones. In Africa, women were more likely to have HIV than men, in part because younger women were more likely to have older sexual partners, including so-called sugar daddies who offered gifts such as cell phone air time or money for school fees.[10] Little of this surprised scientists familiar with the ways the virus spread. But what did prove surprising to many were the overall infection rates, including in some rural areas where traditional African ways still dominated life. Years of estimates by UNAIDS suddenly looked too high, in some cases much too high.

Signs of discrepancies emerged as early as 2001, when a DHS report came out for Zambia. UNAIDS had estimated the adult infection rate in Zambia at 20 percent, but the DHS measurement was 15.6 percent. In the Dominican Republic the following year, the UNAIDS estimated 2.5 percent, while the DHS found a rate of just 1 percent.[11] But perhaps the most startling gap was in Kenya, a much larger country and one with a more extensive epidemic. A DHS survey in 2003 put the Kenyan infection rate at 6.7 percent, less than half of the official UNAIDS rate of 15 percent. This was good news: The difference was equal to more than a million Kenyan lives. But it also was apparent that the UNAIDS reports for years had failed to make clear the path of HIV through the world. Piot and others have said that they made the best possible estimates in an era when the measurement tools were limited, and also that their tracking was far more precise than what they inherited from WHO.[12] But PEPFAR and other major donors based their funding decisions on the UNAIDS estimates of the severity of epidemics. Perhaps equally important, countries themselves relied on those statistics in order to make decisions about where to focus their limited public health resources. The overestimations in several West African nations, such as Liberia, for which the DHS estimate was one quarter the UNAIDS one, also made it harder to detect just how powerfully patterns of male circumcision were affecting the spread of the virus.[13]

UNAIDS did not always move swiftly to incorporate the insights that the ORC Macro studies were revealing. After the DHS study showed that Zambia's rate had been overestimated, UNAIDS actually raised its estimate in its next report, up to 21.5 percent. The agency did accept the newer Kenyan numbers, but it did not take the next logical step of also attempting to mitigate the larger trend of overestimation apparently at work across many nations. When the next wave of DHS surveys came out, the UNAIDS numbers looked even more off track. The 2005 DHS in Ethiopia reported an infection rate of 1.4 percent, not the 4.4 percent UNAIDS had estimated. And the 2006 DHS in Swaziland found that the infection rate was not the UNAIDS number of 38.8 percent; it was 25.9 percent, still horrible but slightly less so.

Armed with the trove of DHS reports, I soon organized a trip to Rwanda, a nation that had been on the leading edge of the AIDS epidemic as HIV moved east into the Great Lakes region. In the 1980s, researchers estimated that close to 30 percent of urban adults had the virus. And many experts feared that the 1994 genocide, with its mass rape, murder, and dislocation, had made the epidemic even worse. But the real story, as was so often the case with HIV, defied the best instincts of many intelligent people. UNAIDS estimated Rwanda's adult infection rate at 13 percent in 1998. In early 2004, when the Bush administration selected Rwanda as a focus country deserving hundreds of millions of dollars of intensive help annually, UNAIDS listed the adult infection rate for Rwanda at 8.9 percent. The agency lowered it to

5.1 percent later that year, but a DHS survey the next year estimated that it was even lower: 3 percent. That was one of the lowest HIV rates in East Africa, and far lower than most experts had expected.

For a story that appeared on the *Post*'s front page in April 2006 I quoted James Chin, the former WHO epidemic tracker, who had been frustrated when the official HIV estimates tacked sharply upward soon after UNAIDS took them over in the mid-1990s. Not only did the numbers seem too high to Chin, the shape of the epidemic also seemed to defy the basic laws of epidemiology. The so-called epidemic curve called for the pace of new infections in outbreaks to rise, then fall, even in the absence of effective interventions. Chin complained in his 2007 book, *The AIDS Pandemic: The Collision of Epidemiology with Political Correctness*, that the UNAIDS models showed HIV rates somehow managing a slow, steady increase that appeared to go on forever. Chin suspected that the imperatives of raising money and political awareness had trumped science at UNAIDS.[14] "It's pure advocacy, really," Chin said in an interview. "Once you get a high number, it's really hard once the data comes in to say, 'Whoops! It's not 100,000. It's 60,000.' . . . They keep cranking out numbers that, when I look at them, you can't defend them."[15]

Chin had one other key point. He predicted that soon UNAIDS would be forced to accept that HIV transmission was falling—and the response of Piot and others would be to claim credit. Chin called it "riding to glory on the down slope of the epidemic curve." If that was coming, it wasn't happening yet. When I called veteran epidemiologist Peter Ghys at UNAIDS he disputed that there was political pressure to inflate numbers, but he acknowledged that the DHS surveys made clear that the agency had "overestimated the epidemic a bit." Ghys also agreed that rates of new HIV infection had stabilized in Africa.

There was one troubling exception to the overall trends in the DHS surveys in Africa. Out of more than a dozen available at the time, the UNAIDS estimates looked too high in all but one. The outlier was Uganda. UNAIDS had put its adult infection rate at 4.1 percent, but the DHS conducted in

2004 estimated it at 6.4 percent. This was one of the first hints that even as the Bush administration had embraced Uganda as a model for its AIDS prevention policies, progress there was beginning to falter.

Raymond Kwesiga was much like the young Ugandan men who had made substantial changes to their sexual behavior in the late 1980s as the fear of death peaked and Philly Lutaaya conducted his crusade.[16] But Kwesiga was in elementary school when all that happened. Uganda had changed in the intervening years, and so had he. Now Kwesiga had grown handsome and charismatic, with dark eyes framed by rectangular, metal-rimmed glasses. He tried to be a good citizen of the moral universe he inhabited as a young man, periodically getting tested for HIV and keeping packs of condoms on hand. Kwesiga used them when he was sober, especially with the prostitutes he could hire for about seventy-five cents a visit. But Kwesiga was drunk many nights. He favored a cheap Ugandan gin called *Waragi*, and the warmth that flowed from the bottles he shared with friends fueled a dangerous sense of invulnerability. Though Kwesiga was shy in some situations, the bottles of *Waragi* helped him get the most out of his reserves of charm. By the time he was in his early twenties his friends had started calling him "Raymond the Great" in honor of his prodigious number of sexual conquests.

It wasn't that Kwesiga didn't fear AIDS. He had watched an aunt and uncle die from the disease when he was still in school. The visions endured of them unable to eat, vomiting, dashing to the bathroom with diarrhea. His uncle grew so thin, Kwesiga recalled, that his head looked small and bony. And in church—Kwesiga was Catholic, an altar boy—he would pray for protection against HIV. "I would believe that God was ever with me."[17] But the AIDS education programs in high school did little to arm Kwesiga against the temptations all around him. These campaigns, in the first few years of the twenty-first century, focused mainly on abstinence. Not once could he recall his teachers talking about staying faithful to a single partner or the dangers of joining the freewheeling sexual networks once again reaching across Kampala. The cultural signals Kwesiga was receiving, meanwhile, were potent. The pecking order among his friends depended on the numbers of women they could bed. "Here girls can be with one partner, but for the boys, it's very, very hard."

As surprised as Kwesiga was to discover he had HIV—a positive test came when he was twenty-three years old, young for a man even in countries with severe epidemics—he was more surprised by the reaction of others. He wasn't shunned. He wasn't shamed. But he also wasn't believed. One former girlfriend kept badgering Kwesiga to get back together with her. Six months after his positive test for the virus, she propositioned him at a bar. He replied sharply, "Are you really sure you want to die?" She took it as merely a ploy: "Raymond, I know you are lying."

But gradually, as time passed, his friends accepted the news, and another friend within the hard-partying gang soon discovered that he had HIV as well. Kwesiga joined an AIDS group that performed at schools and youth centers to warn others against making the same mistakes he had. Yet here too he was astonished by the aggressiveness of some of the young women he met during the programs. Even after Kwesiga talked about his battle with HIV, they would approach him afterward with their cell phone numbers scrawled on a piece of paper, eager to arrange dates with the thoughtful man who looked anything but sick. Kwesiga's diagnosis, however, did bring some belated peace to his life. He laid off the *Waragi*, opting for just a beer or two on nights he went out. He was scrupulous in warning potential girlfriends about his infection. He was disciplined about using condoms for every sexual encounter. Kwesiga started to think about his own future for the first time. Though he knew it would be more difficult to marry and have children, he hoped these remained possibilities. And he also started thinking about his past—the drinking, the casual sex, and the possibility that he was not merely a victim of HIV but one who had spread it to others. "I was enjoying my life, and I thought I wouldn't get the virus," Kwesiga said, speaking slowly, determined to pick the right words, to live by newfound ideals of caution and honesty. "I wasn't very scared. . . . During the night, you don't get scared."

It was also harder, of course, to be scared in a society that has stepped back from its war footing against the epidemic.

President Museveni had made a stirring address at the 2004 International AIDS Conference in Bangkok, reprising many of the themes of his "thin piece of rubber" speech thirteen years earlier in Florence. And even though he had now led his country through the world's most dramatic HIV decline, his speech again was met with some displeasure. But in the years that followed, he talked about AIDS less often. Billboards and radio messages about sexual behavior had grown scarce.[18] And the government response had adapted almost entirely to international—meaning Western-driven—strategies for fighting the epidemic. There were no more war drums on the radio at 6:00 A.M., with the desperate voice of a young girl urging her father toward fidelity. Billboards didn't carry images of skulls and crossbones. And famous Ugandans weren't talking about dying from AIDS anymore. ARVs, thankfully, were increasingly available, courtesy of investments by major international donors such as PEPFAR and the Global Fund to Fight AIDS, Tuberculosis and Malaria. But the fear, the urgency, were gone. Uganda had become Botswana North.

On a reporting trip to Kampala in 2007 I spoke to students at Makerere University, the school where Philly Lutaaya had once delivered his chilling warnings to an earlier generation of students. But that message, like Zero Grazing, was now a distant memory in a bustling capital with a vibrant youth culture. I saw many more signs in downtown Kampala offering easy sex—GET A LOVER, 077-974995 said one typical message—than warning against AIDS. HIV rates had begun to tick back upward in some clinics. Longtime presidential spokesman John Nagenda said that Museveni "has gotten a bit bored with the AIDS story." Measures of risky sexual behavior, meanwhile, had also begun to move in the wrong direction. Rates of multiple partnerships had more than doubled since 1995.[19]

I spent a Friday evening at Makerere watching the comings and goings in a pair of dormitory parking lots with student Cathy Katumba, who had a heart-shaped face and long braids that she tied into a loop at the back of her neck. As we watched, men in their thirties or forties—often in spotless cars carrying the corporate insignias of their employers—would drive into the lots, park, and wait for their college girlfriends to come down for dates. The men, Katumba assured me, typically had wives at home. And the girlfriends had younger college boyfriends for the other nights of the week. But on Friday nights these older men got sexy young companions, and the girlfriends got a night out, maybe new clothes, cell phone air time, or a refrigerator for their dorm rooms. In the ARV era, Katumba said, these young women worried more about getting pregnant than HIV. "They don't look at it as a deadly disease now," she said.

As darkness fell, the men could be seen waiting in their cars, illuminated by the pale green lights of their cell phones. Gradually the young women sauntered out of their concrete dormitories, walked over to the waiting vehicles, and headed out into the Ugandan night.[20]

MAKHWAPHENI UYABULALA

Halperin has a habit of getting tested for HIV in many places that he travels. He does this because the experience offers him a window into the quality of HIV counseling sessions and into how the AIDS epidemic is experienced by those surrounded by it. Early in his time in Swaziland, Halperin walked up to the HIV clinic at Mbabane Government Hospital, the sprawling public health facility on a hillside in the capital, and asked to be tested. It turned out that he had arrived too late for the test itself, but the nurse said he could still get the counseling, so he retreated to a waiting room filled with about a dozen Swazis sitting together nervously on worn-out couches and chairs. Swaziland is a convivial country where conversations tend to break out wherever people come together. Yet this waiting room was striking for its anxious silences. A few bursts of chatter began but faded quickly. Some of the men and women were visibly shaking with fear.

The clinic supposedly was offering "rapid tests," but hours passed like this. Finally, the nurse emerged from her office and began calling people in for what seemed like remarkably brief visits. HIV tests are supposed to be accompanied by in-depth counseling sessions. Nurses are charged with instructing those who don't have the virus in how to keep themselves safe. And those who have HIV are supposed to receive even more time and care. Suicides after positive tests are not unknown. Women who reveal news of their infections to husbands or boyfriends are sometimes beaten or shunned.[1] And everyone with HIV needs guidance on how to maintain their health as long as possible and in how to seek the appropriate care when their condition eventually deteriorates. Some people, meanwhile, simply need tissues and a sympathetic ear.

This was not what was happening at this clinic. Men and women kept emerging after just a few minutes and, in every case, wore broad smiles. Some gave a thumbs-up sign to their fellow worriers in the waiting room. One man leaving the nurse's office went so far as to grin and wink at an attractive young woman who was still waiting for her result. In a nation with staggering rates of HIV, the scene puzzled Halperin, though he began to suspect that much of what he was seeing was for show. Swaziland is a small country, with tightly woven social networks. In any hospital waiting room, there is a good chance of running into a relation or a friend, or at least a friend of a friend. Emerging in tears, or even just wearing a worried expression, could spark rumors.

When Halperin's turn came he asked the nurse, "Wow, those people sure looked happy. Isn't it unusual to get so many negative tests here?"

The nurse corrected him. "Almost every one of them today tested positive."

Halperin's encounters with AIDS programs underwritten by international donors frequently depressed him. In Swaziland and in the other southern African countries he regularly visited, time, energy, and money were being spent on increasingly expensive initiatives that had little apparent impact. All those dollars came with reams of paperwork to complete as well as demands to orient national policies to the priorities of foreign donors.[2] Some ideas, such as the youth-oriented interventions of loveLife and similar programs, got hundreds of millions of dollars even if the evidence of their value was thin. Efforts to improve human rights, gender relations, and economic development also were popular, and, although worthy of support in their own right, often did not have clear connections to stopping HIV.[3]

There also were questions about the role of condom promotion in combating Africa's worst epidemics. In 2002, UNAIDS hired Norman Hearst, a former epidemiologist colleague of Halperin's from the University of California, San Francisco, to examine the effectiveness of such programs. In the resulting report—which UNAIDS never released, but which eventually appeared in the journal *Studies in Family Planning*—Hearst found that condoms rarely failed when used properly by individuals, but he couldn't find any examples of condom promotion campaigns slowing HIV's spread in African societies with widespread epidemics. He acknowledged their role in reducing infection in epidemics such as Thailand's, where transmission was concentrated within the sex industry. But while African men often used condoms in casual hookups or with prostitutes, few did so with their wives or girlfriends, despite years of public health campaigns encouraging the practice. He also raised the unsettling possibility, stimulated by some disturbing findings his research team had made in Uganda, that aggressive condom promotion campaigns, often featuring racy images and double entendres, may make casual sex seem more acceptable, potentially helping to spread HIV.[4]

Condom promotion was just one part of what Halperin began calling the "holy trinity" of prevention strategies, because devotion to them survived even as their scientific rationales gradually weakened over time. The other members of the holy trinity were HIV testing and the treatment of other sexually transmitted infections, such as gonorrhea, syphilis and herpes. Many large, carefully controlled trials seeking to test how these strategies worked in the real world turned up disappointing results.[5] The evidence is clear that people with other sexually transmitted infections are more likely to also have HIV, and vice versa. But it's unclear that treating those other infections will actually prevent this virus from spreading. All these infections result from similar behaviors and the sexual networks they create. If the goal is preventing disease, it would be more effective to break apart those networks through campaigns aimed at changing behavior.

The failure of HIV testing and counseling in trials also surprised many experts. Studies have found that most people who test negative—who learn that they don't have HIV—do not necessarily adopt more careful sexual behaviors. There is even evidence that a negative test can lead to more dangerous choices, a phenomenon called "risk compensation," because a false sense of security undermines the natural caution people might otherwise feel.[6] Those who test positive sometimes do make changes, but often these happen too late to protect their partners, because even frequent testing would not catch most infections during the early, acute phase, when those with HIV are especially contagious and before most standard tests would register a positive result.[7]

All three elements of the holy trinity clearly have important public health benefits of their own, but Halperin favored HIV prevention strategies that more directly addressed the root social and cultural causes of the virus's spread. The key, he was sure, was sexual behavior, and especially concurrent sexual relationships. But it was clear that they were not enough to entirely explain the hyperepidemics of East and southern Africa. Some countries in Central and West Africa, such as Cameroon and Ivory Coast, had even riskier reported sexual behavior than Botswana or South Africa, but their HIV epidemics were much less widespread than in those southern African nations. He believed that male circumcision was crucial to the pattern but, again, not by itself. After all, few European or Chinese men were circumcised but HIV rates were low in those areas.[8] Halperin concluded that it took both factors working together to create the patterns of infection seen around the world, and he came up with a slogan of his own. What existed in Africa's AIDS Belt, and in only a couple of other places on earth, was a "lethal cocktail" of extensive heterosexual networks and low circumcision rates. Changing either factor, on a broad enough level, could cause the pace of new infections to slow dramatically.

In Swaziland, Halperin supported a new initiative that some national AIDS officials hoped would get people to view the disease in a new way. The campaign relied on a provocative word, *makhwapheni*, that translated literally as something a person might hide under his or her arm. But the word was universally understood to mean "secret lover." To Halperin it seemed to capture the same direct tone used during Uganda's Zero Grazing era. If anything, *makhwapheni* had a sharper edge. The Swazis hoped the explicitness would

Images designed to look like text messages appeared on Swazi billboards in July 2006. The message on the left says: I'm dying to have you. *The one on the right says:* Why destroy your family. Your secret lover can kill you.

provoke conversation and help nudge the entire society toward confronting the behavior spreading HIV.

In a land where cell phones were common across nearly all parts of society, text messages became the central motif of the campaign. In July 2006, billboards began to appear across Swaziland that mimicked the face of a familiar Nokia handset, and the words on the screen bore racy invitations in cell-phone pidgin. One said I'M ALL ALONE. CUM 4 A QUICKY. Another said, SHE'S WORKING LATE, CUM WORK ON ME. Beside these invitations were punch lines, usually delivered in the local siSwati language, that hit with the force of a stick in the side. One was WHY DESTROY YOUR FAMILY. Another said AND MORE ORPHANS WERE LEFT BEHIND. The slogan for the entire campaign was *makhwapheni uyabulala*. It meant: "Your secret lover can kill you."[9]

Swaziland's top AIDS official, former cabinet minister Derek von Wissell, regarded the aggressive tactics as necessary. "We have been pussyfooting around sex," he said. "The first thing you must recognize is that sex spreads HIV."[10] The billboards, accompanied by animated television ads that showed attempted romantic liaisons interrupted by concerns about HIV, quickly became the talk of Swaziland. For a few weeks, whenever a cell phone buzzed in the country, there was a pretty good chance someone would deliver the stock joke: "Is that your *makhwapheni*?" The same thing happened when young women and older men were spotted leaving bars together. The *Swazi Observer* newspaper even put the words SAY NO TO MULTIPLE SEX PARTNERS!

in red capital letters on its front page, just below the newspaper's nameplate. "Sometimes you need shock therapy," said editor Musa Ndlangamandla.

But the *makhwapheni* campaign also provoked a furious backlash among some AIDS activists, who thought the ads were accusatory. They enlisted hundreds of people to march on von Wissell's office and chant, "Away with *makhwapheni!*" in one of the largest demonstrations in Swaziland's limited history of democratic protest. One activist, Siphiwe Hlophe, complained that her daughter began asking probing questions about her sex life and HIV infection. "Do you have a *makhwapheni?*" her daughter asked. Another activist, Gcebile Ndlovu, said angrily, "Who wants to be called a *makhwapheni?*" She added, "Don't tell me how many people to have sex with. . . . You can't dictate that to me. I go to bed with someone, that's my choice. Rather, tell me how to be safer."

In the context of a Swazi culture that was profoundly patriarchal, infidelity often was assumed to be the the fault of women rather than the men they slept with. Women had fewer rights than men and less social standing than their own sons. The *makhwapheni* campaign with its racy scripts accidentally touched on these sensitivites, and this controversy eventually spread overseas to activists in Europe and North America. Halperin, who had given input into the ads and billboards, was astonished when friends and colleagues back in the United States started sending him e-mails asking if he had heard of the infamous *makhwapheni* campaign. Many wanted to know: Why were the ads focusing on the sexuality of women?

The accusation puzzled Halperin. The term *makhwapheni* applied equally to both genders, and the secret lovers in the ads were as often men as women. Many Swazi women had told him they appreciated the campaign and were pleased to see such a frank discussion of the issue. Several said that it had inspired them, for the first time, to confront their husbands or boyfriends about the possibility that they had other women. But as the controversy grew internationally, Swazi activists kept pushing as well. The government suspended the program less than two weeks after it had begun. The billboards came down; the commercials went off the air.

Halperin was stunned, and he turned his energies to a USAID-funded effort to assess the program. A survey conducted a couple of months after it ended found that 86 percent of the Swazis surveyed had heard of the campaign, and despite the controversy, 91 percent said they agreed with its message. More important, 78 percent said it made them consider changing their own sexual behavior. Compared to a baseline survey conducted a year earlier, it also showed that the number of people reporting multiple partners had dropped by about 50 percent.[11] Both men and women reported similar declines. One young Swazi woman later told Halperin that the *makhwapheni* program had saved her life. But in the end, von Wissell was enough of a politician to know that a line had been crossed. The government revived the campaign two months later with new cell phone messages and updated commercials. But the word *makhwapheni* was gone, and with it much of the sharp edge.

THE FLOOD

L ike many mothers in Botswana, Chandapiwa Mavundu knew she had HIV. But she also thought that her infant son, Kabelo, would live a long, healthy life.[1] The staff at a government clinic had given Mavundu pills and Kabelo a dose of medicated syrup shortly after his delivery. Together they were supposed to shield him from the virus in his mother's blood. The final measure of protection was to come from avoiding a practice that had always been synonymous with motherhood in Africa. Mavundu was under strict orders to never breast-feed Kabelo. The nurses explained that virus hiding in her milk could infect her baby with HIV.

Mavundu was in her late twenties and had wide-set eyes and hair that she braided into neat rows stretching down to her shoulders. She already had two daughters, so she took special joy in her round-faced son with his playful disposition. In the tiny village of Nkange, just a scattering of a few dozen homesteads on the dusty northern edge of the Kalahari Desert, Mavundu cared for Kabelo in a thatched-roof hut as goats and chickens wandered nearby. Using cans of whitish powder provided by the government clinic, Mavundu made his formula with water from a nearby communal tap and heated it over a wood fire. And so long as the supplies of formula held out, Kabelo seemed to be growing strong. But when the clinic ran out of formula—"sometimes it was there," Mavundu later recalled, "sometimes it was not there"—she had no good choices. Her breasts had stopped producing milk already. The few shops nearby often were out of formula. So Kabelo often sipped cow's milk diluted with water or, when that upset his stomach, flour mixed with water, which offered the color and consistency of formula but few of the nutrients. Kabelo's development slowed during these stretches. The yellowed medical card that Mavundu kept showed worrisome dips in what should have been a steadily upward arc of rising weight. Yet when the formula returned, so did Kabelo's health. He could sit up by six months of age and soon began to crawl around the yard of the sandy homestead, amid the livestock, in a timeless scene of rural African family life.

"He was a happy boy," recalled Kabelo's grandmother, Thaloso Mavundu, who often cared for him when her daughter was away. "He was growing quite fast."

Then the rains came.

The Kalahari does not look like a desert out of an old movie, with nothing but miles of blowing sands stretching on forever. There are flat-topped acacia trees, squat bushes, long grasses, and enough root vegetables and wildlife to have sustained humans here for tens of thousands of years. And when the rains start falling, after months without them, the Kalahari gushes with life. The browns turn green. Springboks and wildebeests drink deeply at water holes. And people celebrate their good fortune. But in late 2005 and early 2006—summertime in Botswana—the rains became torrents. Storms saturated the soil and flooded the streets. Unlike in most African countries, indoor plumbing was fairly common, but pit latrines remained in many poorer communities, and they were in the path of the floodwaters as they surged through the countryside. The murky water, teeming with disease, gradually worked its way into streams and wells.

As the floods spread, Kabelo began vomiting and having diarrhea. His mother was away in the capital, looking for work, so his grandmother took Kabelo to Nkange's village clinic. But he didn't fully recover, and the rains kept coming, as did Kabelo's vomit and diarrhea. When Mavundu returned she found a shrunken, exhausted remnant of the son she remembered. Other children in Nkange were falling ill as well, and it wasn't just there. The outbreak was so terrible that the government of Botswana turned for help from the U.S. Centers for Disease Control and Prevention, which long had kept a major presence in the country to help combat tuberculosis and, more recently, AIDS. The CDC investigators tested village water supplies and stool samples from sick children. They tracked deaths from diarrhea and malnutrition—a malady rarely seen in Botswana in recent decades—and they scrutinized the yellowed health cards that often were the only medical records parents had for their children. The investigators also gathered information on HIV rates and breast-feeding practices. What they found was that despite one of the world's highest rates of childhood HIV, AIDS was not a direct factor in the child deaths. Those who died were no more likely to have HIV than those who survived. Instead, Botswana's children were victims of a dangerous social shift caused by the fear of AIDS.

African children with AIDS had long generated intense concern, and intense calls for international help. The Bush administration's first AIDS initiative there, in 2002, was a major expansion of funding for programs that sought to curb the spread of HIV between mothers and their infants. Babies were the classic "innocent victims" of an epidemic whose spread was often entwined in the public imagination with the twin taboos of illicit sex and drugs. Without some kind of medical intervention infected women passed HIV to their babies about one third of the time.[2] And nearly all babies born with the virus died from it, after a few months or a few years. That grim equation began to change with the discovery of ARVs in the 1990s. Even a single dose to the mother followed by a single dose to the newborn child could block the spread of infection much of the time. But this development also shifted concern to the other main path for spreading the virus between mother and baby—through breast-feeding.

The risk of HIV infection from nursing can reach 1 percent a month for an infant.[3] In the United States, Europe, and other parts of the developed world that made for a simple public health calculation: Replace breast milk with formula and that risk disappeared too. But most of Africa was different. Supplies of formula were erratic. It was much harder to sterilize bottles and rubber nipples over a smoky wood fire than over a gas or electric stove. Few places had ready, consistent sources of the clean water that was essential to preparing safe formula from powdered mixtures. Most babies in wealthy countries, meanwhile, had access to routine health care. Breast milk is valuable for children everywhere, but especially in poorer African nations, where the combination of fluid, nutrients, and antibodies in a mother's milk offers babies their best hope of surviving the continent's biggest killers of children: diarrhea, pneumonia, malnutrition, and malaria. Taken together, these diseases were bigger and more immediate threats than AIDS.

Two camps gradually emerged within the global public health community. One argued for promoting a period of exclusive breast-feeding for all mothers in poor settings, regardless of whether they had HIV or not. But the other—and this included most experts at the time—pushed for exclusive bottle feeding for the babies of women with HIV wherever there was a steady supply of clean water, powdered formula mix, and the means to sterilize bottles and nipples. The debate grew intense, as the two sides lobbied the global health authorities whose policies would impact the lives and deaths of the world's most vulnerable children.[4] Botswana's health minister during the floods, Sheila Tlou, recalled attending a conference in Montreal years earlier in which experts debated whether there should be two sets of policies, one for mothers in developed nations and another for those in places where clean water and other infrastructure for safe formula feeding was scarce. "We saw red!" Tlou recalled. "Why are you sentencing all of our children to death? And why are you sentencing all of us to psychological damage in knowing that we were the ones who infected them?"

But those who advocated for exclusive breast-feeding, though outnumbered, were equally fierce in their conviction that anything that undermined nursing was a mistake. They argued that the loss of breast-feeding's benefits, combined with the dangers of faulty sterilization or contaminated water, made bottle feeding a bad choice for most African mothers, even ones with HIV. "Everyone who has tried formula feeding . . . found that those who formula feed for the first six months really have problems," said Hoosen Coovadia, a pediatrician at the University of KwaZulu-Natal. "They get diarrhea. They get pneumonia. They get malnutrition. And they die." The United Nations eventually crafted a compromise, but one that tilted toward bottle feeding for women with HIV. The official guidelines called for infected women to avoid breast-feeding if formula was culturally appropriate, feasible for the parents to provide, affordable, sustainable, and safe. UNICEF, meanwhile, created a program in which it attempted to provide all of those conditions for poor African women. It eventually distributed 365,000 packs of formula in eight countries, and also provided training and technical assistance to government programs in Botswana and South Africa that offered free formula as a standard service to infected mothers.[5]

But evidence continued to mount on the side of those who favored exclusive breast-feeding, and UNICEF in 2002 announced that it was ending its formula program. An agency report two years later found that there had been substantial "spillover" of formula feeding into the larger population of women who did not have HIV or had not been tested. Only relatively affluent, urban women—who had the means anyway to buy formula if they wished and didn't need UN handouts—had the ability to consistently prepare formula safely. Poorer women, meanwhile, struggled with water supplies and the demands of sterilization. And mothers who skipped breast-feeding were assumed to have HIV, causing some to be shunned. The benefits of formula also turned out to be less than expected. In one major international study conducted in Botswana, breast-fed babies contracted HIV at a slightly higher rate than those fed with formula, but formula-fed babies were more likely to die of other causes.[6] If the goal was avoiding both early death and HIV, the results after eighteen months suggested that formula offered no advantage to babies of infected mothers. HIV-free survival rates were nearly identical for both groups of children. The worst outcomes, studies showed, were for babies who received what experts called "mixed-feeding," which included some breast milk along with formula and often other substances, typically solid foods mixed with water. That created a dangerous combination, in which the solids or formula can cause subtle damage in the baby's gastrointestinal tract, making it less able to block HIV if the infant later consumed infected breast milk.[7] Compared to that, either exclusive breast-feeding or exclusive formula feeding was safer.

The United Nations convened an international conference to examine its approach toward formula feeding and HIV-infected mothers in 2006. The result was new guidelines emphasizing the importance of breast-feeding and stressing that formula feeding was safe only under particular conditions that were rare in most countries with major AIDS epidemics. Yet as the scientific consensus shifted, Botswana kept on course, confident that with its relatively modern infrastructure of roads and piped water, formula remained the best option. The program was small at first, but that changed as Botswana became more successful at using ARVs to keep women with HIV from infecting their babies during pregnancy and birth. That helped create a growing pool of women who knew that they had HIV but that their babies probably did not.[8] Demand for formula soared. And, as happened in other countries, bottle feeding caught on widely in Botswana. Mothers who knew they had HIV got it for free. Many others who feared infection but were uncertain bought formula at stores. And even some who had tested free of the virus began to embrace the convenience of bottle feeding, especially if they traveled regularly or had jobs that made breast-feeding impractical. Breast milk had always been the standard diet for all babies born in Botswana. Now, suddenly, for many it wasn't.

The results of the CDC investigation challenged the idea that even a relatively wealthy African nation could consistently provide the conditions necessary for safe formula feeding. This was a country that had never known war, that had good governance, that was developing faster than almost any nation on the continent. And yet when the CDC tested twenty-six village water systems during the floods they found pathogens in every one. The stools of sick children also turned up a witch's brew of water-borne disease—cryptosporidium, salmonella, *E. coli,* shigella—often working in combination to ravage the digestive system of a single child.

It was clear that the formula shortages that hobbled Kabelo's natural growth were common as well. Records showed that clinics overseen by one of Africa's most competent health ministries still managed to run out of stock often. The program used a powdered formula made by the international food conglomerate Nestlé in neighboring South Africa, and Botswana struggled to keep the distribution system flowing. The story was told in the growth charts of sick children, with dips and surges that left immune systems compromised. In one village CDC investigators visited, 30 percent of the babies fed by formula died, but among those whose mothers had breast-fed them, none did. This pattern held true in villages and cities, among the relatively affluent and the very poor, among those who had HIV and those who didn't.

In the case of Kabelo, he went to the hospital after nearly two months of illness, but the doctors sent him home a week later. Soon Kabelo joined a grim but growing category in Botswana: Children protected from HIV but killed by another disease that, in ordinary circumstances, he probably would have survived. By the time the rains finally subsided in April Kabelo was one of eight dead children in Nkange. One family lost an eight-month-old boy and his female cousin, who was just a month older; both had been fed mainly with formula. A child from another family survived diarrhea only to die soon after from pneumonia. The toll was unlike anything Nkange's elders could recall striking their village. "Since I was a girl, I can't remember a time when we lost so many kids," said Ntselang Swimbo, whose nine-month-old grandson died during the outbreak. "Once a kid gets diarrhea you knew the chance of surviving was almost zero."

The nationwide picture was nearly as grim. Official counts put the toll for child deaths from diarrhea at 532. That was twenty times what was typical for the season, and the real number almost certainly was substantially higher, because many children died at home, away from health facilities and the reach of government record keeping.

In Nkange, about fifty friends, neighbors, and relatives assembled for the funeral of Kabelo. But after losing her son, Mavundu could not stand, could not walk, could not join the procession. Instead she stayed behind, sobbing inconsolably as the others carried Kabelo's body away in a tiny cream-colored casket to a nearby cemetery. They buried him next to his aunt, who had

died years earlier, when the family could afford a metal canopy to mark her grave, as was the custom in this part of Botswana. Kabelo got only the simplest of graves, his time on the earth marked by just a small board protruding upward from the Kalahari sand. Back at the homestead, all Mavundu had left of her son was the health card, a few ratty old clothes, and a battered photograph. It showed mother and son in happier days. Mavundu was standing beside a tree wearing a cream-colored skirt, a dark blouse, and a white

Chandapiwa Mavundu holds up a picture of her son, Kabelo, who died during heavy floods that contaminated water in parts of Botswana.

cloth tied around her head, hiding her braids. She was holding a red-and-blue-plaid umbrella to shade herself and the chubby-cheeked Kabelo, who was tied on her back, in the African style, with a baby-blue blanket.

Even as the CDC reported on what killed those hundreds of children in Botswana—lead investigator Tracy Creek made a presentation at a scientific conference in Los Angeles in February 2007—Mavundu had no idea of the connection between the flood, the formula, and Kabelo's death. A year after his terrible illness she had started taking a combination of ARVs that suppressed her HIV infection. She also was pregnant again. Several studies had recently found that women on such drug regimens posed little risk of spreading the virus to their babies during breast-feeding. But Mavundu didn't know that either. So even after most international authorities had backed away from bottle feeding in such settings, she was planning to again use the formula Botswana's government provided.

Rubbing her tummy, she smiled and said, "I think it's a boy."

MOTHER AND SON

Living in South Africa, the AIDS epidemic was at once invisible and everywhere. The brother of my best South African friend got HIV. So did several close friends of our family's nanny, as well as the maintenance man at our children's nursery school. As the years wore on I kept a worried eye on the shrinking waistline of the gardener who lovingly tended the roses in the yard of the *Post's* house in Johannesburg. Reporting on AIDS gave me a certain advantage in dealing with these issues. When word reached me of a new infection, I had names and numbers of AIDS doctors that I could scrawl down on a piece of paper, offering some hope in a sea of worry and despair.

But for most South Africans, and most Africans in the hardest-hit parts of the continent generally, getting care was not simple, even in the PEPFAR era. South Africa was a land of catchy billboards, for cell phones, for radio shows, for cars, for liquor. Those huge glossy loveLife ones, each as cryptic as the last, kept appearing. But even after the government belatedly accepted that it needed to distribute ARVs, and put serious resources into the program, it took years more before there were simple billboards guiding people with AIDS toward taking what should have been a simple step. The drugs arrived in a scattershot mix of health-care facilities, with no logic that was discernible to those not already in the know. Somebody with advanced HIV could walk into the right clinic and walk out with the drugs necessary for a new life. Or walk into the wrong one and walk out with nothing more than painkillers available at any pharmacy. Meanwhile, South Africans were still getting new HIV infections much faster than the old ones were getting treated. The doctor to whom I most often sent sick people, Francois Venter, once told me: "I just see a never-ending sea of disaster."

That disaster came especially close to my family when we heard that Yvonne was sick.[1] My wife and I had met Yvonne in the first weeks after we moved to Johannesburg, when we were looking for a nanny to care for our children. She was working for some friends of ours, but they were contemplating a change. Yvonne visited our house, smiled broadly, played joyfully with our kids, and generally charmed us all. She was slim but not in a way that seemed sickly. We were ready to offer her the job when our friends changed their minds. They wanted to keep Yvonne but suggested we instead hire her hardworking and wise cousin, Sarah. We did. Yvonne, meanwhile, faded from our minds as the curious rhythms of expatriate family life consumed us.

What we didn't know was that she was indeed sick, and it's easy now to guess when it happened. She had a son in 2001. An HIV test at that time was negative. But in the years that followed Yvonne's relationship with her husband deteriorated. He would disappear for days and sometimes weeks at a time. When she confronted him about his apparent infidelities he would not deny them. Once during a fight he declared cruelly, "You know what, I've got many wives!"

Yvonne reacted as many southern African women do, not by leaving but by vowing to win him back, and by continuing to have sex with him whenever he returned home. To do otherwise would have been to surrender her husband without a fight, to let another woman win him away. "They will say you are a loser," she recalled years later. "That's what kills us."

Yvonne's husband left in 2004, not long after she had visited our home on that job interview. Three years later she grew terribly weak. At this point she was no longer working for our friends but for a South African couple who required Yvonne to call them Sir or Madam in every interaction. She struggled to finish each work day and quarreled with her frustrated bosses. When they urged her to get an HIV test, she refused, believing it impossible that she had the virus.

In February 2008, as Yvonne's energy continued to ebb from her shrinking body, she quit her job. Two weeks later, unable to deny the obvious any longer, she went to a clinic, where a nurse pricked her finger, squeezed a drop of blood onto a plastic stick, and told Yvonne that she had HIV. No counseling or kind words followed. The nurse left her in the room sobbing and bereft, certain of imminent death and terrified about the fate of her two sons, as well as of a young nephew she had adopted. "If there was something there that could have taken my life, I would have done it," Yvonne later said.

The clinic offered no useful medicine. A visit to a second clinic proved no better. Weak and unable to eat, with swollen legs and thinning hair, she made a grueling bus ride to the home of her mother in a remote rural area. Her mother agreed she could stay, offering her a dark, nearly airless back room. There Yvonne prepared to die, miserable and alone, convinced she would never see her boys again.

My wife and I knew none of this until our nanny, Sarah, who had grown suspicious about her cousin's prolonged absence from her apartment in Soweto, called her mother and demanded to know what was happening. Sarah soon told us, and together we began to strategize about how to help. I scribbled down the name and number of a clinic on a piece of paper and gave Sarah money for a bus ticket so that Yvonne could return to Johannesburg quickly.

She did. Two days later the two of them traveled together to the Hillbrow Clinic, a government facility in downtown Johannesburg where I had arranged an appointment. But even then, getting the medicine was not simple. They waited for hours in what they thought was the right ward, but the doctor there simply gave Yvonne some painkillers and told her to go home. Sarah then called me in a panic, and I made a quick call to get two more crucial facts. Yvonne needed to go to Ward 21 and ask for Sister Pat (senior nurses in South Africa are called "sisters"). Even in a government clinic that had plentiful, free AIDS drugs, you needed to find the right ward and ask for the right nurse, or you would die.

They arrived at Ward 21 as it was preparing to close for the day. Sister Pat was waiting for them. She whisked Yvonne inside, managed to extract a few drops of blood from her collapsed veins, and instructed her to return two days later. Her CD-4 count, a standard measure of immune health, was 3. A healthy person typically had a CD-4 count of 800 or more. When the number drops below 250, antiretroviral drugs generally were prescribed immediately. When it got below 10, most people with AIDS were regarded as too sick to save. A CD-4 count of 3 meant Yvonne essentially had no defenses left against any kind of infectious disease. Few people ever reached that level alive.

She returned for her appointment two days later and, with Sister Pat shepherding her through the system, got a standard combination of three ARVs. Her instructions were to take them that night, but Yvonne had absorbed the doubts that Thabo Mbeki and others often voiced about the dangers of the medicine. That night the pills sat on the bedside table of her small apartment, untouched. She was sure that they, rather than AIDS, would kill her. The next day, a Saturday, she refused the medicine again. On Sunday she refused again. And on Monday morning, she refused yet again. Finally, on Monday night, three days after first getting the drugs, Yvonne decided she was ready to die. She downed the medicine, turned out the lights, and pulled the covers over her head.

In the morning, she was amazed to wake up alive.

As Halperin and I began working on this book I decided to interview more young South Africans about their attitudes toward HIV and sex, with an eye toward what might cause them to change the most dangerous behaviors. And I thought that if anyone had something insightful to say, it might be Yvonne's son, Sifiso. He was eighteen—near the beginning of his sexual life—but had recently watched his mother reach the brink of death. If that didn't frighten a young man, I didn't know what could.

Sifiso turned out to be baby-faced and slim. His mother had warned me that Sifiso had become "a bit naughty" in recent years. And in one glance, I knew she was right. It wasn't so much the small, silver hoop earrings or the cut yellow ropes he had tied onto his wrists as a poor man's gold bracelets, though both hinted at a certain attentiveness to a fast-paced style imported straight from American pop culture. What I noticed more was Sifiso's mischievous smile, his easy charm. I liked him instantly, and could see why girls did too.

Sifiso had started drinking, and lost his virginity on the same night, at a party when he was fourteen years old. Now, four years later, he was out of school, unemployed, and dangerously adept at talking girls into bed. His life revolved around friends, parties, and sexual adventures. He dreamed of someday owning a large house, driving a nice car, but had only the vaguest possible plans for finding work or furthering his education. Money was an overriding concern, and his chronic shortage of cash was the only useful brake against acquiring even more girlfriends, and even more risk for HIV.

I asked Sifiso how many girls he had slept with.

"I can't count," he replied with the hint of a boast, his voice thin and metallic, his smile embarrassed. But the number was somewhere between a dozen and twenty. When I pressed him, Sifiso quickly rattled off the names of seven he had slept with in the past few months. During December's Festive Season, when long school breaks, Christmas bonuses, and the southern hemisphere's warm summer nights arrived together, Sifiso and his friends hooked up with two girls apiece some weekends. The key was to scrounge together enough cash to buy the bottles of hard cider many young women in his Soweto neighborhood liked. The nature of the deal—drinks for sex—was straightforward. "Maybe we have to buy them three, but not more than five," he explained.

Each bottle cost one dollar. So the price of sex in Sifiso's world averaged four dollars per act. This was not prostitution. It was simply the rules of the road in many parts of sub-Saharan Africa, where sex often was exchanged for beer, food, clothes, groceries, school fees, or cash. Many women here said, in a notion opposite of the familiar Western idea on such matters, that they would feel cheapened if they had sex and didn't get something tangible in return.

Sifiso and his friends had reduced this equation to a Zulu credo: *Kudliwa mali, kudliwa muntu.*

The literal translation was: You eat my money, I eat you. Practically it meant: I buy drinks, we have sex.[2]

Of all the AIDS-related billboards, radio shows, television ads, and educational programs aimed at young southern Africans like Sifiso over his lifetime, he could recall only one: In high school his class read a book about AIDS. It told of an American tennis star who died from the disease. With a bit of prompting from me, Sifiso recalled the name: Arthur Ashe.

Do you remember anything the book said about how to protect yourself from AIDS?
Sifiso shook his head.

Did your parents ever discuss the subject with you?
Once. Sifiso's mother had urged him to use condoms after overhearing him talking on the phone to a girlfriend. That was the first and last time, he said, that either parent had talked about AIDS prevention.

Do you talk about AIDS with anybody else?
His friends occasionally discussed it. They agreed that condoms were a good idea.

Do they use them?
Sometimes. Sifiso often used condoms when he was sober. When he had too much of his favored concoction of Klipdrift Brandy and Coke, he often didn't. "If I'm drunk, I don't mind anything, because I just tell myself everybody will die."

Do you think you are at high risk for getting AIDS?
Sifiso shook his head.

Have you ever been tested for HIV?

He shook his head again. "If you go and test, you'll have more stress, and you'll get sick."

Can you imagine having only one girlfriend at a time?

"My friends tell me that if you have one girlfriend, you are not man enough. They mustn't see you with one girlfriend. You must have many."

Do you know any men who are faithful to a single girlfriend?

Sifiso quickly rattled off the names of a dozen of his friends. Of that group, two were attempting to stay faithful at the moment. Both had pregnant girlfriends. Sifiso himself had no such ambitions for monogamy— even if he some day got married. "I'll just tell [my wife] that I'm going to sleep with my friends, and go get other girls."

At about this point in the interview, I opened my notebook to a blank page and started to draw a diagram on a piece of paper. In the middle I put a circle. I told Sifiso that this represented him. Around that circle I drew several others and connected them with lines. These were his girlfriends. Then I drew still more circles. These were the boyfriends of his girlfriends. Then I drew another set of even more circles. These were the girlfriends of the boyfriends of his girlfriends. I'd sketched such sexual networks many times with men and women I interviewed in Africa. It never failed to provoke alarm. Sifiso's eyes grew wide. His mouth pulled into a tight circle. He made a high-pitched exclamation of concern, common in southern Africa, that is a mix between "Oh!" and "Wow!" Then I explained how a virus infecting one person in my diagram could spread to the others. I dragged my pen through a string of the circles, ending in the middle, with the one representing Sifiso.

Do you see the danger?

He nodded.

Is there any possibility that you could change, or that the sexual culture you live within could change?

He shook his head.

Is there any hope that we might find a way to control AIDS here?

"No," Sifiso said softly but firmly, with another shake of his head. "There's no hope."

After our conversation, I drove away in a foul mood. On the way back to the house where I was staying I had a minor accident that left the driver-side mirror on my rental car smashed. In the shards left behind I caught a glimpse of myself in shattered reflection. Then I climbed into bed for the rest of the afternoon. The pillow over my head shielded me from the sunlight but not from my deepening sense of doom.

WHAT SHALL WE DO? PART II

ifiso's declaration, "There's no hope," clung to me long afterward. Although delivered with a young man's fatalistic bravado, it echoed things I had been told for years by Americans, Europeans, and Africans, by men and women, by experts viewing the epidemic from forty thousand feet and by those living at ground level, wrestling with the terrors of AIDS day by day. Most of them weren't disputing the idea that sexual behavior was driving HIV, or that changing that behavior might reverse the epidemic. They just couldn't imagine how that might come about. "It's such a difficult thing," Angela, the grandmother in Botswana with HIV and an extramarital boyfriend, had told me years earlier. "I take it that it's just human nature."

There have been many examples of sexual behavior shifting in ways that slowed HIV's spread. Uganda and Zimbabwe are among the best documented, but their circumstances were particular and may be difficult to replicate in their entirety. Swaziland's *makhwapheni* program had a promising beginning but was weakened by the political backlash it provoked. New York's gay men in the 1980s did make important shifts in sexual behavior, saving many lives, but that was at a time of tremendous terror, before the arrival of ARVs. Then there is the question of simple fatigue. President Museveni eventually lost focus, especially on the core concept of Zero Grazing, as did Ugandans generally. When I asked Peter Piot if there was any way to re-create that nation's successes from the late 1980s, he said, "I wish I knew the answer."

He followed with this observation: "Working on AIDS made me far more humble about what I know, or thought I knew. . . . The paradox is that there has clearly been success, or achievement, but it's not been clear what to do" to inspire similar successes elsewhere.

South African physician Francois Venter and I used to debate this conundrum in conversations fueled by coffee or red wine. He had been struggling for years to get South Africa's balky public health system to treat as many AIDS patients as possible, but he also worried about the waves of new infection he couldn't yet see but clearly were building in the distance. Venter and I are about the same age, but I am happily married and he is happily single, and he sometimes teased that I was attempting to impose my lifestyle choice on others. Such critiques have instinctive appeal. At a debate on AIDS prevention at the World Bank in 2010, researcher Susan Allen said, "I don't think as Americans we're in any position to tell Africans they're promiscuous."[1]

The line generated a smattering of knowing laughter. And in many ways, Allen was correct. We have little right to instruct on matters of sexual morality across religious and cultural lines. Yet it was also an echo of the time in the 1990s when WHO researchers looked at dramatic findings about sexual behavior in the parts of the world where HIV was spreading fastest and balked at addressing the issue forcefully. It was soon after that when members of Collins Okendi's family started dying in such numbers, when Raymond Kwesiga went on his gin-fueled escapades, when Yvonne got infected while married to a man who thought there was nothing wrong with keeping, as he put it, "many wives" in a land rife with HIV. For many global health authorities, the subject was too awkward, the science elusive.

A 2010 article in the *Journal of the International AIDS Society*, by American economists Larry Sawers and Eileen Stillwaggon, blasted Halperin and several other researchers for a litany of alleged shortcomings in their work on concurrency. Central to their critique was the observation that measurements of sexual behavior in the DHS and other traditional surveys offered little evidence that Africans had higher rates of concurrency or other sexual behaviors key to HIV's spread. This was a familiar argument, and one made by some AIDS experts as well.[3] But it failed to acknowledge that surveys conducted in people's homes, while invaluable in measuring HIV rates and some other health metrics, were demonstrably flawed at estimating risky sexual behavior because they relied on self-reported accounts, given in settings often less than entirely discreet. Such limitations inspired some researchers to develop techniques offering subjects greater privacy, and scientists more illuminating insights. University of Pennsylvania demographers Stephane Helleringer and Hans-Peter Kohler used computerized interviewing techniques in a 2007 study to ensure confidentiality to villagers on an island in Lake Malawi and they cross-referenced answers looking for undisclosed connections. The result was a detailed map of densely interwoven sexual networks and the additional revelation that men reported only two-thirds of their non-marital relationships and women less than half of theirs. In 2009, researchers in Kenya found that women were an astounding ten times more likely to report ongoing relationships with two or more men when providing secret answers through an anonymous "ballot box" compared with face-to-face interviews.[4]

Despite such complexities in the data about African sexual behavior, Sawers and Stillwaggon condemned the entire line of inquiry as fatally flawed. Their article concluded with this remarkable proclamation: "It is customary to end the presentation of research with calls for still more research. This paper, however, calls for an end (or at least a moratorium) to research on sexual behaviour in Africa of the kind discussed in this article." Similar calls had been made before—except the subject wasn't sexual behavior. It was male circumcision. For years some opponents of the procedure sought to block inquiry into the subject and accused researchers such as Halperin of harbor-

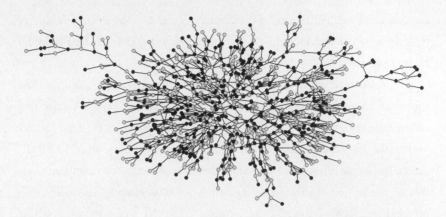

In a community in Malawi, half of all adults were linked in the same sexual network over the previous three years.

ing a bizarre obsession. This had a chilling effect on the normal progress of science and helped delay by at least a decade one of the most promising developments in fighting the epidemic.[5]

Not long after the circumcision trials announced their results, most of the major players in global public health began promoting the procedure. The WHO established clinical guidelines, UNAIDS organized a series of workshops, and some large donors made major funding commitments.[6] Questions, of course, remained: How do you perform enough circumcisions to make a difference? Do you start with baby boys, for whom the procedure is cheaper and simpler, and then wait for the impact to begin twenty years later? Or would it be better to launch mass campaigns among adults, an approach with more involved logistics but more immediate benefits as well? The only precedent for such a dramatic expansion of circumcision services was in Israel, after the fall of the Berlin Wall, when hundreds of thousands of Eastern European Jews belatedly sought to perform a ritual they were denied under communism.[7] But that was in an advanced nation with plenty of operating rooms and physicians, not in African countries with weaker public health systems. Yet with clear trial results from three African nations, the nature of the questions had shifted. It was no longer: Should we offer safe, voluntary male circumcision services in places with high HIV rates? It had become: How do we do it?

The most ambitious effort began in western Kenya's Luoland, on the shores of Lake Victoria, not far from where Collins Okendi's family had suffered waves of devastation that routine circumcision might have largely prevented. The funding for this program, meanwhile, came from PEPFAR, the Gates foundation, and other international organizations that had once treated the subject warily. Some community elders initially objected, arguing that the lack of circumcision among Luo men was a defining ethnic characteristic.[8] But Robert Bailey's lead Kenyan counterpart on the circumcision trial was a Luo woman, social scientist Kawango Agot, who saw the practice as culturally neutral, something Luos didn't traditionally do but could if the reason was compelling enough. "What I care about is people dying," she said.

In 2009 I rode along with a mobile circumcision team that gathered each morning on the second floor of a mall in Kisumu before departing into the countryside. The team had a clinical officer—roughly equivalent to an American nurse practitioner, with advanced medical training that stopped short of the level of physician—along with a nurse, a hygiene officer, and a driver. They headed out in a Toyota van stuffed with surgical equipment, as the Sade song "Smooth Operator" came on the radio. They arrived about an hour later at the Sango Rota Health Centre, a collection of white concrete buildings set in what appeared to be an open field outside a Lake Victoria fishing village. The rudimentary operating theater was in a small white tent. A few feet away stood a second white tent, no bigger than the first, that served as the recovery room. Ten patients arrived soon after in a truck from the village, and the men waited in the shade of a nearby tree, chatting nervously to each other, as the medical team prepared for surgery.

The hygiene officer set out buckets for cleaning. The clinical officer, Godfrey Owino Ochieng, and the nurse, Zephania Akello Opiyo—both men in their late twenties—donned white smocks, scrubbed their hands, and arranged their clamps and scissors in the cramped, stuffy operating theater.

Ochieng, who wore a plaid shirt and dark slacks beneath the surgical smock, had become a circumcision specialist. Once a man had been cleaned and given several shots of anaesthesia in the base of his penis, Ochieng typically needed just fifteen minutes to clamp off the foreskin, cut it away, and suture up the wound. The men got pills to ease the pain as the anaesthesia wore off but most reported feeling better in just a couple of days. The ranks of patients tended heavily toward young men, although many parents brought their sons as well. One man, seventy, brought in all six of his sons and told Ochieng, "No one is going to stay in this home who is uncircumcised."

The program also portrayed circumcision as a way to avoid sexually transmitted infections and improve hygiene. But a secret weapon in the campaign to get Luo men circumcised was the enthusiasm of Luo women. Word had spread that without the protection of a foreskin, a man's penis grew a bit less sensitive, and as a result, he became slower to reach orgasm. Many women believed that made for longer, better sex. A female friend told Ochieng, "Unless a man is circumcised, I can't go to bed with him. He'll leave me hanging."[9]

The surgeries began late in the morning and lasted all afternoon, with just a short break for lunch. As dusk approached, all the newly circumcised men had stable blood pressures and clean sutures, and they all had been warned to avoid sex for six weeks to make sure the wound had healed completely.[10] Ochieng and the others packed the surgical tools and soiled towels into the back of the van and closed up the two white tents. In daily outings like this, a dozen or so men at a time, such crews were making all of Luoland more resistant to HIV. After two years the program had circumcised more than a quarter of a million men.[11]

But as circumcision programs finally started moving forward in some places, the other half of Halperin's lethal cocktail was still generating controversy.[12] The critiques from Sawers, Stillwaggon, and others provoked a backlash against programs addressing sexual behavior generally, including some newer efforts targeting concurrency.

The impact was profound, especially at a time when the faltering fiscal strength of the U.S. government and other major donors made clear that AIDS funding could not keep growing indefinitely—or at least not at such a remarkable clip. Piot, whatever the flaws in his tenure, helped lead an tremendous surge in investments in fighting AIDS, with the total funding going from $292 million in 1996 to $16 billion by 2010. The cost was modest compared to the wars the United States fought in this era, or the bailout organized to save the financial industry from collapse. But by the standards of international public health, the steep trajectory for new funding was historic and, arguably, unsustainable.

Putting someone on ARVs amounts to a moral commitment that should not be taken lightly. Stop these drugs, and most patients will soon die. Current trends make this a terrifying possibility for millions of people, especially at a time when foreign donors are scrambling to close deficits at home. With at least two Africans contracting HIV for every one who receives access to ARVs, treatment costs threaten to escalate indefinitely.[13] Poor adherence to treatment also can cause the virus to rebound and mutate, creating new variants that can be resistant to the most widely used and least expensive drugs.[14] There can be other troubling consequences if the response to the epidemic becomes dominated by treating millions of people but fails to make meaningful inroads on prevention. Among them is the entrenchment of a lopsided relationship between the big, mostly Western donor nations and the hard-hit, poorer African ones. To those with a sense of history there is an unpleasant whiff of an old dynamic lurking here. As long as money and expertise flows in a single direction, south into Africa, many of that continent's professionals will spend their lives answering to foreign decision makers and bending to the imperatives of foreign priorities and sensibilities. That is enervating and potentially infantilizing for those trapped in what threatens to become a permanent state of dependency.[15]

Large amounts of AIDS money, meanwhile, goes to funding the increasingly vast conglomeration of government contractors, university departments, and international agencies selected to lead the fight against AIDS. Some of this, no doubt, is necessary. But recall the point by Malcolm

Potts, who once headed a charter member of this group, when he labeled it the "AIDS Industrial Complex." The homage to President Dwight Eisenhower's cautionary words about the U.S. military is an important one. However laudable the goals and the benefits it sometimes produces, the creation of such an extensive industry creates its own political and financial imperatives. These can sometimes drain resources away from other worthy approaches that might be outside the natural range of what the industry already does.

Is there room in today's multibillion-dollar AIDS responses for the evocative metaphors of a Yoweri Museveni? The potent lyrics of a Franco?

In July 2011, when AIDS experts and activists gathered at a conference in Rome, word spread about the latest research into the power of ARVs to not only treat HIV but to prevent it in some cases. One University of North Carolina trial of discordant couples—meaning those in which one partner had HIV and the other didn't—found that giving ARVs to the infected person before their immune systems deteriorated lowered the risk to their partners by an astonishing 96 percent. Other trials, meanwhile, were showing that when uninfected people took a daily dose of a single ARV as a prophylaxis, they were less likely to contract the virus. The official catchphrase for such approaches, treatment-as-prevention, appeared on the surface to marry the two sides of the AIDS fight. But it was nothing like a marriage of equals. Some proponents went one step further with a new slogan: "treatment-is-prevention." At the Rome conference, when a panelist asked a crowded room of experts who believed there were now enough prevention tools to bring the AIDS epidemic to a halt, every hand went up.

But as so often has happened in the three-decade-old war on AIDS, enthusiasm about such new scientific discoveries gradually became more subdued in the face of the realities of the epidemic. The world already has some prevention strategies that work in highly controlled research settings. After all, condoms block HIV when used properly and consistently. Where they falter—and where many experts fear treatment-as-prevention might falter—

was in more ordinary circumstances, where even the best public health advice can be difficult to follow, especially if it requires making wise decisions every single day. Would people who had HIV but were not yet feeling sick be willing to start ARVs early, and potentially endure serious side effects over many years, if the main benefit was to keep others from becoming infected?[16] Even Myron Cohen, the North Carolina researcher who was the principal investigator of one of the studies that excited researchers in Rome, continued to worry about the limitations of the strategy outside of the controlled environment of clinical trials. "We don't have this all worked out," Cohen told *Science*.[17]

For treatment-as-prevention to have a major impact on the epidemic, testing initiatives would need to be expanded massively in order to identify people with HIV before they develop symptoms. And it would be nearly impossible to catch the many infected during the acute phase, that initial period of just a couple of months when the virus is unusually contagious but typically cannot be detected by standard tests. Expanding treatment regimes surely had medical benefits, and should also prevent some cases of HIV. But years of routine access to ARVs in the United States and other countries such as Australia and Holland had not led to much lower infection rates, largely because people sometimes engage in riskier behavior when they don't see AIDS as a death sentence.[18] There may also be some nearly insurmountable logistical barriers when it comes to keeping enough people on treatment to impact the spread of the virus.[19] It would be vastly more expensive as well, and require much more infrastructural support than other strategies. The funding needed to put many millions of additional people on ARVs for decades could threaten to crowd out other worthy global health initiatives, such as campaigns to provide clean water, family planning services, and treatments for a range of other diseases. Among the most vulnerable programs were those targeting the sexual behaviors that spread the virus. An editorial in *The Lancet* put the bull's-eye on these efforts: "Findings now need to be translated into policy and action. Agencies such as the President's Emergency Plan for AIDS Relief and the Global Fund to Fight AIDS, Tuberculosis and Malaria need to reassess their prevention portfolios and consider diverting funds from programmes with poor evidence (such as behavioural change communication) to treatment for prevention."[20]

There was nothing wrong, of course, with desiring more evidence. But there is subtle bias at work in favor of strategies that are fundamentally biomedical, such as treatment-as-prevention, and against those that might be called anthropological. Such prevention approaches occur organically within cultures and are more easily observed than tested. Male circumcision is an ancient cultural practice but one that eventually was able to prove its worth in randomized controlled trials.[21] Yet it is not clear that the same will ever be true of initiatives aimed at changing norms of sexual behavior.[22] How would scientists feasibly construct a randomized controlled trial for programs targeting sexual networks? There was no way to have an experimental group receive the "treatment"—in this case intensive radio campaigns, grassroots community action, and the cautionary words of political and cultural leaders—while the other group received none of this. Information simply spills out in every direction in the modern world. If the Ugandans had first demanded a randomized controlled trial for Zero Grazing, the campaign never would have happened.[23]

During the years that scientists and policy makers debated the relationship between male circumcision and HIV, it already had the effect of preventing millions of infections worldwide because so many cultures already practiced it.[24] The evidence for the ability of changes in sexual behavior to slow HIV's spread is in some important ways even stronger today than the evidence for male circumcision was in the 1990s. This evidence does not come from randomized controlled trials but from analyzing natural experiments, in which the course of history reveals truths that other research can struggle to discern. And as the modeling work of Martina Morris and other epidemiologists has suggested, even relatively small changes in overall rates of concurrency can make big changes in the speed of HIV's spread. If men and women living amid severe epidemics dial back, even modestly, their average number of sexual relationships, everyone in the community would be safer.

All of this brings us back to a question once put so lyrically by the Zimbabwean singer Oliver Mtukudzi: What shall we do?

Or put another way: How can we turn the world's tinderboxes into wet moss?

First, it is time to let die the myth that Asia, Europe, and the Americas will develop African-style epidemics, with infections raging across entire societies. Instead, in these places we must target prevention efforts at the groups facing the most risk. Campaigns promoting condom use with sex workers have proven successful in places where prostitution is the main source of the epidemic. Programs to prevent addicts from sharing needles, meanwhile, have made inroads from New York to China to Iran, and it is also important to provide drug treatment programs that can help people free themselves from addiction. Addressing the severe epidemics among men who have sex with men, including the many who are not openly gay, is especially challenging. Strategies should seek to improve access to condoms and lubricants, warn about the dramatically higher risk of anal sex, and promote approaches such as partner reduction and perhaps male circumcision.[25]

Within Africa, which still accounts for about 70 percent of new HIV infections and AIDS deaths each year, condoms have a role but do not appear to be the key to reversing the most serious epidemics. Effective vaginal microbicides and one-a-day prevention pills might help protect the most vulnerable people, including people whose regular sex partners are known to have HIV.[26] ARV treatment programs, meanwhile, should continue expanding—so long as prevention and other important initiatives are also protected—and this likely will also help block some new infections from beginning. And safe, affordable, and voluntary circumcision services should be made widely available to men, especially in Africa, where the practice is so ancient and the health benefits so profound. ARV programs that help pregnant women from giving HIV to their babies should continue to grow. And mothers without reliable access to clean water should be encouraged to breast-feed exclusively, regardless of their HIV status, and appropriate steps should be taken to make transmission of the virus as rare as possible during nursing. In the long run, however, the most important initiatives in preventing HIV should be those aimed directly at the main cause of its spread: sexual behavior.

History makes clear that changing behavior is a natural response to a spreading AIDS epidemic, especially when the conversation about its causes is straightforward, as it was in Uganda and Zimbabwe, and among gay men in the United States and Europe. There also have been substantial HIV declines in Kenya, Malawi, Zambia, Rwanda, Ethiopia, India, Cambodia, Haiti, and the Dominican Republic. In each case these have coincided with reported declines in risky sexual behavior, and most notably with declines in multiple partnerships.[27] These changes in norms of behavior already have resulted in millions of averted HIV infections. Even Thailand, well known for the "100 percent condom" campaigns that slowed the spread of HIV at brothels, also experienced shifts in its broader sexual culture. Mechai Viravaidya, whose first name became slang for "condom," sometimes told halls full of male military recruits to hold up one finger for each of the women they had slept with in the last three months. Then he would say, "Okay, now cut that in half."[28]

The instinct toward greater caution amid a sexually transmitted epidemic should be forcefully supported, with well-crafted campaigns led by society's natural leaders—pastors, artists, traditional healers, politicians.[29] Such programs should be direct enough to cut through the clutter of modern life, as Swaziland's blunt *makhwapheni* program did, while also displaying understanding that such initiatives are fraught with gender and other cultural issues that must be handled sensitively. And Uganda's experience shows how these efforts must mix a sense of urgency about preventing new infections with compassion for those already infected.

In a 2007 study conducted across ten southern African nations by the nonprofit health education group Soul City, an overwhelming majority of the more than two thousand people interviewed said concurrency was common within their communities.[30] Soul City soon after created a soap opera called *One Love*, which dealt frankly with sex and HIV. The thirteen half-hour episodes tracked how the virus spread through a family of shopkeepers—the Molois, including parents Zimele and Lebo, and two school-age children— because of a succession of risky sexual decisions. Lebo Moloi learned she was pregnant but almost immediately had to rebuff the advances of a male friend. Her husband, Zimele, meanwhile, had an affair with a nineteen-year-old girl go sour. As the episode ended Lebo headed to a clinic to get a HIV test, apparently unconcerned that she could be infected. . . .

One Love had the melodramatic tone of soap operas everywhere—the close-ups, the ominous music, the exaggerated facial expressions—but the characters felt sympathetic and real.[31] South Africans gave the series some of the nation's best television ratings in 2009; viewers jammed lines to radio shows to talk about the issues of sexual culture raised by the show. If one of the challenges of AIDS prevention was delivering smart, hard-hitting messages in rapidly modernizing societies, here was a model that seemed promising. There have been others, including an approach developed by Population Services International in Mozambique in which small groups of men and women gathered in a room to explore the networks of relationships in their own communities by sketching the connections on sheets of butcher paper. Also in Mozambique, a foundation led by fomer First Lady Graca Machel helped design a national campaign that used the slang term "walk outside" to warn against having multiple partners. The slogan was: *Andar Fora é Maningue Arriscado!* meaning "To Walk Outside Is Very Risky!"[32] And in western Kenya, Martina Morris helped develop a program called Know Your Network. It used a series of anonymous questions sent as cell phone text messages to help people understand the sexual networks in their communities and to spark open discussions about addressing the danger of concurrency.

Done properly, preventing new infections should cost pennies on the dollar compared to lifelong treatment with ARVs, and the outcomes are much better for those who get to live their days free of HIV. Putting millions more people on medicine, while an urgent humanitarian priority, should not distract from the imperative of preventing new infections. Millions of lives arguably could have been saved had the world acted more quickly on the evidence for male circumcision.[33] Will we repeat the mistake as experts continue debating the importance of addressing sexual behavior?

In *The Ghost Map*, published in 2006, author Steven Johnson tells how a physician traced the origins of a cholera outbreak in Victorian London to the water flowing from a single infected water pump. The book is a study in classic epidemiology near the birth of the discipline, just four years after the creation of the London Epidemiological Society. Dr. John Snow took time away from his patients to follow a hunch that cholera was spread not by noxious gases, as most scientists at the time supposed, but by tainted water. His groundbreaking investigation got little help from the primitive biomedical tools of the time; instead it relied on the observation that those who were using the Broad Street pump, long renowned for its clear, cool water, were getting sick and dying, and those who got their water elsewhere were not. Snow also discovered that people who consumed water in conjunction with alcohol, as did a group of men working at a beer factory that paid wages in the brewery's product, were much more resistant to cholera than those who took their water straight.

In this metaphor, the alcohol would be male circumcision—something that has a strong protective effect, limiting the ability of the pathogen to start an infection. But the source of the spread, the infected water pump, is a pattern of sexual behavior more common in parts of Africa than almost anywhere else in the world. The parallel is important not only because of the insights it offers into epidemiology but because of the insights it offers into the nature of scientific evidence, and especially about who has the authority to designate some evidence as worthy and other evidence unworthy. Snow delivered his argument to the city fathers, who reluctantly agreed to remove the handle from the Broad Street pump, ending its flow. The decision generated immediate protest and, in some cases, outright ridicule. And the evidence that Snow produced, a detailed map connecting cholera deaths with proximity to the infected pump, was nowhere near as rigorous as what public health authorities would demand today. Yet in Johnson's telling, that one action likely saved hundreds of lives and became a key building block in the world's understanding of disease transmission—and, more important, how to prevent it.

We believe something similar is possible for AIDS. We sympathize with the frustrations expressed by Piot, near the end of a long career in fighting AIDS, about the difficulty of trying to change sexual behavior in foreign cultures. But humility can't be an excuse for resignation.

Halperin not long ago made a two-week trip to Tanzania. The Kagera region near Uganda once had its own version of Zero Grazing that had achieved similarly dramatic results through homegrown, inexpensive measures. But Tanzania overall had not made major strides against HIV. I received e-mails almost every day he was in Tanzania about money he believed was being used inefficiently while worthier efforts struggled for funding. He was especially incensed at the shocking inequities that resulted when big international donors poured money into one important cause, such as AIDS treatment, while neglecting others. On a visit to Muhimbili National Hospital in the capital, Dar es Salaam, Halperin saw overwhelmed wards where patients were crammed into every corner, sometimes screaming in pain, while nearby a modern outpatient AIDS ward gleamed with state-of-the-art equipment, thanks to PEPFAR dollars.[34] And he grew particularly alarmed when it became clear that, with funding from donors likely to stagnate amid budgetary pressures, biomedical approaches were winning the bureaucratic fight for resources, while behavior change efforts were facing the possibility of severe downsizing. High-tech strategies for fighting AIDS were threatening to finally crush those based more on targeting the root causes of HIV's spread and informed by an understanding of how culture and history shaped the epidemic.

Halperin was growing frustrated, and was even tempted to focus on other long-standing passions, such as international family planning or global climate change, when a Tanzanian man working for USAID pulled him back. The man, Duncan Onditi, was a devout Christian—something Halperin decidedly was not—but expressed a thought that resonated with his deepest convictions. During a heated exchange among some U.S. government officials debating the prevention approach in Tanzania, Onditi pleaded that however challenging it can be to address the sexual behavior driving HIV, we should not stop trying. The cause is too important, the cost of failure too high: "If Jesus hasn't given up on man, then how can we give up on him?"

EPILOGUE

The arrival of antiretroviral drugs in Kenya stabilized the wave of death crashing through the family of Collins Okendi, and he was pleased to see circumcision programs growing in his homeland. As the new patriarch of the Audi clan, he oversees the fortunes of an array of surviving relatives, many of whom live in his small concrete apartment building in Nairobi at any given time. Another group live at his parents' house—the one his father never quite completed before dying—on Homa Mountain. Collins, though, has found a good job as an aide to a member of parliament representing a large section of Luoland. He and his girlfriend had a daughter in November 2009. They named her Karin, after his late mother.

Late in 2007 Peter Piot oversaw a wholesale overhaul of HIV estimates worldwide, acknowledging that UNAIDS had veered beyond the mark by seven million infections. The global number was not forty million and rising, as the agency had said for years; it was thirty-three million and stable. The agency also went back and recalculated some of the numbers in its previous reports. In 2002, when the agency announced that forty-two million people had HIV, the actual number that year was thirty million, UNAIDS now said. The most authoritative, trusted tracker of the epidemic had been off by a remarkable 40 percent. Even more pronounced were the revisions in the estimated number of people becoming infected each year, which were also recalibrated downward. And the estimated numbers of deaths continued to decline, mainly because of growing access to ARVs.[1]

Piot defended his handling of HIV statistics to the end, saying that UNAIDS was right to increase the estimates in the mid-1990s, and that, as newer studies made clear that overestimates had crept in, the agency responded as swiftly as practical. He retired from UNAIDS the following year, at the end of 2008, and eventually became head of the London School of Hygiene and Tropical Medicine.

By the end of his tenure Piot had come to have doubts about how he had handled the research into male circumcision. He maintained that the science was unconvincing until the three randomized controlled trials proved its effectiveness against HIV but in retrospect, he wished he had pushed years earlier for such rigorous research so that programs could have started sooner. "I accept the criticism," Piot said. "It could have saved lives."[2]

Piot's replacement at UNAIDS was his gregarious deputy, Michel Sidibé, who initially made some efforts to reach out to scientists who had felt shut out during his predecessor's reign. Among Sidibé's new hires, in 2009, was Rand Stoneburner, who became a senior adviser. He retired in 2011 and returned to working as a consultant, but not before getting frustrated once again with UNAIDS. "There's a lot of talk, empty talk. The colors of reports and the font are way more important than any of the content," he said on one of his final days at the agency.[3] In this later era, Stoneburner said, UNAIDS officials acknowledged that partner reduction was key to Uganda's historic HIV decline but said they were unsure how to encourage such changes in other places. "Prevention is now taking a lot of drugs, and fifteen other things . . . It's just kind of sad."

Michel Caraël, meanwhile, left UNAIDS but continued doing consulting work for the agency. For two years, he was, by chance, living in the same French village as Stoneburner. They once ran into each other in a market and exchanged pleasantries. They spoke in general terms about sharing a drink together sometime. They never did.

· · ·

South Africa's Thabo Mbeki, the flawed prince of the African National Congress, gradually lost control of the party to his onetime deputy, Jacob Zuma. Mbeki had fired Zuma in 2005 over allegations of corruption, and Zuma appeared on the verge of political oblivion when prosecutors also charged him with raping a family friend who was decades younger. He offered a vehement defense that the encounter was consensual, but he outraged the AIDS community with his acknowledgement that he knew the woman had HIV but did not use a condom. For protection, he told the court, he had taken a shower shortly after sex—a move, he believed, that lowered his chances of contracting HIV.

Yet Zuma had a political touch, a warmth, that Mbeki never could muster. Zuma beat both the corruption and rape charges. Along the way he rallied South Africa's labor unions and other parts of the ANC's political left in a powerful challenge that forced Mbeki out of office in September 2008, more than seven months before his second five-year term was supposed to end. Mbeki's defeat did not result directly from his mishandling of AIDS, but his approach to the issue contributed to his aura of remoteness from the most urgent needs of his citizens. It also remained a signature policy failure of a political career that, on other matters, was generally admired.[4] Mbeki's reputation never entirely recovered from those few hectic months in 2000 when he seemed to legitimize a group of denialists whose views on the epidemic were irresponsible, if not outright loony. There were some intriguing insights lurking amid Mbeki's critique of the global response to AIDS, which, as he argued, tended to fixate on one epidemic to the exclusion of other diseases that were as common but generated far less political interest among rich nations. Had he been a college professor, or a newspaper columnist, Mbeki's eagerness to challenge the conventional wisdom on AIDS might have been remembered more sympathetically. But as the leader of a nation with more HIV infections than any other, Mbeki had veered far from the most basic goal of giving South Africans the best chance of surviving a historic plague.

In his role as president of South Africa,[5] Zuma quickly exceeded the low expectations of AIDS activists, embracing science in some cases when it was not popular, as his administration did in August 2011 when it reversed course on infant feeding policies and began promoting exclusive breast-feeding.[6] Zuma also supported male circumcision as a way to slow the spread of HIV, and he publicly acknowledged that he himself had been circumcised, which was unusual but not unheard of among Zulu men. The Zulu king also endorsed the practice, as did the Treatment Action Campaign. And the Orange Farm trial, which had transformed itself into a program offering any man in the community free circumcisions, continued to report successes. Seven years after the first procedure men were still lowering their risk of HIV infection—by more than 75 percent—and the community as a whole was also safer. An analysis of the impact of the program on the township estimated that if no circumcisions had been performed, the number of men younger than thirty-five who became infected would have been about 60 percent higher.[7]

But elsewhere in South Africa, the national effort did not move ahead nearly as swiftly as ones in Kenya or Tanzania, despite Zuma's official support. This was a major lost opportunity in a nation that stood to prevent far more infections than any other from expanding these services.[8] Zuma, meanwhile, balked at challenging the sexual behavior that spread HIV. Zuma had several wives and also acknowledged sexual relationships outside of his marriages. He had even fathered a daughter with a girlfriend in October 2009—dubbed by the South African press Zuma's "love child"—just a few months into his presidency. This was his twentieth child with at least five different women.

George W. Bush's presidency ended a few months after Mbeki's did, and though he wasn't ousted by a rival, his popularity in the final phase of his presidency sunk to Mbeki-like levels that hobbled Republican chances of keeping the White House when elections came in November 2008. Yet as Bush's party got pounded at the polls, one of the few elements of his legacy embraced by both parties was his historic campaign of mercy to African

AIDS victims. Congress tripled the initial $15 billion allocation to the effort by approving an expansion of the program to $48 billion.

PEPFAR's great success was to lead the way in dramatically expanding access to ARVs. By the end of his administration Bush's initial goal of putting two million people on antiretroviral drugs appeared to have been met, though the reality turned out to be more complex than it seemed during his 2003 State of the Union address. PEPFAR funded a variety of efforts—involving training, supply chains, and monitoring efforts, along with many other programs—much more often than it actually bought drugs for people with AIDS. Bush's big bet on antiretroviral drugs altered the global debate and accelerated the moment when important medicine reached sick Africans. But the second half of Bush's promise—to prevent seven million new infections—was quietly superceded by rough estimates about how many people saw, heard, or otherwise might have benefited from U.S.-funded prevention campaigns. Yet even with this attempted recasting, there wasn't much to celebrate. The closest thing to a clear success was the expansion of services to keep mothers from spreading HIV to their babies during pregnancy and birth, but that program lagged far behind expectations; only a relatively small percentage of infected African women took the antiretroviral drugs that inhibited transmission in 2009, the year Bush left office. More infections could have been prevented, and for less money, by simply making sure that contraception was easily available in poor countries.[9] Women who learn they are infected with HIV often don't want to have more children, but that choice became more difficult during the Bush years as the U.S. government shifted attention away from international family planning efforts.

Overall, the trajectory of the virus's spread appeared little changed in the nations where Bush focused his energies. The HIV levels in several countries, including Kenya and Ethiopia, fell during his administration, but there is little persuasive evidence that it had much to do with Bush's initiative. The timing made a cause-and-effect link unlikely, as most of the decline in new infections took place before the PEPFAR programs got rolling.[10] The African nation that had made the most decisive progress against the spread of HIV during Bush's administration was one that got relatively little funding from donors, Zimbabwe. And the one that the Bush administration celebrated as the HIV prevention model for the world, Uganda, had clear signs in these same years of slipping backward toward a dangerous past. Museveni's days as a beacon of hope on AIDS prevention continued to fade into the distance at the same time as rising numbers of Ugandans came to regard him as a dictator unwilling to relinquish power.[11] As of this writing, he is still president, and his tenure has pushed past the quarter-century mark.

The Global Fund to Fight AIDS, Tuberculosis and Malaria stopped funding loveLife in 2005, citing its poor record. A spokesman said, "This is a fund with limited resources. . . . loveLife is extremely costly—there are programs that have been very effective which cost a fraction of what loveLife costs. It would be irresponsible of the Global Fund to spend almost $40 million without seeing results."[12] But other funders, including the South African government and the Henry J. Kaiser Family Foundation, continued to underwrite the program, which bills itself as South Africa's "largest national HIV prevention initiative for young people."

In 2010, the CDC's Global AIDS Program cut its support for the *One Love* campaign developed by Soul City by 75 percent. The agency told the South Africa program that CDC wanted to focus on other issues, such as male circumcision and preventing HIV transmission from mothers to their babies. Soul City's requests for funding to study the impact of the campaign generated little interest from donors.

Francois Venter became an increasingly prominent voice on AIDS policy. He continued to lobby for male circumcision programs but maintained his doubts about how to effectively curb the sexual behavior that drives HIV. Venter, meanwhile, became an advocate of treatment-as-prevention. Venter, Halperin, and many others continued to debate the issue passionately through epic e-mail chains spanning the globe.

. . .

After Halperin and his family moved to the Boston area, where he took a job at Harvard in late 2006, he often longed to return to Africa and eagerly accepted consulting work that allowed for visits. During his first return trip to Swaziland, he stopped by a Zionist indigenous church where he had become something of a regular during the years he was posted in the country. The tiny concrete building was perched on a hilltop near his former rental home in Mbabane's arresting Pine Valley, named for the trees that, from a distance, gave the region's craggy mountainsides a verdant sheen. As drums pounded and the parishioners began chanting, some entered into trances; their eyelids were closed but their faces still somehow appeared alert. The minister welcomed Halperin back and then began a lecture on the necessity of fighting against the terrible plague that was consuming the nation and had recently taken the lives of several congregants. He counseled the men and women sitting on hard wooden benches on opposite sides of the church to avoid straying from their spouses, and he reported that some of the church's young men had recently gone to town seeking circumcision.

To Halperin, it seemed that the best hope for preventing HIV could be found in places such as this, on the front lines of the epidemic, where values were defined and often powerfully expressed. He thought, as he often had before, that initiative coming from those most directly affected, working in their own communities, could be the most potent force to finally overcome AIDS.

On my own first trip back to Africa after leaving Johannesburg in 2008, I went looking for Yvonne and found her at a modest concrete home in Soweto. As I pulled up in my rental car she appeared before me, chubby and exuberant, with a red ribbon on her chest and a new diamond ring on her left hand. As her strength had returned with the help of ARVs, she had met a new man and married. She was still unemployed, as huge numbers of black South Africans were, but she was delighted simply to be alive, to be the woman she was before AIDS.

"It's just me! I'm so big! I'm so healthy!" Yvonne gushed.

She remained deeply worried, however, about keeping her son Sifiso from becoming infected with HIV. When it became clear that he had no interest in changing his sexual behavior—or even to get tested for HIV—I suggested (with his mother's approval) that he at least get himself circumcised. With the procedure easily available at Orange Farm, all Sifiso needed was the courage to show up. So one warm summer morning, Sifiso and I drove down to Orange Farm, through the parched South African highveld and into the dense and dingy township known more for its crime and violence than cutting-edge HIV prevention services. When the nurse called his name Sifiso grabbed my hand and dragged me into the surgical ward behind him. I tried to explain to the genial doctor what I was doing there, how I knew the young man about to get surgery. The best I could do was, "Here in South Africa, I guess I'm his uncle." The doctor laughed and agreed, "Yes, you're his uncle."

The operation went smoothly. It is possible Sifiso already had HIV, but if he didn't, he was now much less likely to become infected.[13] And if he never gets infected, he can never give the virus to anybody else. One possible link in Africa's chain of infection will have been broken.

APPENDIX

How the AIDS Epidemic Can Be Overcome[1]

The parenthetical numbers in this appendix and in the corresponding endnotes are references to related scientific publications. These references begin on page 388.

We have avoided the familiar term "pandemic" in writing *Tinderbox* because it could suggest that the pattern of HIV's spread has been similar across the planet. What we see instead is a series of epidemics that share a birth and early life in the Congo River basin during the colonial era but otherwise have evolved in strikingly different ways in different places. Variations in HIV infection rates hint at this, and scientists and policy makers often group epidemics as either "generalized," meaning that the overall population faces substantial risk for infection, or "concentrated," meaning risk is confined mainly to particular high-risk groups. There has been some debate on what range of infection rates should correspond to each category, (2) but the rates of infection are less important than the predominant ways that the virus is spread. These differences dictate what kinds of public health responses will be most effective in attempting to overcome a particular epidemic.[2]

The epidemic in San Francisco is concentrated because there is one key driver: sex between men. There is a clear geographic footprint in the city's gay enclaves, but the more important factor, especially during the period of HIV's most rapid spread in the early 1980s, was involvement in sexual networks where both unprotected anal sex and encounters with many partners were common. The situation is similar among epidemics driven by users of injecting drugs in parts of Eastern Europe, Southeast Asia, and some other regions, in that there is also a single main driver of transmission: sharing of infected needles. Controlling such concentrated epidemics requires carefully targeted approaches. (2, 3) Telling all San Franciscans that they are in danger of HIV, while technically true, would do little to slow the spread of the virus through that city. Instead, prevention messages must be aimed at men facing the highest risk, and the remedies—condoms, lubricants, and a reduction in multiple sexual partnerships—need to be tailored to the main source of transmission.[3] Likewise, prevention messages in epidemics where injecting drug use is the main factor spreading HIV must also be on target, focusing on access to clean needles and efforts to help people kick their addiction. A famous example of precise targeting was the intensive "100 percent condom use" campaign of Thailand. HIV was concentrated among sex workers and their customers, and transmission was confined mainly to brothels. In this kind of epidemic, such targeted condom promotion worked remarkably well.[4]

The HIV epidemics in most of the world, including virtually everywhere in Asia, Europe, and the Americas, are of this variety: They are concentrated and, as a result, they require prevention strategies that are targeted to curtail the most likely sources of new infections. A number of nations, such as Russia, Brazil, and India, have a combination of concentrated epidemics. These involve considerable numbers of men who have sex with men, injecting drug users, and sex workers and their clients.[5] That can result in a higher infection rate overall, at least in some parts of these countries. But in most places, the virus has been present for decades without putting the majority of the population at substantial risk of infection (see map on page 168).[6] The main exception is within the generalized epidemics of Africa—and especially those in parts of East and southern Africa—home to only about 2 percent of the global population yet burdened with a tragically disproportionate share of the world's total number of infections. In most of these places, there is little injecting drug use and little open homosexuality. That's not to say that sex between men and hypodermic needles have no role in HIV's spread in these areas. Unsafe medical practices have been identified in some places, and recent research has documented the high vulnerability of homosexual men and injecting drug users in parts of Africa. But these groups, while growing, remain relatively small in number. The preponderance of HIV transmission in Africa results from heterosexual contact. (2–5) There are additional elements that help the virus move more widely, including significant age disparities as some older men have sex with much younger women. These and other relationships often have what experts call a "transactional" element, meaning sex is implicitly traded for rent money, school fees, cell phone air time, and other kinds of support. (5–8) Rates of rape and other kinds of sexual coercion are also distressingly high in many of the societies where HIV rates are highest.[7] Yet in a generalized epidemic, it is often a major risk factor simply to be sexually active. Prevention efforts in such regions must necessarily be aimed broadly at most members of society. Only in this way, as the experience of places such as Uganda shows, can generalized epidemics be turned around.

What follows are short discussions of the most common prevention strategies and explanations of where they might be effective in fighting HIV.[8]

Abstinence: This approach, of course, completely prevents sexual transmission of HIV or other sexually transmitted infections, and there is consensus that young people should be encouraged to delay the initiation of sexual relations until they are physically, psychologically, and otherwise prepared. (15) But while there are many benefits, especially for lowering unintended pregnancy among adolescent women, this strategy generally will only delay the age of HIV infection rather than actually reduce lifetime risk.[9] Most HIV infections occur among individuals in their twenties or older, when few people will abstain entirely from sex. As a result, abstinence campaigns are unlikely to have major impacts on the overall epidemic. (1, 4, 5, 16, 17)

Condoms: When used consistently, male condoms are estimated to be somewhere between 80 and 90 percent effective against infection from HIV.[10] Condoms also substantially block several other sexually transmitted infections, such as gonorrhea and chlamydia.[11] (19) Condom promotion has been most effective in epidemics where HIV is spread mainly through sex work, as has been documented in a number of places, including Thailand, Cambodia, the Dominican Republic, Senegal, and parts of India. (1–3, 14, 20–32)

Condom programs have helped as well among some other high-risk populations such as men who have sex with men. Although condom use may also
have contributed to the decline in HIV infection in some generalized African
epidemics, such as in Zimbabwe (33, 34), there is not yet evidence of a decisive role. (1, 4, 5, 20, 21) This is probably because consistent condom use
has not reached a high enough level—even after many years of widespread
and often aggressive promotion—to produce a measurable slowing of new
infections in these places. In these hardest-hit regions of the continent, most
HIV transmission typically occurs within more regular, concurrent sexual
partnerships. Numerous studies have made clear that consistent condom use
is difficult to achieve in these kinds of relationships. (1, 5, 20, 21, 35–38)
Even when couples use condoms during their first several sexual encounters,
use tends to diminish over time, as relationships become steadier.

   HIV testing and counseling: Rigorous reviews of many studies have shown
no consistent reduction in risky behavior for people who undergo counseling
and who test HIV-negative. (1, 4, 39–43) Some risk reduction, such as increased condom use, has been reported by people who test positive for HIV
and among couples in which only one partner is infected. (39–46) And a
growing focus of AIDS programs is an area called "Prevention for Positives,"
though the potential of this approach to impact the overall epidemic remains
unclear (please see chapter 4, endnotes 1 and 2). Some individuals do make
positive changes in their behavior after learning their HIV status or being
counseled. But in studies that specifically measured rates of infection for HIV
or other sexually transmitted infections, (46–49) there has been little solid
evidence of widespread beneficial impact from HIV counseling and testing
interventions.[12] One trial in the United States found a modest decrease in
new cases of sexually transmitted infections following an intensive counseling
and testing intervention. (50) But the most rigorous study so far was set in
Zimbabwe and, unique among trials of such programs, specifically tracked
new HIV infections. (49) Troublingly, it may even have found an increased
risk of infection, apparently as a result of "risk compensation," in which
people behave less carefully because the intervention itself makes them feel
safer.[13] This can also occur with other public health approaches, including

sunscreen lotions, seat belts, and condoms. (51–53) Several other studies of HIV testing, particularly among high-risk populations such as men who have sex with men, have found evidence of risk compensation. (48, 54–57) Another limitation of standard HIV tests are their inability to detect infection during the "acute" phase, during which a newly infected person is far more contagious than a month or two later, (58–60) when the immune system usually begins developing antibodies to the virus. Furthermore, there is no evidence of a pattern of HIV rates declining in those countries or regions where testing services have been made more available. So while HIV testing serves as a crucial entry point for infected people to access ARV medications and other important care, it appears unlikely to substantially alter the epidemic's course.[14]

Male circumcision: Findings from more than fifty studies since the 1980s have indicated that circumcision significantly reduces the risk of heterosexual HIV infection in men. (1, 67–77) These include randomized controlled trials in Africa, all three of which were stopped early for ethical reasons when initial findings demonstrated a risk reduction between 50 and 60 percent. (68, 73–77) Different analyses of the trial data—taking into account various other factors—suggested an even higher level of protection, of up to 76 percent. (78, and see chapter 24, endnote 19). Longer-term follow-up of the participants of these trials eventually corroborated a similarly strong effect of this one-time, permanent procedure. (79, 80) Taking into account so-called herd immunity—the collective effect of many members of society having a lowered risk of infection—the impact could be substantial if many men become circumcised.[15] (81–87) Preliminary modeling by UN and other experts suggests that a large-scale expansion of circumcision services could avert up to 5.7 million new HIV infections and 3 million deaths over twenty years in sub-Saharan Africa. (82) Many of these would be among women, protected indirectly by lower infection rates among their sex partners. Cost savings from widespread adoption of circumcision could eventually run into the billions of dollars by avoiding the long-term expenses of ARVs and other medical care. (83–87) Male circumcision also reduces the risk, in both partners, of some other sexually transmitted infections (67, 68, 90–98), including herpes, chancroid, and HPV, which results in fewer cases of cervical cancer in the partners of circumcised men. (67–69, 72, 95–98) In addition to the indirect effect of male circumcision on reducing HIV infections among women, observational studies from Africa suggest they may also be less likely to contract the virus from infected partners who are circumcised.[16] Nearly all the studies that inquired into Africans' attitudes have found a majority of uncircumcised men and their sex partners would welcome the procedure.[17] Behavioral risk compensation, as with nearly all prevention interventions, remains a concern, (51, 68–71) though there is no evidence yet of this having occurred with circumcision. (68, 75–77, 9, 80, 112–14) Among the likely reasons is that so far men have sought the procedure mainly for reasons such as to improve hygiene rather than specifically to reduce HIV risk.[18] (71, 103) However, experts agree that provision of male circumcision must be combined with promotion of behavior change, including partner reduction and consistent condom use. (1, 15, 68–71, 112–114)

Microbicides: In a South African trial, a vaginal gel that used an ARV as its active ingredient reduced risk of HIV to women by almost 40 percent and of HSV-2 by about 50 percent. (129) This came after years of disappointing results with a dozen other microbicide candidates, including one using the spermicide Nonoxynyl-9, which appears to have actually increased the risk of getting HIV in some settings. (130) While we hope that ongoing trials will confirm the finding of the ARV-based microbicide, it is widely agreed that nearly any topical agent brought to market would be less effective—on a per-act basis—than some existing approaches, including condoms. Such approaches are likely to eventually become an important method to help some women avoid infection, especially in cases where they desire more personal control, but microbicides appear unlikely to substantially alter the overall epidemic. This is in part because these methods, like condoms, would probably need to be reapplied prior to each sex act, making consistent use a challenge.[19] (131, 132) There is also the risk of negative consequences if high-risk populations, such as sex workers, were to switch from using condoms to microbicides in the belief that they are similarly effective. (48, 132) Attempts to develop other female-controlled prevention methods have so far been discour-

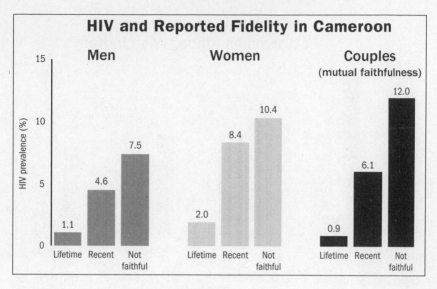

Faithfulness helps protect against HIV infection.

aging. A trial of the cervical diaphragm failed to reduce HIV incidence, though this may have been due to low adherence. (139; chapter 9, endnote 14)

Multiple sexual partnership reduction: Studies from around the world have consistently found an association between multiple sexual partnerships and increased risk of acquiring HIV and other sexually transmitted infections. Reductions in such multiple partnerships appear to have had a considerable effect in many countries already, (1, 2, 4, 5, 14, 15, 20, 140–61) and this approach could have even greater impact if it was more widely and assertively promoted. After the intensive Zero Grazing campaign was initiated in Uganda in 1987, WHO surveys conducted in 1989 and 1995 found a more than 50 percent reduction in the number of both men and women reporting casual partnerships. (1, 5, 14, 140–142, 151–155) The decrease in the number of men reporting three or more non-regular partners during the previous year was even more dramatic, falling from 15 to 3 percent. (140, 152, 155) In Kenya, partner reduction and fidelity have evidently been the main changes associated with the HIV decline there. (1, 21, 149, 150, 156) Similar forms of behavior change have been reported in large surveys in Zimbabwe, where HIV has also fallen considerably, (33, 34, 157, 158) along with

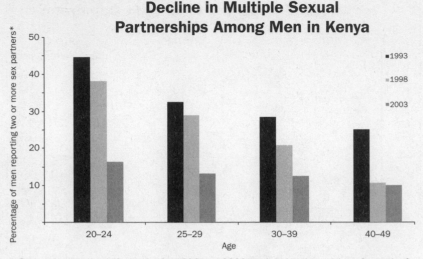

Decline in Multiple Sexual Partnerships Among Men in Kenya

* During the previous 12 months for 1998 and 2003, and during the previous 6 months for 1993 (which would understate the actual reduction in partners between 1993 and 1998).

Men reported sizable reductions in multiple partnerships between 1993 and 2003, roughly corresponding to the main period of HIV decline in Kenya.

a number of other countries including Ethiopia, (1, 149, 159) Malawi, (1, 149, 160) Zambia, (1, 155, 161) Ivory Coast, (1, 149) Haiti, (26, 149, 150) and the Dominican Republic. (26) Epidemiological modeling indicates that even modest reductions in multiple partnerships, if adopted by many people across a society, can have a considerable impact on infection rates over time. (5, 162, 163; and see chart on page 146) Few campaigns to reduce multiple sexual partnerships have been rigorously documented and evaluated, largely because such initiatives have been few in number and generally informal in nature. Some promising examples, however, have begun to emerge. In Swaziland, the number of people reporting multiple partnerships in the previous month fell by half after the aggressive *makhwapheni* campaign in 2006 targeted the danger of having a concurrent partner, or "secret lover." (141, 164) In 2009 the "One Love" multimedia campaign in South Africa, which was highlighted in a television soap opera, also generated a high degree of interest and public approval. (165) Other programs have been launched to educate local communities in several hard-hit countries, including Kenya, Botswana, and Mozambique, about the risk of multiple and concurrent partners and the danger of being in a risky sexual network. As described in chapter 14, the experience in places such as Uganda in the 1980s and early 1990s suggests that both partner reduction and combating stigma against people living with HIV can be successfully addressed together. (5, 14, 151–154, 166)

Pre-exposure prophylaxis: Many experts and others are increasingly interested in using ARVs to help prevent HIV infections (as also discussed in chapter 30 and the "Treatment-as-Prevention" section of this appendix). For a number of years, a short dose of ARVs has often been administered to recent rape victims or health workers accidentally stuck by needles.[20] And some recent trials have found that uninfected people taking ARVs as prophylaxis on an ongoing basis are about 45 to 70 percent less likely to contract HIV.[21] (167–69) For some high-risk individuals, including the sexual partners of people living with HIV, such strategies likely would be helpful. Yet as with treatment-as-prevention, some experts are skeptical about the ability of these approaches to have a broad, population-level impact, particularly in countries that have generalized epidemics. Concerns have been raised about cost and the ability of health systems to provide such services on a massive scale, as well as the possibility of risk compensation if taking ARVs in this manner becomes increasingly perceived as a substitute for other prevention measures such as condom use and partner reduction. (51, 64, 171–173)

Preventing mother-to-child transmission: While addressing only a relatively small proportion of overall HIV infections, interventions to prevent mothers from transmitting HIV to their babies are clearly important. Programs to provide ARVs to mothers and their newborn children have been highly effective at preventing HIV transmission to babies in some more-developed countries, such as Botswana, where remarkably few infants now end up infected. (174) However, the WHO recommends a four-pronged approach to PMTCT, of which this is only one. The other aspects include general prevention efforts—through promotion of behavior change and other strategies—to help women from becoming infected in the first place. It is also important to address infant feeding practices, particularly through promotion of exclusive breast-feeding during the first six months of a baby's life. As noted in chapter

28, a number of studies have found that this practice considerably reduces the risk of transmitting HIV to babies when compared to feeding them both breast milk and other substances. (174–79) Experience suggests this is best accomplished through warning mothers and others about the dangers of mixed feeding, which also poses other serious health risks yet is still the main way of feeding infants in most of Africa and many poorer regions of the world. Making family planning services more widely available also is fundamental, so that women who know they have HIV can choose whether to become pregnant, which could result in having children who also end up infected. Making these services available to the many women who are unaware of their HIV status also would result in fewer unintended pregnancies overall and, consequently, fewer babies born with HIV.[22] Modeling has estimated that providing widespread access to family planning services would eventually result in much less mother-to-child transmission than the use of ARVs alone during pregnancy and birth. (184–86) It also would result in fewer AIDS orphans left behind when some parents die.[23]

Regulatory, legal, gender-based, and economic approaches: A number of such strategies, often called "structural" approaches, have been proposed over the years, including even the idea of paying people to remain HIV-negative. Some recent studies have suggested that financial interventions such as buying school uniforms for female students can lead to outcomes like reduced teen pregnancy, (188) and one trial resulted in lower HIV infection among young women who were given cash incentives to avoid having relations with older "sugar daddies." (189, 190) However, this research was conducted in an extremely poor and culturally conservative rural area of Malawi. There are indications that such approaches may not work as well, and may even backfire, in regions where the cash economy is already more established and where social norms are less traditional. (5–7, 154, 165, 191, 192) A number of ethical, sustainability and other issues also have been raised regarding such strategies. Meanwhile, some policy and regulatory approaches, such as eliminating or restricting access to subsidized alcohol for uniformed personnel and other high-risk populations, have been found to be effective in some settings. In 2010, Botswana instituted a substantial tax on alcohol products and reportedly sales have declined considerably, which might also help reduce rates of sexually transmitted infections including HIV. (193, 194) Legal approaches such as establishing and enforcing curfews for young people have been used in places like Uganda during the Zero Grazing years to bolster the impact of behavior change campaigns. (5, 152, 153)

Screening the blood supply and preventing other nonsexual transmission: Biological interventions such as screening of the blood supply are highly effective, but at this stage in the epidemic address only a small proportion of HIV infections. Some have argued that unsafe medical practices, such as unsterile injections, cause a large portion of HIV cases, (195–98) but rigorous analyses of the available evidence show this is highly unlikely.[24] (199–202)

Treatment of other sexually transmitted infections: Nine randomized controlled trials measuring the impact of treating other sexually transmitted infections on HIV transmission have been conducted. The first study, conducted in Mwanza, Tanzania, in the early 1990s, found a nearly 40 percent reduction in HIV when sexually transmitted infections were treated.[25] But despite this initially encouraging finding, all the subsequent trials found no effect on HIV acquisition or transmission. (1, 205, 206) These trials included three studies aimed at preventing HIV infection by treating HSV-2, the virus that causes genital herpes. HSV-2 likely is a risk factor for HIV, and it's clear that people with this or other sexually transmitted infections are much more likely to also have HIV, and vice versa. But the evidence from these trials showed that treating these infections with medications did not affect HIV transmission. Because the same risky behaviors that cause HIV also cause other sexually transmitted infections, the association between these two may not necessarily be mainly a causal one. And there are no patterns across time or geography suggesting a direct relationship between treatment of sexually transmitted infections and reductions in HIV. In many parts of Africa, levels of bacterial infections such as syphilis, gonorrhea, and chlamydia now have a nearly inverse correlation with HIV prevalence. There are extremely high rates of these infections in some places, including many parts of Central Africa, where HIV prevalence remains relatively low.[26] (89) Conversely, HIV rates have reached exceptionally high levels in other areas, particularly parts of southern Africa, where such infections were greatly reduced many years ago following the establishment of effective control programs. (208–210) So while treating sexually transmitted infections remains critical for broader public health reasons—to prevent acquisition of syphilis in babies, for example—there is little evidence of an impact on the overall HIV epidemic.[27]

Treatment-as-prevention: Some experts regard this as the most exciting area of HIV prevention today, and it's easy to see why. Many observational studies have shown that an infected person taking effective, consistent doses of ARVs is much less contagious than an untreated person (214–17), and in 2011 a randomized trial showed a striking 96 percent reduction in risk among people who had regular sex with an infected person taking ARVs. (218) For some high-risk populations, such as the spouses of people with HIV, this approach indeed shows promise.[28] There should be benefits, in terms of both improving individual health and preventing transmission, as treatment guidelines are shifted for people to start ARVs before significant deterioration of their immune systems. What is less clear, however, is the capability of "treatment-as-prevention" to more broadly impact the AIDS epidemic, particularly in countries where it already has spread widely in the general population. (62–65, 171–173, 222–225) While evidence is emerging that expanded access to treatment may have had some promising population-level impact in places such as Vancouver and San Francisco, (226, 227) it remains uncertain whether a substantial decline in new HIV infections has occurred as a result. An important factor is risk compensation, as AIDS is no longer seen as a death sentence. The prevention impact of treatment—which does lower viral load at the population level—appears to have been at least partially canceled out by the negative consequence of lowered risk perception and, as a result, a greater propensity by some gay men and others to engage in risky behavior. (51, 65, 172, 173, 223–225) The effects of this can be observed in the relatively stable HIV rates in communities such as San Francisco at a time when rates of other sexually transmitted infections, such as rectal syphilis, have been increasing sharply. (223–225) A further limitation of treatment-as-prevention is that a third or more of all infections appear to occur from sex during the "acute infection" period, when the virus is unusually contagious but generally undetectable by standard HIV tests. (58–60, 163, 229, 230) It may also prove difficult to convince large numbers of people to take pills for many years mainly to prevent infection to others, as opposed to alleviating the symptoms of disease.[29] These drug regimens can have serious side effects, and adherence rates tend to be lower outside the strictures of rigorously controlled clinical trials (chapter 30, endnote 14). Condoms, for example, long have offered a highly effective way to slow transmission, and yet use has often proven to be inconsistent. In addition, many experts worry that it could prove exorbitantly expensive and impractical, for various logistical reasons, to try placing tens of millions of otherwise healthy people on ARVs for the rest of their lives. (62–65, 171–173, 232, and see chapter 30)

Vaccines: Most work on AIDS vaccine development, despite investments of many billions of dollars over three decades, has been disappointing, with several trials having been stopped prematurely because of lack of impact or possibly even harm. (1, 233–238) In 2009, a trial of one candidate appeared to find a mild protective effect of between about 25 and 30 percent, (236) but some experts have raised doubts about the interpretation of this result.[30] Additional trials would be necessary to confirm this finding. Vaccine researchers are investigating other promising candidates. Overall, however, there is little realistic prospect of an effective vaccine in the near future. (235) We certainly hope that this changes because vaccines are among the most powerful tools in public health. Yet HIV, partly because of its ability to evade the body's immune response with unusual speed, has proven to be an exceedingly difficult target for vaccine developers.

ACKNOWLEDGMENTS

Tinderbox was born in large part from conversations the authors had over many years with scholars, scientists, doctors, nurses, religious leaders, educators, politicians, journalists, community activists, and, most of all, thousands of ordinary people from Africa and elsewhere whose insights and generosity never failed to astound us. Only a small percentage of these people are named in this book, but the cumulative power of the contributions of those unnamed are at least as profound. Few asked for anything in return, other than the faint hope of having their experience understood by the wider world. We particularly wish to thank Faruk Maunge, Serara Selelo-Mogwe, Tsetsele Fantan, Oliver Mtukudzi, Brian Khumalo, Collins Omondi Okendi and his relatives, Raymond Kwisega, Cathy Katumba, Thandi Mzukwa, Gcebile Ndlovu, Siphiwe Hlophe, Gordon Mthembu, and Chandapiwa Mavundu, as well as Sarah, Yvonne, and Sifiso, whose last names will remain unpublished out of respect for their privacy.

For the historical, cultural and epidemiological information in *Tinderbox*, vital perspectives were offered by, among others, Phil Burnham, Swizen Kyomuhendo, Tamara Giles-Vernick, Robert Bailey, Helen Epstein, Jim Shelton, Nancy Rose Hunt, Ludo Margaret Mosojane, Michael Worobey, Beatrice Hahn, Martine Peeters, Kawango Agot, Edward Green, Ronald Gray, Norman Hearst, Khalidah Bello, Allison Herling Ruark, Peter Kilmarx, Kate Crawford, Tim Hallett, and Steve Hodgins. Valuable remembrances on the global response to the AIDS epidemic came from Jesse Kagimba, Michel Carael, Rand Stoneburner, Fred Wabwire-Mangen, Peter Piot, Elizabeth Pisani, David Serwadda, Anne Peterson, Helen Jackson, Malcolm Potts, Sam Okware, Kent Hill, Elly Katabira, John Cleland, and many others.

Timberg's time reporting and writing *Tinderbox* was supported in part by The John Templeton Foundation, which provided a grant in the crucial early days of this project. This work also was made possible by the generosity and patience of many of Timberg's current and former colleagues at *The Washington Post*, including Marcus Brauchli, Liz Spayd, Raju Narisetti, Kevin Merida, Cameron Barr, Emilio Garcia-Ruiz, Vernon Loeb, Nick Anderson, Leonard Downie, Phil Bennett, David Hoffman, Scott Wilson, Tiffany Harness, Mary Hadar, Vanessa Williams, Shirley Carswell, Peter Perl, Donald Graham, and Katharine Weymouth. Also providing wisdom and unfailing hospitality were many friends and colleagues in Africa, including Lebogang Montjane, Robyn Dixon, Stephanie Nolen, Karima Brown, Eddie Mleya, Aminu Abubakar, Charles Ngereza, Joy Brady, Shakeman Mugari, Darlington Majonga, and Musa Ndlangamandla.

Halperin wishes to recognize the enduring contributions, intellectual and otherwise, of his former coworkers at USAID, including Jim Shelton, Glenn Post, David Stanton, Kate Crawford, Kendra Phillips, Jeff Spieler, Nomi Fuchs, Peter Halpert, Michael Cassell, Jeff Ashley, Shanti Conly, and Paul Delay. In addition to those colleagues already mentioned, he is grateful to the many others who have directly and indirectly enriched this endeavor, including David Alnwick, Clemens Benedikt, Doug Call, Ward Cates, Eduarda Cipriano, Antonio de Moya, Ken French, Exnevia Gomo, Simon Gregson, Poloko Kebaabetswe, Jeff Klausner, Bongani Langa, Jay Levy, Shahin Lockman, Khanya Mabuza, Tim Mah, Ray Martin, Willi McFarland, Marc Mitchell, Elisa Ruiz, Roger Shapiro, Roger Short, Ann Swidler, Richard Wamai, Helen Weiss, David Wilson, Derek von Wissell, and Godfrey Woelk. He wishes to pay homage to his dear friend and colleague Doug Kirby—scientist, humanitarian, adventurer—who dedicated his life to the pursuit of truth and excellence and was a beacon of decency and integrity.

Tinderbox could not have reached fruition without the work and devotion of our terrific agent, Gillian MacKenzie, and the excellent team at Penguin Press, most notably Laura Stickney, Mally Anderson, and Ann Godoff. We also would like to thank our friends and colleagues who graciously took the time to read and comment on various drafts, including Arthur Allen, Kelley Andrews, Robert Bailey, Khalidah Bello, Alan Brody, Debby Cornelius, Justin Gillis, Marelize Gorgens, Neil Gutman, Tim Hallett, Samuel Halperin, Stephanie Hanes, Allison Herling Ruark, Steve Hodgins, Helen Jackson, Roy Jacobstein, Judith Kneeter, Todd Koppenhaver, Suzanne Leclerc-Madlala, Lebogang Montjane, Stephanie Nolen, Glenn Post, Lee Ross, Roger Short, Carol Stack, Markus Steiner, Mark Tilton, Robert Timberg, Scott Timberg, Francois Venter, Jake Waskett, and Alan Whiteside. Several friends within the publishing industry also were generous with their time, including Flip Brophy, Steve Wasserman, David Patterson, Scott Moyers, and Peter Osnos. All errors and shortcomings of *Tinderbox* are, of course, the responsibility of the authors.

This book, like many others, took a heavy toll on the family time of both of its authors. Timberg would like to express his profound affection and gratitude to his wife, Ruey, and their children Cecilia, Andrew, and Natalie, and also to the many friends and family members who supported him along the way with love and kindness, including functioning Internet connections on those many occasions when his own failed. Halperin is indebted to the loving support and patience of his wife, Moira, and their daughters Leila and Ariel, and to his dear mother Tam and siblings Dina and Jonathan. He would also like to honor the memory of his father Irving, who instilled a deep respect for the craft of writing.

NOTES

Prologue

1. The term "Pygmies" is regarded by some as pejorative.
2. Scientists once thought simian immunodeficiency virus (SIV) was largely benign, but more recent research by scientists, particularly a team led by Beatrice Hahn of the University of Alabama at Birmingham, has indicated that it does kill many chimps.
3. The evidence regarding the exact nature of the initial transmission event from chimp to human remains circumstantial and open to some debate. It is possible, for example, that an infected chimp bit a human, which occasionally occurs in Central Africa. However that first infection happened, it appears unlikely that a major HIV epidemic could have emerged in Africa during this era without other key elements, including the impact of colonialism discussed in this book (see chapter 5, endnote 14).
4. Infections from HIV-1, Group M have been the cause of nearly all deaths from AIDS, but there have been some infections from other types of HIV, including HIV-2, a much less deadly virus confined mainly to parts of West Africa. There have also been some other HIV-1 types, such as Groups N and O, which are still found almost exclusively in Cameroon. In the historical recounting of this book, when we refer to the birth and global epidemic of "HIV," we are referring specifically to the HIV-1 Group M virus.
5. According to the 2010 *UNAIDS Global Report* (Geneva), people living with HIV in Africa represent 70 percent of the total number of infected people worldwide. It is possible that this proportion is even greater because the estimates for HIV prevalence in other regions have tended to be less precise than those in Africa. The newer types of measurement methodologies, based upon Demographic and Health Survey–type household surveys, which are discussed in later chapters, have only been conducted in very few countries outside of sub-Saharan Africa.
6. UNAIDS, *Report on the Global HIV/AIDS Epidemic* (Geneva: UNAIDS, June 2000):

pp. 16–19, www.thebody.com/content/art31144.html#africa. The cumulative risk of infection for individuals over the course of their lifetimes is much higher than the prevailing HIV rate at any given moment, as the risk for contracting HIV continues over many decades of sexual activity. Simon Gregson and Geoff P. Garnett, "Contrasting Gender Differentials in HIV-1 Prevalence and Associated Mortality Increase in Eastern and Southern Africa: Artefact of Data or Natural Course of Epidemics?" *AIDS* 14 (2000) (Supplement 3): S85–S99; Wambura Mwita, Mark Urassa, Raphael Isingo, et al. "HIV Prevalence and Incidence in Rural Tanzania: Results from 10 Years of Follow-up in an Open Cohort Study." *J Acquir Immune Defic Syndr* 46 (2007): 616–23.

7. Civil war and conflict often are cited as major causes of HIV's spread in Africa, but careful analysis of the evidence suggests this is unlikely. Paul B. Spiegel, Anne R. Bennedsen, Johanna Claass, et al., "Prevalence of HIV Infection in Conflict-Affected and Displaced People in Seven Sub-Saharan African Countries: A Systematic Review." *Lancet* 369 (2007): 2187–95. (Also see Malcolm Potts, Daniel T. Halperin, Douglas Kirby, et al., "Reassessing HIV Prevention." *Science* 320 [2008]: 749–50, including the Supporting Online Supplemental Material.) Furthermore, HIV rates in Africa have tended to be higher among people with higher income: Mishra Vinod, Simona Bignami-Van Assche, Robert Greener, et al., "HIV Infection Does Not Disproportionately Affect the Poorer in Sub-Saharan Africa." *AIDS* 21, suppl. 7(2007): S17–28; James D. Shelton, Michael M. Cassell, and Jacob Adetunji, "Is Poverty or Wealth at the Root of HIV?" *Lancet* 366 (2005): 1057–58. Daniel Halperin, "Old Ways and New Spread AIDS in Africa." *San Francisco Chronicle*, Nov. 30, 2000: A31 (sfgate.com/cgi-bin/article.cgi?file=/chronicle/archive/2000/11/30/ED113453 .DTL). In Tanzania, for example, HIV rates among men in the highest wealth quintile were about two and a half times higher than among men in the lowest quintile, and the corresponding difference among women was about fivefold. There and in most other African countries, more educated people also have higher HIV rates. (Tanzania HIV/AIDS Indicator Survey 2003/04; Tanzania Commission for AIDS/National Bureau of Statistics, Dar es Salaam, Tanzania/ORC Macro, Calverton, MD.) However, in recent years this tendency has begun to reverse in a few countries. In Zimbabwe, for example, rates of HIV infection have become similar for poorer, less-educated people and wealthier, more-educated ones. James R. Hargreaves, Christopher P. Bonell, Tania Boler, et al., "Systematic Review Exploring Time Trends in the Association Between Educational Attainment and Risk of HIV Infection in Sub-Saharan Africa." *AIDS* 22 (2008): 403–14.

8. It is common when describing the path of HIV through the world to use the term "pandemic." As discussed in the appendix, we avoid using the term in this book because we feel this could suggest that the virus has infected relatively similar percentages of the population in much of the world. We prefer the terms "AIDS epidemic" or "HIV epidemic." When talking about two or more communities or countries, we refer to them as "epidemics." This better reflects, we believe, the reality that the virus has spread in particular ways through particular communities, and often at remarkably disparate rates.

9. The impact of male circumcision on the homosexual spread of HIV is more complex and less profound, mainly because most transmission among men who have sex with men occurs from being the receptive—not the insertive—partner in anal sex, in which case circumcision

status is irrelevant. However, in some regions, including much of Latin America, Asia, and Africa, there are substantial numbers of men who report being exclusively the insertive partners when having sex with other men. In such instances, circumcision appears to be protective to a similar degree as it is for men having vaginal sex, based on findings from recent studies conducted in South Africa, Peru, and Australia. David J. Templeton, Fengyia Jin, Liminc Mao, Garrett P. Prestage, et al., "Circumcision and Risk of HIV Infection in Australian Homosexual Men." *AIDS* 23 (2009); Charles S. Wiysonge, Eugene J. Kongnyuy, Muki Shey, et al., "Male Circumcision for Prevention of Homosexual Acquisition of HIV in Men." *Cochrane Database Syst Rev* 6 (2011): CD007496; Jorge Sánchez, Victor G. Sal y Rosas, James P. Hughes, et al., "Male Circumcision and Risk of HIV Acquisition Among MSM." *AIDS* 25, no. 4 (Feb. 20, 2011): 519–23; Simeon Bennett, "Circumcision Reduces HIV Risk for Some Gay Men, Research Shows." Bloomberg News Service, July 20, 2009.

10. See the writings of Geoffrey Rose ("Sick Individuals and Sick Populations." *Int J Epid* 14 [1985]: 32–38), Roy Anderson (*Infectious Diseases of Humans: Dynamics and Control* [New York: Oxford University Press, 1991]), Douglas Weed ("A Radical Future for Public Health." *Int J Epid* 30 [2001]: 440–41), and Steven Johnson (*The Ghost Map: The Story of London's Most Terrifying Epidemic—And How It Changed Science, Cities, and the Modern World* [New York: Riverhead, 2006]).

11. Estimates for HIV prevalence in adults vary tremendously around the world, ranging from less than 0.1 percent in the Philippines, Bangladesh, and South Korea, to 0.2 percent in Bolivia, Belgium, and Madagascar, and up to about 25 percent in the southern African nations of Swaziland, Lesotho, and Botswana (UNAIDS Global Report, Geneva, 2010). Also, in most of the world male circumcision is not nearly as crucial a factor as it is in sub-Saharan Africa, in part because of the much larger proportion of infections in those places due to sex between men and injecting drug use.

12. Some new studies suggest that the acute infection period for the subtype C of the HIV virus, which is found predominately in southern Africa and parts of Asia, might be of longer duration. Vladimir Novitsky, Rui Wang, Hermann Bussmann, et al., "HIV-1 Subtype C-Infected Individuals Maintaining High Viral Load as Potential Targets for the 'Test-and-Treat' Approach to Reduce HIV Transmission." *PLoS One* 5 (2010): e10148. While this may help explain the unusually high HIV rates in the southern African region, it is clear that severe epidemics can also take off in the absence of subtype C. For example, very high rates have occurred in places such as Uganda and western Kenya, where other HIV subtypes predominate.

13. In this book, for the sake of readability, we often use "Africa" when actually we mean "sub-Saharan Africa."

14. Even if prevention efforts were to succeed—that is, if the number of people becoming newly infected fell significantly—it could still take many years for HIV prevalence rates to also decline substantially. This will increasingly be the case if many people infected with HIV are taking and adhering to ARV medications, allowing them to live longer. As that happens, the total number of people infected with HIV would tend to stay elevated because many more of them would continue to stay alive and be counted. That is why HIV rates alone are an incomplete measure of progress against the epidemic. The public health goal should be

to prevent new infections while extending and improving the lives of those already living with HIV.

Chapter 1: Francistown

1. Several elements of this chapter appeared in *The Washington Post*, on March 2, 2007, in an article by Timberg headlined "Speeding HIV's Deadly Spread: Multiple, Concurrent Partners Drive Disease in Southern Africa."

2. For the sake of simplicity we use the term "HIV rate" when technically we mean "HIV prevalence." Prevalence is the more formal epidemiological term for describing the level of infections in a population, meaning that, if one hundred adults out of one thousand total are infected with the virus, this would translate to a prevalence "rate" of 10 percent.

3. Although many experts and others prefer the term "sex worker," in this book we often use the colloquial term "prostitute."

4. Please see appendix, endnote 7 for an explanation of why rape does not appear to be a major factor for HIV transmission.

5. In 1996 almost 85 percent of women ages fifteen to twenty-four in Botswana reported using a condom during their most recent sexual encounter with a nonregular partner. At the time, this percentage was much higher than similar measures reported anywhere else in Africa, or probably anywhere else globally. A similar figure was still being reported eight years later (UNAIDS/WHO Epidemiological Fact Sheets, 2000 and 2004, Geneva).

6. Municipalities such as Francistown display various similarities to some American cities or suburbs, but of course they are not completely identical, especially in areas farther from the downtown centers. In addition, it is worth noting that in countries such as Botswana the level of income inequality is much higher than in Europe or North America. (However, standard measures of inequality, including the Gini Co-Efficient, normally do not take into account the provision of free or heavily subsidized health care, education, housing, and other social services in some nations such as Botswana.)

7. We have, at Angela's request, not used her real name.

8. Studies in Africa have found that in nearly half of all serodiscordant couples, meaning where only one person in the couple is infected, it is the woman who has HIV, not the man. Oghenowede Eyawo, Damien de Walque, Nathan Ford, Gloria Gakii, Richard T. Lester, and Edward J. Mills, "HIV Status in Discordant Couples in Sub-Saharan Africa: A Systematic Review and Meta-Analysis." *Lancet Infect Dis* 10 (2010): 770–77; Damien de Walque, "Sero-Discordant Couples in Five African Countries: Implications for Prevention Strategies." *Population Devel Rev* 33 (2007): 501–23; James D. Shelton, "Ten Myths and One Truth About Generalised HIV Epidemics." *Lancet* 370 (2007): 1809–11; "The Not-So-Fair Sex." *The Economist*, June 28, 2007, www.economist.com/science/displaystory.cfm?story_id=9401560. A number of these studies are discussed in the chapter on gender, marriage, and HIV in Edward Green and Allison Herling Ruark, *AIDS, Behavior, and Culture: Understanding Evidence-Based Prevention* (Walnut Creek, CA: Left Coast Press, 2011).

9. For a description of the nuances of infidelity and other aspects of sexual culture in Botswana, see Isaac Schapera, *Married Life in an African Tribe* (Evanston, Ill.: Northwestern University Press, 1966). For example, Schapera recounts how a groom on his wedding day was attracted to two young women attending his wedding and asked his bride if he could take them on

as mistresses. She agreed, provided that he do so with discretion and that he promise to father children with her (pp. 206–7). On a 2000 visit to Botswana, Halperin interviewed the traditional leader, Kgosi Linchwe, of the same Mochudi area where Schapera conducted most of his research (Sandy Grant, *People of Mochudi* [Mochudi: Phuthadikobo Museum, 2001]). Chief Linchwe, now deceased, confirmed the main findings of Schapera's work, which has occasionally been controversial. For example, he agreed that while premarital sex, including even preliminary experimentation by young children, was generally tolerated, there were also clear cultural rules regarding matters such as pregnancy outside of marriage.

10. Kenyan anthropologist Jomo Kenyatta, who later became that nation's president, wrote that some premarital sex play traditionally was permitted among his Gikuyu ethnic group. But a man was never supposed to ejaculate inside a woman who was not his wife: "Any intercourse which may result in pregnancy before marriage is strictly forbidden. Any young man who may render a girl pregnant (*kohira moiretu ihu*) is severely punished by the *kiama* (tribal council). The fine for this is nine sheep or goats and three big, fat sheep (*ndorome*) as the *kiama* fees. Besides this, the man is made a social outcast or 'sent to Coventry' (*kohingwo*) by all the young men and girls of his own age-group. Punishment is also extended to the girl. She pays a fine by providing a feast to the men and the girls of her age-group. She is also liable to ridicule (*kohingwo* and *gocambio*)." Jomo Kenyatta, *Facing Mt. Kenya: The Tribal Life of the Gikuyu* (London: Secker and Warburg, 1938).

11. Among males, these Setswana coming-of-age ceremonies included circumcision. This too would have made HIV's spread much less likely.

Chapter 2: Searching for the Beginning

1. An early example of this gulf in perceptions came after the publication of the first studies suggesting that HIV likely was born in the Congo River Basin from human contact with primates. Once scientists discovered, in the mid-1980s, that HIV closely resembled a simian virus (SIV), it was fairly clear that the epidemic started in Africa. The gorillas, monkeys, and chimpanzees that harbored SIV were indigenous to the continent. Yet many Africans bristled at the suggestion that the supposed "Gay Plague"—already entwined in the public imagination with behavior they regarded as perverse—somehow emanated from their own continent. Some Westerners, meanwhile, revived old stereotypes, born of colonial dispatches from missionaries and explorers, suggesting that Africans were hypersexual as well as indiscriminate in their choices of partners. At the first International AIDS Conference, in Atlanta in 1985, an American journalist was tasteless enough to ask a group of doctors visiting from Zaire, "Is it true that Africans have sex with monkeys?" Laurie Garrett, *The Coming Plague: Newly Emerging Disease in a World Out of Balance* (New York: Penguin Books, 1994).

2. Suspicions that HIV began as a Western bioweapon also have traveled widely in the African American community in the United States; see Seth Kalichman, *Denying AIDS: Conspiracy Theories, Pseudoscience, and Human Tragedy* (New York: Springer/Copernicus, 2009).

3. Joby Warrick, "Biotoxins Fall Into Private Hands." *The Washington Post*, April 21, 2003.

4. Robert Cooke, "Scientists: AIDS, Polio Vaccine Are Unrelated." *Newsday*, September 12, 2000.

5. Worobey and Hamilton believed that any chimpanzees used in the Kisangani lab would have been caught in this area rather than shipped in from somewhere else. The varieties of SIV

found among chimp communities are based heavily on geography, so the SIV identified in one part of central Congo would closely resemble that found in other communities in central Congo. This allowed the scientists, they believed, to test the SIV in the region without finding the precise community of chimps that might have been used in polio vaccine production. Michael Worobey, Mario L. Santiago, Brandon F. Keele, et al., "Contaminated Polio Vaccine Theory Refuted." *Nature* 428 (2004): 820.

Chapter 3: One Tiny Speck of Truth

1. Sooty mangabeys have been found to be the source of the epidemic of HIV-2, a related but much less virulent and deadly virus concentrated in parts of West Africa. In the past some researchers postulated that the HIV-2 epidemic may have had a dampening effect on the spread of HIV-1 in that region. Today this theory has largely been abandoned, partly because the near-universality of male circumcision in West Africa provides a more compelling explanation.
2. Brandon F. Keele, Fran V. Heuverswyn, Li Yingying, et al., "Chimpanzee Reservoirs of Pandemic and Nonpandemic HIV-1." *Science* 28 (2006): 523–26.
3. André J. Nahmias, Jeffrey Weiss, Xi Yao, et al., "Evidence for Human Infection with an HTLV III/LAV-like Virus in Central Africa, 1959." *Lancet* 327 (1986): 1279–80.
4. Michael Worobey, Marlea Gemmel, Dirk E. Teuwen, et al., "Direct Evidence of Extensive Diversity of HIV-1 in Kinshasa by 1960." *Nature* 455 (2008): 661–64.

Chapter 4: A Tale of Two Viruses

1. The approximate risk of HIV infection from vaginal sex is estimated mainly from discordant couple studies, in which initially only one partner is already infected and the other partner is followed for a period of time, often for many years. In these studies, as well as in national surveys conducted in various African countries, there are nearly always many more discordant couples than ones in which both partners have the virus. This suggests that, although it is important to address the risk of infection in discordant couples, the majority of them will remain discordant, even in the absence of interventions such as condom use. Ronald H. Gray, Victor Ssempijaa, James Shelton, David Serwadda, Fred Nalugoda, and Maria J. Wawer, "The Contribution of HIV-Discordant Relationships to New HIV Infections." *AIDS* 25 (2011):1343–44; Green and Ruark, *AIDS, Behavior, and Culture* pp. 211–12; Eyawo et al., "HIV Status in Discordant Couples in Sub-Saharan Africa"; Shelton, "Ten Myths and One Truth About Generalised HIV Epidemics"; Nancy Padian, Stephen Shiboski, Sarah Glass, and Eric Vittinghoff, "Heterosexual Transmission of HIV in Northern California: Results from a Ten-Year Study." *Am J Epidemiol* 146 (1997): 350–57; Rebecca Bunnell, Alex Opio, Joshua Musinguzi, et al., "HIV Transmission Risk Behavior Among HIV-Infected Adults in Uganda: Results of a Nationally Representative Survey." *AIDS* 22 (2008): 617–24; James D. Shelton, "A Tale of Two-Component Generalised HIV Epidemics." *Lancet* 375 (2010): 964–66. The rough estimate of one infection per every one thousand unprotected acts obscures that certain cofactors can alter the risk substantially. These include the "acute infection period" discussed in later chapters, or whether the man is circumcised. Regarding the risk of HIV infection from oral sex, some studies among men who have sex with men have detected a very small risk, but most have found none. In a study of women in Spain

whose partners had HIV, researchers documented more than nineteen thousand acts of unprotected fellatio but not a single case of HIV transmission. Jorge del Romero, Beatriz Marinocovich, Jesús Castilla, et al., "Evaluating the Risk of HIV Transmission Through Unprotected Orogential Sex." *AIDS* 16 (2002): 1296–97. However, there is certainly a risk of contracting some other sexually transmitted infections, such as gonorrhea or herpes, from oral sex.

2. In one trial published in 2011, for example, among the couples where one of the partners became infected nearly 30 percent of those new infections appeared to occur from sex outside the relationship, even though at the beginning of the study only 5 percent of participants reported having more than one partner. Myron S. Cohen, Ying Chen, Marybeth McCauley, et. al., "Prevention of HIV-1 Infection with Early Antiretroviral Therapy." *N Engl J Med* 365 (2011): 493–505. Also see Gray et al., "The Contribution of HIV-Discordant Relationships"; Connie Celum, Anna Wald, Jairam R. Lingappa, et al., "Acyclovir and Transmission of HIV-1 from Persons Infected with HIV-1 and HSV-2." *N Engl J Med* 362 (2010): 427–39; Oliver Manigart, Claudine Kraft, Nathan Makombe, et al., "Co- and Superinfection of Partners in a Cohort of Couples Previously Infected by Genotypically Different Viruses in Kigali." *Retrovirology* 6 (2009): 14; Herling Green and Ruark, *AIDS, Behavior, and Culture*; Allison Herling Ruark, James D. Shelton, and Daniel T. Halperin, "Acyclovir and Transmission of HIV-1 from Persons Infected with HIV-1 and HSV-2." *N Engl J Med* 362 (2010): 1741; Ronald Gray, "Discordant Couples and HIV Transmission." Presentation for "Emerging Issues in Today's HIV Response: A Debate Co-hosted by the World Bank and USAID," Washington, D.C., August 26, 2010, http://go.worldbank.org/OPC5GJ5RD).

3. "Zimbabwe Demographic and Health Survey, 2005–6." Harare/Calverton, MD: ORC MACRO International. The underreporting of risky sexual behavior common to such household-based surveys is discussed in greater detail in chapter 30. In some of these surveys, including the one in Zimbabwe, the interview respondents were also invited to be tested for HIV. Note that not all of these women would necessarily have been sexually active; some could have been victims of early childhood sexual abuse or originally infected from their mothers before or during birth, or during breast-feeding. Some might also have been infected through other forms of nonsexual transmission, such as from unsafe medical injections. A few writers have postulated that such data is evidence of nonsexual transmission being the main source of HIV in Zimbabwe and other African countries with severe HIV epidemics, but as noted in the appendix, rigorous analysis of the totality of the evidence suggests this is unlikely.

4. In a very small percentage of cases, some individuals became infected with HIV many years ago but have not yet developed AIDS. Scientists think this probably results from a genetic immunity.

5. Condoms have been shown to provide a high level of protection against infection from HIV and some other sexually transmitted infections (see appendix). However, in a CDC study conducted among women in Brazil, researchers detected "moderate-to-high" levels of prostate-specific antigen (PSA), which is only produced in men, in the vaginal tracts of 3.5 percent of users of male condoms and 4.5 percent of users of female condoms immediately after sex. (Considerably higher numbers of women had "low" levels of PSA.) Maurizio Macaluso, Richard Blackwell, Denise J. Jamieson, et al., "Efficacy of the Male Latex Condom

and of the Female Polyurethane Condom as Barriers to Semen During Intercourse: A Randomized Clinical Trial." *Am J Epidemiol* 166 (2007): 88–96. This finding suggests there may be occasional leakage of semen during condom use. (PSA is much smaller than sperm and about the same size as HIV.) In addition to known "user failures" associated with condom breakage and slippage, this could help explain why condoms are less than 100 percent effective for both contraception and HIV prevention. In the United States, rates of male condom breakage during sexual intercourse and withdrawal are approximately two broken condoms per hundred condoms used. (Susan Weller and Karen Davis-Beaty, "Condom Effectiveness in Reducing Heterosexual HIV Transmission." *Cochrane Database Syst Rev* [2001]: CD003255; Lee Warner and Markus J. Steiner, "Male Condoms." In *Contraceptive Technology*, 20th revised ed., Robert A. Hatcher, James Trussell, Anita L. Nelson, et al., eds. [New York: Ardent Media, 2011]). However, studies in some other countries have found higher rates of breakage and slippage (Léonard Mukenge-Tshibaka, Michael Alary, Nassirou Geraldo, and Catherine M. Lowndes, "Incorrect Condom Use and Frequent Breakage Among Female Sex Workers and Their Clients." *Int J STD AIDS* 16 [2005]: 345–47; Pauline Russell-Brown, Carla Piedrahita, Robin Foldesy, et al., "Comparison of Condom Breakage During Human Use with Performance in Laboratory Testing." *Contraception* 45 [1992]: 429–37). In a Family Health International–led study conducted in Jamaica, male participants reported breaking 18 percent of condoms used during the week prior to the interview and having 3.5 percent of condoms slip off completely during sex. A brief counseling session on correct use led to a reported reduction in such mishaps. Markus J. Steiner, Doug Taylor, Tina Hylton-Kong, et al., "Decreased Condom Breakage and Slippage Rates after Counseling Men at a Sexually Transmitted Infection Clinic in Jamaica." *Contraception* 75 (2007): 289–93.

6. Women occasionally do become infected through the vaginal tissue, but most research suggests that the cervix is likely the main entry point of HIV infection in women.

7. In many parts of Africa there is a common practice of "dry sex," in which women insert powders, herbs, or other substances to dry or "tighten" their vaginas prior to intercourse. Judith E. Brown and Richard C. Brown, "Traditional Intravaginal Practices and the Heterosexual Transmission of Disease: A Review." *Sex Transm Dis* 27 (2000): 183–88; Mags E. Beksinska, Helen V. Rees, Immo Kleinschmidt, and James McIntyre. "The Practice and Prevalence of Dry Sex Among Men and Women in South Africa: A Risk Factor for Sexually Transmitted Infections?" *Sex Transm Infect* 5 (1999): 178–80. Dry sex is also commonly practiced in parts of the Caribbean region. (Daniel T. Halperin, "Dry Sex Practices and HIV Infection in the Dominican Republic and Haiti." *Sex Trans Infect* 75 [1999]: 445–46.) As noted in the appendix, endnote 19, some studies have found that such practices, which can cause trauma and tearing of vaginal tissue during sex, may increase women's risk of HIV infection.

8. Epidemiologists refer to this concept as R0=1, with the R0 representing the rate of transmission. So long as R0>1, meaning that ongoing transmission is happening faster than existing victims die or are cured, the epidemic continues to spread. When it falls below the crucial threshold, such that R0<1, the epidemic goes into reverse and begins to fade away. Along these lines, William D. Hamilton conducted some interesting theoretical research, before the discovery of AIDS, showing how viruses that are not very aggressive could end up infecting many more people than more immediately virulent ones. And see Cristoff Fraser, T. Déirdre

Hollingsworth, Ruth Chapman, Frank de Wolf, and William P. Hanage, "Variation in HIV-1 Set-Point Viral Load: Epidemiological Analysis and an Evolutionary Hypothesis." *P Natl Acad Sci USA* 104 (2007): 17441–46.

9. There have been other instances of SIV crossing into the human population, such as those resulting in the much more minor epidemics of HIV-2 and of HIV-1, Groups N and O (prologue, endnote 4).

Chapter 5: The Lion and Dr. Livingstone

1. David Livingstone, *Travels and Researches in South Africa* (Philadelphia: G. G. Evans, 1859).

2. Livingstone named this Victoria Falls.

3. George Seaver, *David Livingstone: His Life and Letters* (New York; Harper & Brothers, 1957), p. 79.

4. The most comprehensive account of this historical period can be found in Thomas Pankenham's *The Scramble for Africa* (London: George Weidenfeld & Nicolson, 1991).

5. For more on Belgian brutality, see Adam Hochschild's *King Leopold's Ghost: A Story of Greed, Terror, and Heroism in Colonial Africa* (New York: Houghton Mifflin Company, 1998).

6. Sometimes the ivory was found on the ground.

7. The same is true today. Most surveys make it clear that during years of civil war it is the diseases that produce most of the body counts, not the bullets. For more information, see the excellent mortality surveys produced by the International Crisis Group, www.crisis group.org.

8. W. O. Henderson, *The German Colonial Empire: 1884–1919* (London: Routledge, 1993), p. 84.

9. Helmuth Stoecker, ed., *German Imperialism in Africa* (London: Humanities Press International, 1986), p. 84.

10. Harry Rudin, *Germans in the Cameroon, 1884–1914* (New Haven: Yale University Press, 1938), p. 92.

11. Anthropologist Philip Burnham of University College of London, in an interview with Timberg.

12. Peter Geschiere and James J. Ravell (trans.), *Village Communities and the State: Changing Relations among the Maka of South-Eastern Cameroon Since the Colonial Conquest* (Boston: Kegan Paul International, 1982), p. 141.

13. A key role could also have been played by Brazzaville, on the other side of the Congo River. The cities are so close that they are almost indistinguishable from the point of view of tracking an epidemic.

14. More recently this part of Cameroon has seen a heavy influx of logging. Combined with the bush meat trade it is possible that—even in the absence of the colonial history—HIV might have eventually emerged in the more recent era. If this happened it would likely have produced a significantly different kind of epidemic than the one that in fact started spreading near the beginning of the twentieth century. Jacques Pepin (*The Origins of AIDS* [New York: Cambridge University Press, 2011]) suggests that HIV-1 groups N and O, which appear to have emerged more recently in Cameroon, may represent the kinds of small, localized epidemics that could have developed without the influence of colonialism in Central Africa. These epidemics, each of which also originated from a single contact with

chimps infected with SIV, sparked limited outbreaks that never traveled far. Even in present-day Cameroon most HIV infections are from Group M, probably originally by way of Congo.

Chapter 6: *Femmes Vivant Théoriquement Seules*

1. E. J. Glave, *In Savage Africa, or, Six Years of Adventure in Congo-Land* (New York: R. H. Russell & Son, 1892), p. i.
2. Ibid., p. 26.
3. Ibid., p. 31.
4. Ibid., pp. 33–34.
5. Catherine Coquery-Vidrovitch, "The Process of Urbanization in Africa (From the Origins to the Beginning of Independence)." *Afr Stud Rev*, 34 (1991): 24.
6. Anthropologists and historians who have investigated the precolonial sexual culture of the region do not always agree on its character, including the extent to which such rules regarding sexual behavior were adhered to. There are ethnographic accounts, for example, of fathers offering the sexual favors of their daughters to visitors in exchange for goods. John and Pat Caldwell, the anthropologist team who have worked for many years in Nigeria, have explored this in a number of papers, including: John C. and Pat Caldwell, "The African AIDS Epidemic." *Scientific American* (March 1996). They wrote that many sub-Saharan African cultures put more emphasis on the importance of fertility than on preventing sex outside of marriage: "Infidelity might occasionally spark fights, punishment and, more rarely, marital dissolution, but it was never equated with sin and excoriated in the way that it was in traditional Western and Asian societies. Much good flowed from this permissive attitude: women were not suppressed and hidden, and girls had survival chances as great as their brothers." Yet eventually these traditions, as the Caldwells noted, did make these societies more vulnerable to the spread of sexually transmitted diseases. Also see John C. Caldwell and Pat Caldwell, "High Fertility in Sub-Saharan Africa." *Scientific American* (May 1990).
7. Johannesburg was founded in 1886, Nairobi in 1899, and Yaoudé in 1888.
8. Rudin, *Germans in the Cameroon,* p. 269.
9. Quoted from "Rapport sur l'hygiène du 3ème trimestre." *Archives Africaines* H 834 (1906) by Nancy Rose Hunt, "STDs, Suffering, and Their Derivatives in Congo/Zaire: Notes Towards an Historical Ethnography of Disease," in Charles Becker et al., eds., *Vivre et Penser le Sida en Afrique / Experiencing and Understanding AIDS in Africa* (Paris: Codesria, IRD, Karthala, 1999), pp. 111–31.
10. Jean Comhaire, "Some Aspects of Urbanization in the Belgian Congo." *Am J Sociol* 62 (1956): 10.
11. Hunt, "STDs," p. 115.
12. Nancy Rose Hunt, "Noise Over Camouflaged Polygamy, Colonial Morality Taxation, and a Woman-Naming Crisis in Belgian Africa." *J AfrHist* 32 (1991): 476.
13. Ibid., p. 484.
14. From Charles Lodewijckx, "Note sur la dénatalité de la cuvette centrale et sur les résultats obtenus jusqu'à présent par la propagande anti-malthusienne menée à Bolingo et aux environs," April 23, 1948. Unpublished manuscript in Fonds Hulstaert, 7.4, Centre Æquatoria, Bamanya, Zaire, quoted in Hunt, "STDs," p. 117.

Chapter 7: The Gift

1. Demographers generally included Brazzaville, just across the Congo River from Leopoldville, in their calculations of the metropolitan area.
2. William A. Hance and Irene S. van Dongen, "Matadi, Focus of Belgian African Transport." *Ann Assoc Am Geogr* 48 (1958): 41–72.
3. George Lardner, "Did Ike Authorize a Murder? Memo says Eisenhower Wanted Congolese Premier Dead." *The Washington Post*, August 8, 2000.
4. Kevin Whitelaw, "A Killing in Congo." *U.S. News & World Report*, July 24, 2000.
5. Some researchers have argued that unsafe medical practices such as the reuse of syringes contributed significantly to the earlier spread of HIV (see Pepin, *Origins of AIDS*).

Chapter 8: The Big Bang

1. Adam Hochschild, "An Assassination's Long Shadow." *The New York Times*, January 16, 2011.
2. Significant parts of this account of the Ebola crisis are based on Garrett, *The Coming Plague*.
3. The space capsule, in the end, was not used.
4. Garrett, p. 118–19.
5. "Ebola Haemorrhagic Fever in Zaire, 1976." *B World Health Organ* 56, no. 2 (1978): 271–93.
6. Ibid., p. 273.
7. Ibid.
8. Garrett, p. 124.
9. Subtype A also made its way to East Africa; in some parts of that region it now appears to be nearly as common as subtype D.
10. Interview with Timberg.
11. M. Thomas P. Gilbert, Andrew Rambaut, Gabriela Wlasiuk, Thomas J. Spira, Arthur E. Pitchenik, and Michael Worobey, "The Emergence of HIV/AIDS in the Americas and Beyond." *P Natl Acad Sci USA* 104, no. 47 (2007): 18566–70.
12. According to Worobey, until 1975 many of the commercial blood products used in the United States came from Haiti. An employee at an airport in Miami once e-mailed him with recollections of cargo planes arriving from Port au Prince loaded with blood. Pepin (*Origins of AIDS*, pp. 197–208) also discusses the role of the blood and especially the plasma industries in potentially facilitating the spread of HIV in Haiti and elsewhere.
13. Some writers such as Paul Farmer have postulated that HIV was first spread to Haiti from the United States, most likely by a gay man, but the genetic research of Worobey's group shows this is unlikely (Gilbert et al., "The Emergence of HIV/AIDS in the Americas"). However, as Worobey himself has stressed, it ultimately does not make much difference—in practical terms—in which direction the transmission event occurred.

Chapter 9: Americanizing AIDS

1. The AIDS tragedy in the United States received more attention after it became known that some famous figures, such as movie star Rock Hudson and tennis champion Arthur Ashe, had AIDS.
2. Margaret Garrard Warner, "The Savior on the Right?" *Newsweek*, January 19, 1987.

3. Sue Cross, "Falwell Says Government Must Control Homosexuals to Stop AIDS." Associated Press, July 4, 1983.

4. Such estimates are, of course, based on retrospective modeling, because there was virtually no record keeping of AIDS deaths anywhere in Africa at that time.

5. Robert H. Byers Jr., W. Meade Morgan, William W. Darrow, et al., "Estimating AIDS Infection Rates in the San Francisco Cohort." *AIDS* 2 (1988): 207–10. These men were recruited from a clinic that treated sexually transmitted diseases, so they may have had riskier behavior than other gay men in San Francisco. Yet even among a cohort of lower-risk gay men in the city (those initially uninfected with hepatitis B), 28 percent already had HIV by 1981, and by 1988 more than half had become infected (Nancy A. Hessol, Alan R. Lifson, Paul M. O' Malley, Lynda S. Doll, Harold W. Jaffe, and George W. Rutherford, "Prevalence, Incidence, and Progression of Human Immunodeficiency Virus Infection in Homosexual and Bisexual Men in Hepatitis B Vaccine Trials, 1978–1988." *Am J Epidemiol* 130 [1989]: 1167-75.)

6. A similar transmission pattern appeared in some other regions as well, including the southern cone countries of South America. In most regions outside of Africa, including in most of Latin America and Asia, such "concentrated epidemics" persist to the present day, including in many places where commercial sex work is also a major factor.

7. Harry Nelson, "Advances in AIDS Study Foreshadow Long Battle." *Los Angeles Times*, April 21, 1985.

8. Transfusions are even more efficient than needle sharing for spreading HIV, which resulted in very high infection rates in hemophiliacs before the supply of plasma was made safe from HIV.

9. Rebecca F. Baggaley, Richard G. White, and Marie-Claude Boily, "HIV Transmission Risk Through Anal Intercourse: Systematic Review, Meta-analysis and Implications for HIV Prevention." *Int J Epidemiol* 39 (2010): 1048–63. Daniel Halperin, Stephen Shiboski, Joel Palefsky, and Nancy Padian, "High Level of HIV-1 Infection from Anal Intercourse: A Neglected Risk Factor in Heterosexual AIDS Prevention." Poster presented at the International AIDS Conference, Barcelona, 2002.

10. As explained in chapter 18, African American and (to a lesser degree) Hispanic women face higher risk than white women. While HIV rates in these ethnic groups have never approached those in the severely affected parts of Africa, they are much higher than among white American women; rates for this group remain extremely low. (See chapter 18, endnote 2.)

11. Joseph A. Catania, Thomas J. Coates, Ron Stall, et al., "Changes in Condom Use Among Homosexual Men in San Francisco." *Health Psychol* 10 (1991): 190–99.

12. Many unwanted pregnancies have also been avoided through the use of condoms, though many contraceptive experts consider them a "second-rate" method, due to relatively high failure rates. Most of these failures result from inconsistent or incorrect use. (See appendix and chapter 4, endnote 5.)

13. According to the Cochrane review of public health interventions, consistent condom use is estimated to be approximately 80 percent effective at reducing HIV transmission (see appendix). In the 1990s, a review panel for the Centers for Disease Control and Prevention and the National Institutes of Health estimated male condoms to be approximately 86 percent effective for HIV prevention. A subsequent review commissioned by UNAIDS es-

timated them to be about 90 percent effective with consistent use. (Norman Hearst and Sanny Chen, "Condom Promotion for AIDS Prevention in the Developing World: Is It Working?" *Stud Family Plann* 35 [2004]: 39–47.) Condoms are likely not the only contraceptive methods that can help prevent some sexually transmitted infections, including perhaps HIV. The diaphragm and cervical cap also effectively block entry into the cervix. Although a randomized trial examining the effect of the diaphragm to prevent HIV infection was not successful, this may have been largely due to inconsistent use. (Nancy S. Padian, Ariane van der Straten, Gita Ramjee, et al., "Diaphragm and Lubricant Gel for Prevention of HIV Acquisition in Southern African Women: A Randomized Controlled Trial." *Lancet* 370 [2007]: 251–61.) The lead investigator, Nancy Padian, has told Halperin that if she were to conduct this trial over again, she might have used the cervical cap instead, as it may have a higher acceptance rate among users. The cap can be left in the vagina for many days, or even weeks, unlike the diaphragm, which must be inserted prior to each sex act. Rebecca Chalker, *The Complete Cervical Cap Guide*. (New York: Harper and Row, 1987.) Ariane van der Straten, Nuriye Sahin-Hodoglugil, Kate Clouse, Sibongile Mtetwa, Mike Chirenje, "Feasibility and Potential Acceptability of Three Cervical Barriers Among Vulnerable Young Women in Zimbabwe," *J Fam Plann Reprod Health Care 36* (2010): 13–19.

14. The U.S. response overseas, especially in Africa, would eventually include more focus on abstinence and fidelity-based approaches, in large part because of the influence of cultural and political conservative forces in the United States. The United States, meanwhile, was not the only country whose foreign assistance approach to HIV prevention focused on condom promotion. So did those of the British, German, and most other Western governments. For that reason it would also be true to say that the world's response to the epidemic had become "Westernized."

Chapter 10: It Can't Be Here Already!

1. Nathan Clumeck, Jean Sonnet, Henri Taelman, et al., "Acquired Immune Deficiency Syndrome in Belgium and Its Relation to Central Africa." *Ann NY Acad Sci* 437 (December 1984): 264–59. The initial incursions of HIV into Europe from African immigrants did not spread extensively into the wider European population. Research by Worobey and his colleagues has made clear that the epidemics among gay men and injection drug users that emerged in Europe had their origins in North America. Nearly all cases of HIV infection in the Americas and in Europe derived from a single virus transmitted from Congo to Haiti sometime in the 1960s, according to their research. The descendants of this virus spread to the United States and, soon thereafter, to Europe.

2. Garrett, *The Coming Plague*, p. 111.

3. Larry Altman, "Battle-Scarred Veteran Is General in Global War on AIDS." *The New York Times*, July 21, 1998.

4. AIDS may have appeared to have become more severe in the early 1980s in part because of the deterioration of the health care system in Kinshasa. With few functioning hospitals, any surge of new patients would have exacerbated capacity limitations.

5. This account draws in part from Garrett's book.

6. Peter Piot, Henri Taelman, Kapita B. Minlangu, et al., "Acquired Immunodeficiency Syndrome in a Heterosexual Population in Zaire." *Lancet* 324 (1984): 65–69. San Francisco's

level of AIDS cases, due to its large male homosexual population, was in fact many times higher than that of Kinshasa.

7. Ibid.

8. Interview with Timberg.

9. Interview with Timberg.

10. Piot et al., "Acquired Immunodeficiency Syndrome in a Heterosexual Population in Zaire"; Philippe van de Perre et al., "Acquired Immunodeficiency Syndrome in Rwanda." *Lancet* 324 (1984): 62–65.

11. These estimates for HIV rates in Kinshasa and Kigali were based on relatively small numbers of patients. As explained in chapter 26, with the more accurate Demographic and Health Survey methodology now available, it is clear that the rates were probably not as high as initially estimated in those years. In addition, it is very likely that HIV prevalence has declined substantially in Rwanda, due to mortality, and probably changes in behavior, particularly partner reduction. Levels of multiple sexual partnerships are now among the lowest in the region; in the 2010 Demographic and Health Survey less than 2 percent of Rwandan men reported multiple partners in the past year, while in most other sub-Saharan African countries more than 15 percent of men reported multiple partners. In Cameroon, over 30 percent did. JB Bingenheimer, "Men's Multiple Sexual Partnerships in 15 Sub-Saharan African Countries: Sociodemographic Patterns and Implications," *Studies in Family Planning* 41 (2010): 1–7. Vinod Mishra, Praween Agrawal, Soumya Alva, Yuan Gu, and Shanxiao Wang, *Changes in HIV-Related Knowledge and Behaviors in Sub-Saharan Africa*. DHS Comparative Reports No. 24. Calverton, MD: ICF Macro., 2009, table 6.2. (www.measuredhs .com/pubs/pdf/CR24/CR24.pdf).

Chapter 11: Attention na SIDA

1. Loren K. Clarke and Malcolm Potts, eds., *The AIDS Reader: Documentary History of a Modern Epidemic*, vol.1 (Boston: Branden Publishing, 1988), p. 288.

2. Cesar Nkuku Khonde, "An Oral History of HIV/AIDS in the Congo," in *The HIV/AIDS Epidemic in Sub-Saharan Africa in a Historical Perspective,* Philippe Denis and Charles Becker, eds. (Dakar, Senegal: Law, Ethics, Health, 2006).

3. Interview with Timberg.

4. Jon Cohen, "The Rise and Fall of Projet SIDA." *Science* 278 (1997): 1565–68. Several other leading AIDS experts, such as Jonathan Mann, helped establish and conducted research at Projet SIDA.

5. Graeme Ewens, *Congo Colossus: The Life and Legacy of Franco & OK Jazz* (Norfolk, UK: Buku Press, 1994).

6. John Nimis, "Literary Listening: Readings in Congolese Popular Music." NYU doctoral dissertation, unpublished, 2010.

7. Ewens, p. 85.

8. Ibid., p. 30.

9. This is a mix of French and the local Lingala language.

10. Ewens, p. 248.

11. No definitive evidence has ever emerged that Franco had AIDS. His symptoms, his chilling song, and the widespread assumption that he had the disease may make the question moot

for the purposes of discussing the impact on people in Congo, and indeed throughout much of the rest of Africa, where he was also widely popular.

12. Interview with Timberg.

13. Peter Kilmarx, "AIDS Prevention and Control in Rural Africa: Risk Behaviors, Knowledge and Attitudes in a Group of Rural Zaire Men." Unpublished study, 1987.

14. Jane T. Bertrand, Bakutuvwidi Makani, Balowa Djunghu, and Kinavwidi L. Niwembo, "Sexual Behavior and Condom Use in 10 Sites of Zaire." *J Sex Res* 28 (1991): 347–64. Also see Jane T. Bertrand, Bakutuvwidi Makani, Susan E. Hassig, et al., "AIDS-Related Knowledge, Sexual Behavior and Condom Use Among Men and Women in Kinshasa, Zaire." *Am J Public Health* 81 (1991): 53–58.

15. Melinda Moore, Susan E. Hassig, Namagosi Lusakulira, William E. Bertrand, and Tumba D. Kashala, "Sexual Behavior and Perceived AIDS Risk Among 3,500 Zairian Health Workers." Presentation at the 5th International Conference on AIDS, Montreal, Canada, June 1989, Abstract # MDO 17.

16. Cohen, "The Rise and Fall of Project SIDA."

17. Interview with Timberg.

18. HIV rates in Congo appear to have declined in recent years, at least in Kinshasa, where the CDC has been conducting and overseeing routine testing of pregnant women since 1987. In that year the estimate was about 6 percent; by 2007 it had declined to about 3 percent. In the 2007 Demographic and Health Survey, the estimated prevalence for the country was estimated to be 1.4 percent.

Chapter 12: You Won't Believe

1. The causal relationship between HIV and other sexually transmitted infections is unclear (see appendix), but it is likely that chancroid, especially in the earlier stages of the African epidemic, was an important facilitator of HIV transmission. The extremely high rate of HIV among these Pumwani prostitutes, meanwhile, is not unusual for Africa. Even in West African countries, where national HIV rates are much lower, rates among sex workers are typically very high. In Accra, Ghana, 75 percent of brothel-based prostitutes are infected, compared to only 2.2 percent of pregnant women in the community overall. Anne-Marie Côté, François Sobela, Agnes Dzokoto, et al., "Transactional Sex Is the Driving Force in the Dynamics of HIV in Accra, Ghana." *AIDS* 18 (2004): 917–25.

2. Ritual scarring once was considered a risk factor for the spread of HIV, but there has been little evidence of transmission this way, even when blades are shared. However, diseases that are much more contagious, such as hepatitis and tetanus, can be spread in this manner.

3. D. William Cameron, J. Neil Simonsen, Lourdes J. D'Costa, et al., "Female to Male Transmission of Human Immunodeficiency Virus Type 1: Risk Factors for Seroconversion in Men." *Lancet* 3 (1989): 403–7. As detailed in later chapters, the appendix, and subsequent review papers, this would be only the first of dozens of epidemiological studies to indicate a strong protective effect of male circumcision: Aaron A. Tobian, Ronald H. Gray, and Thomas C. Quinn, "Male Circumcision for the Prevention of Acquisition and Transmission of Sexually Transmitted Infections." *Arch Pediatr Adolesc Med* 164, (2010): 78–84, http://arch pedi.ama-assn.org/cgi/reprint/164/1/78.pdf; Helen A. Weiss, Daniel Halperin, Robert C. Bailey, Richard J. Hayes, George Schmid, and Catherine A. Hankins, "Male Circumcision

for HIV Prevention: From Evidence to Action?" *AIDS* 22 (2008): 567–74; Richard G. Wamai, Helen A. Weiss, Catherine Hankins, et al., "Male Circumcision Is an Efficacious, Lasting and Cost-Effective Strategy for Combating HIV in High-Prevalence AIDS Epidemics: Time to Move Beyond Debating the Science." *Futur HIV Ther* 2 (2008): 399–405; USAID, "Male Circumcision: Current Epidemiological and Field Evidence; Program and Policy Implications for HIV Prevention and Reproductive Health." USAID Office of HIV-AIDS/AIDSMARK (2002), www.dec.org/pdf_docs/PNACS892.pdf; Helen Weiss, Jonny Polonsky, Robert Bailey, Catherine Hankins, Daniel Halperin, and George Schmid, "Male Circumcision: Global Trends and Determinants of Prevalence, Safety and Acceptability." Geneva: WHO/UNAIDS (2007), www.malecircumcision.org/media/documents/MC_Global_Trends_Determinants.pdf; Jon Cohen, "Male Circumcision Thwarts HIV Infection." *Science* 309 (2005): 860; Jeffrey Klausner, Richard Wamai, Kasonde Bowa, Kawango Agot, Jesse Kagimba, and Daniel T. Halperin, "Is Male Circumcision as Good as the HIV Vaccine We've Been Waiting For?" *Futur HIV Ther* 2 (2008): 1–7; WHO/UNAIDS/AIDS Vaccine Advocacy Coalition/Family Health International: Clearinghouse on Male Circumcision for HIV Prevention: www.malecircumcision.org.

4. The anthropologist Jeff Marck (personal communication) notes that the genes of Bantu speakers offer clues about their complex history. Their Y chromosomes show signs of descent from the Bantus' ancestral homeland, near the Cross River region of Nigeria and the Grassfields region of Cameroon. But the mitochondrial DNA of these Bantu speakers reflects the genetic history of the areas where they eventually settled. This suggests that what is often called the Bantu "migration" resulted from men gradually moving into new areas and finding wives among the indigenous communities they found. That is why anthropologists prefer the term "Bantu encroachment." See Brigitte Pakendorf, Koen Bostoen, and Cesare de Filippo, "Molecular Perspectives on the Bantu Expansion: A Synthesis." *Language Dynamics and Change* 1 (2011):50–88; Jeff Marck, "Aspects of Male Circumcision in Subequatorial African Culture History." *Health Transition Review* supplement to volume 7 (1997): 337–59.

5. The near-universal practice of male circumcision in most other West and Central African countries helps explain why HIV in those regions is also much lower than in most African nations with low circumcision rates (see chart on page 179).

6. HIV acts like a Trojan horse seeking to infiltrate the very same immune cells, such as Langerhans, which normally protect the body from foreign infections. For more information on the biology of male circumcision and HIV infection, see Brian J. Morris and Richard G. Wamai, "Biological Basis for the Protective Effect Conferred by Male Circumcision Against HIV Infection." *Int J STD AIDS* (2011; in press); Min H. Dinh, Kelly M. Fahrbach, and Thomas J. Hope, "The Role of the Foreskin in Male Circumcision: An Evidence-Based Review." *Am J Reprod Immunol* 65 (2011): 279–83; Taha Hirbod, Robert C. Bailey, Kawango Agot, et al., "Abundant Expression of HIV Target Cells and C-type Lectin Receptors in the Foreskin Tissue of Young Kenyan Men." *Am J Pathol* 176 (2010): 2798–805; Ronald H. Gray, Robert C. Bailey, and Brian J. Morris, "Keratinization of the Adult Male Foreskin and Implications for Male Circumcision." *AIDS* 24 (2010): 1381–82; Zhicheng Zhou, N. Barry de Longchamps, Alain Schmitt, et al., "HIV-1 Efficient Entry in Inner Foreskin Is Mediated By Elevated CCL5/RANTES That Recruits T Cells and Fuels Conjugate Formation with Langerhans Cells." *PLoS Pathog* 7 (2011): e1002100, epub June 30, 2011; Bruce K. Pat-

terson, Alan Landay, Joan N. Siegel, et al., "Susceptibility to Human Immunodeficiency Virus-1 Infection of Human Foreskin and Cervical Tissue Grown in Explant Culture." *Am J Pathol* 161 (2002): 867–73; Betty A. Donoval, Alan L. Landay, Stephen Moses, et al., "HIV-1 Target Cells in Foreskins of African Men with Varying Histories of Sexually Transmitted Infections." *Am J Clin Pathol* 125 (2006): 386–91; Scott G. McCoombe and Roger V. Short, "Potential HIV-1 Target Cells in the Human Penis." *AIDS* 20: (2006): 1491.

7. Christopher Ehret, *The Civilizations of Africa: A History to 1800* (Charlottesville: University of Virginia Press, 2002). It is widely believed in Ethiopia that early Jewish traditions also influenced when and how Ethiopian boys are circumcised.

8. David Shingirai Chanaiwa, "The Zulu Revolution: State Formation in a Pastoralist Society." *African Studies Review* 23 (1980): 12.

9. John Bongaarts, Priscilla Reining, Peter Way, and Francis Conant, "The Relationship Between Male Circumcision and HIV Infection in African Populations." *AIDS* 3 (1989): 373–77. Because the paper was based on outdated cultural information regarding circumcision taken from George P. Murdock's *Ethnographic Atlas*, it contained some serious errors for certain countries, but the overall pattern is remarkably consistent with today's data.

10. Male circumcision also helps prevent various other diseases and conditions, most notably cervical cancer, in women. And, as discussed in the appendix (endnote 16), there is observational evidence suggesting a direct HIV prevention benefit for women if their partner is circumcised. (A clinical trial in Uganda that sought to confirm this effect was terminated early because an insufficient number of study participants were enrolled.)

Chapter 13: Fear Worked

1. Interview with Timberg.

2. Peter Mugyeni, *Genocide by Denial: How Profiteering from HIV/AIDS Killed Millions* (Kampala, Uganda: Fountain Publishers, 2008), p. 2.

3. David Serwadda, Nelson K. Sewankambo, and J. Wilson Carswell, "Slim Disease: A New Disease in Uganda and Its Association with HTLV-III Infection." *Lancet* 326 (1995): 849–52.

4. This anecdote has been published in several places. Jesse Kagimba recounted it to Timberg in an interview. Elements of this chapter appeared in *The Washington Post* in an article by Timberg on March 29, 2007, headlined "Uganda's Early Gains Against HIV Eroding."

5. Interview with Timberg.

6. Museveni stopped considering himself a born-again Christian in 1966, according to his autobiography, *Sowing the Mustard Seed* (New York: Macmillan, 1977). His wife Janet Museveni, however, continues to to be actively involved in evangelical Christian causes, including abstinence-only campaigns for youth. (Helen Epstein, "God and the Fight Against AIDS." *The New York Review of Books* [April 28, 2005].) The influence of some American evangelicals in Uganda in recent years has been troubling. In 2009 one group supported proposed legislation which threatened to impose the death penalty for homosexuality. (Jeffrey Gettleman, "Americans' Role Seen in Uganda Anti-Gay Push." *The New York Times*, January 3, 2010.) The law ultimately was not enacted.

7. Blaine Harden, "Uganda Battles AIDS Epidemic," *The Washington Post*, June 2, 1986.

8. Ibid.

9. Interview with Timberg.

10. Kim Witte and Mike Allen, "A Meta-Analysis of Fear Appeals: Implications for Effective Public Health Campaigns." *Health Educ Behav* 27 (2000): 608–32; Edward C. Green and Kim Witte, "Can Fear Arousal in Public Health Campaigns Contribute to the Decline of HIV Prevalence?" *J Health Commun* 11 (2006): 245–59.

11. This traditional moral order probably was in part idealized, but that did not keep the Ugandans from invoking it to combat HIV.

12. Museveni's parable about the snake and the termite mound is still widely remembered in Uganda. Halperin and behavioral scientist Doug Kirby often heard it mentioned in the focus groups and interviews they conducted and oversaw in 2003 for a USAID-funded research project on the history of behavior change in Uganda. Douglas Kirby and Daniel T. Halperin, "Success in Uganda: An Analysis of Behavior Changes That Led to Declines in HIV Prevalence in the Early 1990s." Scotts Valley, CA: ETR Associates, 2008, http://programservices.etr.org/base/documents/Uganda-BehaviorsFormat.pdf.

13. Interview with Timberg.

14. The key role of religious institutions was frequently mentioned during the USAID-funded fieldwork conducted by the Kirby/Halperin-led research team. (Douglas Kirby, "Presentation by ABC Study Director Douglas Kirby to USAID," October 23, 2003.) Although in the anecdote recounted here it was a man who had died, of course many women were also dying. However, a common recollection was that "the big men were the first to go," meaning that it was often the relatively older, wealthier, more educated and more urban males who seemed to die first. The USAID study and other research would later suggest that these same men were also the earliest and most likely people to change their sexual behavior at that time. (And see Elizabeth Pisani, "We Can't Wait for Equality: The Focus on Female Empowerment Has Made Us Forget the Key Role of Men in AIDS Prevention." *The Guardian* December 1, 2007; www.guardian.co.uk/commentisfree/2007/dec/01/comment.gender.)

15. Green and Witte, p. 251.

16. S. Mulinde Musoke, *New Vision*, Oct. 23, 1987, cited in James D. Shelton, Daniel T. Halperin, Vinand Nantulya, Malcolm Potts, Helene D. Gayle, and King K. Holmes, "Partner Reduction Is Crucial for Balanced 'ABC' Approach to HIV Prevention." *Brit Med J* 328 (2004): 891–94. Much of this *New Vision* coverage is also cited in Kirby and Halperin, "Success in Uganda."

Chapter 14: Born in Africa

1. Fela Kuti, by nearly every account but his own, died of AIDS.

2. Senegal is another African country that had a successful approach to reducing both HIV transmission and stigma against people with AIDS. Because most infections there stemmed from commercial sex work, early in the epidemic the government adopted effective measures to provide condoms and reproductive health services for prostitutes. (UNAIDS, "Acting Early to Prevent AIDS: The Case of Senegal," June 1999.) These interventions, along with the universal practice of male circumcision, have helped keep the adult HIV rate at below 1 percent. Although comparisons have occasionally been made between Senegal and much more severely affected countries elsewhere in Africa, a more appropriate comparison would be to its neighbor Gambia, which is similar culturally but until recently did not adopt such

aggressive prevention measures. The HIV rate in that country has been about double that of Senegal, suggesting that Senegal's approach was effective.

3. Helen Epstein, *The Invisible Cure: Why We Are Losing the Fight Against AIDS in Africa* (New York: Picador, 2008); Edward C. Green, Daniel T. Halperin, Vinand Nantulya, and Janice A. Hogle, "Uganda's HIV Prevention Success: The Role of Sexual Behavior Change and the National Response." *AIDS Behav* 10 (2006): 335–46; David Wilson, "Partner Reduction and the Prevention of HIV/AIDS." *Brit Med J* 328 (2004): 848–49; Daniel T. Halperin, "The Controversy over Fear Arousal in AIDS Prevention and Lessons from Uganda." *J Health Commun* 11 (2006): 266–67.

4. Joseph Lugalla, Maria Emmelin, Aldin Mutembe, et al., "Social, Cultural, and Sexual Behavioral Determinants of Observed Decline in HIV Infection Trends: Lessons from the Kagera Region, Tanzania." *Soc Sci Med* 59 (2004):185–98. The researchers concluded that "multi-partnered sexual relationships, polygamy, and extramarital affairs that were very common in the past are declining. . . . The findings suggest that the number of people who are 'zero-grazing' (sticking to one partner) . . . is increasing."

5. Ibid., p. 188.

6. The information on Uganda in this paragraph is based on research conducted in 2003 by the USAID team led by Kirby and Halperin (chapter 13, endnote 12).

7. Museveni's government did issue condoms to soldiers, on the presumption that young men far from home would have sexual encounters with prostitutes and other casual sex partners. When it came to promoting condoms more generally, however, Museveni expressed various concerns, including that they could deteriorate more easily in the African heat than in cooler climates. More recent manufacturing techniques have largely eliminated that problem.

8. See Gary Slutkin, Sam Okware, Warren Naamara, et al., "How Uganda Reversed Its HIV Epidemic." *AIDS Behav* 10, no. 4 (2006): 351–60; and Green et al., "Uganda's HIV Prevention Success."

9. The section of Museveni's speech that raised questions about the usefulness of condoms in fighting HIV generated a critical response among some activists and experts.

10. Interview with Timberg.

11. Rand Stoneburner and Daniel Low-Beer, "Population-Level HIV Declines and Behavioral Risk Avoidance in Uganda." *Science* 304 (2004): 714–18; Green et al., "Uganda's HIV Prevention Success"; Epstein, *The Invisible Cure*; Ruth Bessinger, Priscilla Akwara, Daniel T. Halperin, "Sexual Behavior, HIV and Fertility Trends: A Comparative Analysis of Six Countries; Phase I of the ABC Study" (Washington, D.C.: Measure Evaluation/USAID [2003]), www.cpc.unc.edu/measure/publications/pdf/sr-03-21b.pdf; Slutkin et al., "How Uganda Reversed Its HIV Epidemic."

Chapter 15: The Condom Code

1. International organizations such as UNAIDS did praise the unusually early level of commitment and openness in fighting AIDS that Uganda displayed.

2. To understand the longer-term effects of prevention programs and their failures, it may be useful to imagine a bathtub with a tap remaining fully open. Even if some of the water is draining out of the bottom of the tub (i.e., if people are dying of AIDS), the overall water level will not drop much if the tap is not closed, at least partially. The same is true for the

numbers of people infected with HIV: only meaningful declines in the numbers of new infections can produce, over the long term, substantially lower levels of infection within a population.

3. The team led by Imperial College of London researcher Timothy Hallett, who has helped perform much of the modeling work for UNAIDS over the years, has shown how careful analysis of changes in HIV in various countries can indicate which declines were due mainly to mortality and other "natural dynamics" and which were largely caused by changes in sexual behavior. Timothy B. Hallett, John Aberle-Grasse, Gonzalo Bello, et al., "Declines in HIV Prevalence Can Be Associated with Changing Sexual Behaviour in Uganda, Urban Kenya, Zimbabwe and Urban Haiti." *Sex Transm Infect* 82 (2006): i1–i8. doi:10.1136/sti.2005.0160; Albert Killian, Simon Gregson, Bannet Ndyanabangi, et al., "Reductions in Risk Behaviour Provide the Most Consistent Explanation of Declining HIV Prevalence in Uganda." *AIDS* 13 (1998): 391–98.

4. Edward C. Green, *Rethinking AIDS Prevention: Learning from Successes in Developing Countries* (Westport CT: Praeger, 2003), p. 160. (Italics in quote appear in original text.)

5. Wiwat Rojanapithayakorn and Robert Hanenberg, "The 100% Condom Program in Thailand." *AIDS* 10 (1996): 1–7; Jon Cohen, "Asia—The Next Frontier for HIV/AIDS: Two Hard-Hit Countries Offer Rare Success Stories: Thailand and Cambodia." *Science* 301 (2003): 1658–62; Charles P. Wallace, "Miracle Man of Thailand," *Los Angeles Times*, January 22, 1990.

6. Paul Blustein and Mary Kay Magistad, "Red Lights and Traffic Jams as Bankers Meet in Bangkok." *The Washington Post*, October 19, 1991.

7. Shelton et al., "Partner Reduction Is Crucial for Balanced 'ABC' Approach to HIV Prevention"; Mark VanLandingham and Leo Trujillo, "Recent Changes in Heterosexual Attitudes, Norms and Behaviors Among Unmarried Thai Men: A Qualitative Analysis." *Int Fam Plann Perspect* 28 (2002): 6–15. There may have been some human rights violations during the Thai campaign. Rumors spread, for example, that some sex workers may have been arrested for not using a condom with clients, but these reports have not been confirmed.

8. Cohen, "Asia—The Next Frontier for HIV/AIDS"; Smarajit Jana, Nandinee Bandyopadhyay, Sampa Mukherjee, et al., "STD/HIV Intervention with Sex Workers in West Bengal, India." *AIDS* 12, suppl. B (1998): S101–8. Sushena Reza-Paul, Tarab Beattie, Hafeez U. Syed, et al., "Declines in Risk Behaviour and Sexually Transmitted Infection Prevalence Following a Community-Led HIV Preventive Intervention Among Female Sex Workers in Mysore, India." *AIDS* 22, suppl. 5 (2008): S91–100; Peter D. Ghys, Mamadou O. Diallo, Virginie Ettiègne-Traoré, et al. "Increase in Condom Use and Decline in HIV and Sexually Transmitted Diseases Among Female Sex Workers in Abidjan, Côte d'Ivoire, 1991–1998." *AIDS* 16 (2002): 251–58; Diana Kerrigan, Luis Moreno, Santo Rosario, and Michael Sweat, "Adapting the Thai 100% Condom Programme: Developing a Culturally Appropriate Model for the Dominican Republic." *Cult Health Sex* 3 (2001): 221–40; Daniel T. Halperin, Antonio de Moya, Eddy Perez-Then, Gregory Pappas, and Jesus M. Garcia Calleja, "Understanding the HIV Epidemic in the Dominican Republic: A Prevention Success Story in the Caribbean?" *JAIDS* 51 (2009): S52–59.

9. The years after the Stonewall Riots also saw an increase in the popularity of anal sex. In addition, men became more likely to alternate in sexual roles, being at different times the insertive and receptive partner.

10. Gabriel Rotello, *Sexual Ecology: AIDS and the Destiny of Gay Men* (New York: Plume, 1997), pp. 148–49.

11. Interview with Timberg.

12. The summary of the 1989 WHO survey was written by American social scientist Maxine Ankrah. For a more detailed examination of Ankrah's role in this research, see Epstein, *The Invisible Cure*. Our account is drawn from interviews Timberg conducted with Stoneburner and Carael, and also from direct experience Halperin later had with this matter when he was working for USAID. In the interviews, Carael firmly disputed Stoneburner's suggestion that he was withholding any of the 1989 data on Ugandan sexual behavior, that night or ever. And Carael further said that even he did not have immediate access to all of the data on the night in question. Stoneburner, meanwhile, acknowledged that he did not directly ask for it that night, but he said that in subsequent years, when it became clear that the 1989 data was key to understanding the behavior change in Uganda, he requested it on several occasions. It was not made available until Halperin intervened in 2002. For more on this, see endnote 6 for chapter 26.

13. Interview with Timberg.

14. Godwil Asiimwe-Okiror, Alex A. Opio, Joshua Musinguzi, et al., "Change in Sexual Behaviour and Decline in HIV Infection Among Young Pregnant Women in Urban Uganda," *AIDS* 11 (1997) 11: 1757–63. It is crucial to note that consistent condom use would have been far lower than the 55 percent of Ugandan men who reported ever having used a condom. In a more recent survey conducted in the Rakai region of the country, less than 5 percent of study participants reported using condoms consistently, even after two and half years of intensive promotion efforts; an additional 16 percent reported inconsistent use. (The inconsistent condom users had similar rates of HIV acquisition, and even higher ones of other sexually transmitted infections, compared to nonusers.) Saifuddin Ahmed, Tom Lutalo, Maria Wawer, et al., "HIV Incidence and Sexually Transmitted Disease Prevalence Associated with Condom Use: A Population Study in Rakai, Uganda." *AIDS* 15 (2001): 2171–79. In no African country have more than 5 percent of married women reported using condoms in their marriage (Demographic and Health Surveys, Calverton, MD: ORC MACRO International). Also see Hearst and Chen, "Condom Promotion for AIDS Prevention in the Developing World"; Thomas A. Peterman, Lin H. Tian, Lee Warner, "Condom Use in the Year Following a Sexually Transmitted Disease Clinic Visit." *Int J STD AIDS* 20 (2009): 9–13; Nelly Westercamp, Christine Mattson, Michelle Madonia, et al., "Determinants of Consistent Condom Use Vary by Partner Type Among Young Men in Kisumu, Kenya: A Multi-Level Data Analysis." *AIDS Behav* 14 (2010): 949–59; Slutkin et al., "How Uganda Reversed Its HIV Epidemic"; James D. Shelton, "Confessions of a Condom Lover." *Lancet* 368 (2006): 1947–49; Iddo Tavory and Ann Swidler, "Condom Semiotics: Meaning and Condom Use in Rural Malawi." *Am Sociol Rev* (2009) 74:171–89.

15. Stoneburner and Low-Beer, "Population-Level HIV Declines and Behavioral Risk Avoidance"; Green et al., "Uganda's HIV Prevention Success"; Bessinger et al., "Sexual Behavior, HIV and Fertility Trends"; Shelton et al., "Partner Reduction Is Crucial for Balanced 'ABC' Approach to HIV Prevention." In the WHO surveys, the number of people who reported having a sex partner they were not married to or living with decreased by about 50 percent. Although the researchers did not inquire specifically into numbers of concurrent partner-

ships, various indications suggest that this behavior probably also declined considerably. (See Epstein, *The Invisible Cure*; Helen Epstein, "Why is AIDS Worse in Africa?" *Discover*, February 2004.)

16. Uganda Demographic and Health Survey, 1988/1989, Kampala: Uganda Ministry of Health/Calverton, MD: ORC MACRO International.

Chapter 16: The Beat-up

1. The International AIDS Conferences took place once a year from 1985 to 1994, then every other year after that.

2. www.actupny.org/diva/van1syn.html. During this era before the widespread availability of effective medications, many AIDS activists were radicalized by the terrible death rates among friends, colleagues, and lovers, along with the conviction that the government and the pharmaceutical industry were moving too slowly to offer relief. Whatever unease their tactics caused, it's clear that the activist community succeeded in keeping the disease in the international spotlight. In addition, they successfully pushed to accelerate testing of the promising drug therapies that would eventually improve and prolong the lives of those with AIDS.

3. Lindsay Knight, *UNAIDS: The First 10 Years* (Geneva: UNAIDS. 2008), p. 60.

4. Lawrence Altman, "AIDS Meeting: Signs of Hope, and Obstacles." *The New York Times*, July 7, 1996.

5. Peter Piot, "AIDS: A Global Response." *Science*, June 28, 1996.

6. The Lancet, "Spain's Unwelcome Distinction." *Lancet* 248 (1996): 1578.

7. Interview with Timberg.

8. The female sex partners of men who injected drugs or had sex with other men eventually made up a substantial portion of the epidemic as well.

9. In some countries, such as Swaziland, South Africa, and Lesotho, HIV rates actually rose higher and faster than many experts had predicted. This shows that there were also some dangers in potentially underestimating the course of the epidemic.

10. An example of such vague billboards was one widely used by UNAIDS: DON'T TURN YOUR BACK ON AIDS.

11. Jeremy Seabrook, *Love in a Different Climate—Men Who Have Sex with Men in India* (London: Verso Press, 1999). In a 2010 study commissioned for UNAIDS in Tanzania, similar reasons for having anal sex with other men were given. Edward S. Maswanya, Prince P. Mutalemwa, Elizabeth H. Shayo, et al., "Drivers of HIV/AIDS Epidemics in Tanzania Mainland: 'Case Study of Makete, Temeke, Geita, Lindi, Kigoma & Meru Districts.' " Dar Es Salaam, Tanzania: National Institute for Medical Research, July 2010. And see Juliet Richters et al., "Condom Use and 'Withdrawal': Exploring Gay Men's Practice of Anal Intercourse," *Int J STD AIDS* 11 (2000): 96–104.

12. In Piot's 1996 editorial in *Science*, even the use of the term "adults" was problematic. At the time, the only group for which there was reliable data was pregnant women. They tended to have higher HIV rates for a variety of reasons, including being more likely to be sexually active and to have had unprotected sex; in Africa women generally have higher rates than men as well.

13. The high rates of HIV found among young army recruits in northern Thailand were an exception to the usual pattern of low levels of infection in the general population in Asia.

In the early nineties as many as 12 percent of this group, most of whom reported frequently visiting prostitutes, were infected. After the aggressive "100 percent condom" program was instituted in the brothels, this rate fell dramatically. Tawesak Nopkesorn, Tim D. Mastro, Suebpong Sangkharomya, et al., "HIV-1 Infection in Young Men in Northern Thailand." *AIDS* 7 (1993): 1233–39; Tim D. Mastro and Kanchit Limpakarnjanarat, "Condom Use in Thailand: How Much Is It Slowing the HIV/AIDS Epidemic?" *AIDS* 9 (1995): 523–25; Rojanapithayakorn and Hanenberg, "The 100% Condom Program in Thailand"; Peter H. Kilmarx, Somsak Supawitkul, Mayuree Wankrairoj, et al., "Explosive Spread and Effective Control of Human Immunodeficiency Virus in Northernmost Thailand: The Epidemic in Chiang Rai Province, 1988–99." *AIDS* 14 (2000): 2731–40.

14. James Chin, *The AIDS Pandemic: The Collision of Epidemiology with Political Correctness* (London, Radcliffe Publishing Ltd, 2006). Piot, in an interview with Timberg, portrayed Chin's modeling as flawed and said that the UNAIDS revisions beginning in the mid-1990s represented major improvements. Chin, both in his book and in interviews with Timberg, gives an almost opposite account, saying that his models took into account the uncertainty of the surveillance systems better, and that the revisions by UNAIDS created significant overestimates in many places.

15. The 2007 UNAIDS Global Report (Geneva: UNAIDS) showed, for the first time in any UNAIDS/WHO report, a decline in the number of new HIV infections, beginning in the mid- to late nineties. (The most recent [2010] UNAIDS Global Report can be found at: www.unaids.org/globalreport/Global_report.htm.)

16. James D. Shelton, Daniel T. Halperin, and David Wilson, "Has Global HIV Incidence Peaked?" *Lancet* 367 (2006): 1120–22. There were exceptions to the generally downward trend in new HIV infections that began around the middle to late 1990s. Some countries, such as Mozambique, experienced rising numbers of infections for several more years.

17. Interview with Timberg. Some critics have suggested that the national origin of HIV experts may influence how they approach questions related to sexual behavior, with the presumption that Europeans tend to be less judgmental than, for example, Americans. This theory is untestable but intriguing. In her book *Lust in Translation: The Rules of Infidelity from Tokyo to Tennessee* (New York: The Penguin Press, 2007), journalist Pamela Druckerman draws upon national survey data to show that although continental Europeans do not necessarily have higher rates of infidelity, they are typically more accepting of its occasional occurrence. Some researchers also have expressed that singling out African sexual behavior as different would amount to an inappropriate moral judgment on African societies.

18. John Cleland et al., *Sexual Behavior and AIDS in the Developing World* (Geneva: World Health Organization, 1995). Regarding the kind of "consensus-dominated" views on AIDS prevention that have often prevailed, see Tachi Yamada, "In Search of New Ideas for Global Health." *N Engl J Med* 358 (2008): 1324–25.

19. In these WHO surveys, the percentages of people reporting such behaviors did not refer to all adults. Instead, these statistics referred to all people who were either married, had a regular partner, or were widowed, divorced, or separated. Charts containing data from the various countries surveyed are included in Daniel T. Halperin and Helen Epstein, "Concurrent Sexual Partnerships Help to Explain Africa's High HIV Prevalence: Implications for Prevention." *Lancet* 364 (2004): 4–6. In addition to the central importance of male sexual behavior,

the less often discussed but also crucial role of female behavior must not be overlooked. Modeling and empirical research indicates that the behaviors of both genders are key to understanding the breadth and depth of the sexual networks that result from multiple partnerships, especially concurrent ones, and consequently the rate at which HIV can spread in a society. Martina Morris and Mirjam Kretzschmar, "Concurrent Partnerships and the Spread of HIV." *AIDS* 11 (1997): 641–48; Mark N. Lurie, Brian G. Williams, Khangelania Zuma, et al., "Who Infects Whom? HIV-1 Concordance and Discordance Among Migrant and Non-migrant Couples in South Africa." *AIDS* 17 (2003): 2245–52; Stéphane Helleringer and Hans-Peter Kohler, "Sexual Network Structure and the Spread of HIV in Africa: Evidence from Likoma Island, Malawi." *AIDS* 21 (2007): 2323–32; Epstein, "Why Is AIDS Worse in Africa?"; Epstein, *The Invisible Cure*; Green and Herling Ruark, *AIDS, Behavior, and Culture*, pp. 211–12; Eyawo et al., "HIV Status in Discordant Couples in Sub-Saharan Africa"; Abigail Harrison, John Cleland, and Janet Frohlich, "Young People's Sexual Partnerships in KwaZulu-Natal, South Africa: Patterns, Contextual Influences, and HIV Risk." *Stud Fam Plann* 39 (2008): 295–308; Potts et al., "Reassessing HIV Prevention"; Shelton, "Ten Myths and One Truth About Generalised HIV Epidemics."

   Damien de Walque, in "Sero-Discordant Couples in Five African Countries" (see also chapter 1, endnote 8) concludes that: "The finding that a substantial proportion of HIV-infected couples are sero-discordant couples in which only the woman is infected . . . contradicts women's low self-reported levels of extramarital sex and . . . is extremely difficult to explain without extramarital sex among married women." Women's sexual behavior is frequently associated with the importance of what experts refer to as "transactional sex"; see Suzanne Leclerc-Madlala, "Transactional Sex and the Pursuit of Modernity." *Social Dynamics* 29 (2004): 1–21; Mark Hunter, "The Materiality of Everyday Sex: Thinking Beyond Prostitution." *Afr Stud* 61 (2002): 99–119; Ann Swidler and Susan C. Watkins, "Ties of Dependence: AIDS and Transactional Sex in Rural Malawi." *Stud Fam Plann* 38 (2007): 147–62; Ann M. Moore, Ann E. Biddlecom, and Eliya M. Zulu, "Prevalence and Meanings of Exchange of Money or Gifts for Sex in Unmarried Adolescent Sexual Relationships in Sub-Saharan Africa." *Afr J Reprod Health* 11 (2007): 44–61; Epstein, "Why Is AIDS Worse in Africa?"; Epstein, *The Invisible Cure*. The urgent need to accurately assess the extent of risky sexual behavior is, however, severely challenged—especially when it comes to women's behavior—by the limitations of self-reported data (see chapter 30).

20. In a national survey in 2005, about 40 percent of South African males aged fifteen to twenty-four and almost 25 percent of young females reported having more than one current sexual partner (Olive Shisana, Thomas Rehle, Leickness C. Simbayi, et al., "South African National HIV Prevalence, HIV Incidence, Behaviour and Communication Survey." Cape Town: HSRC Press, 2005). And women who reported having more than one partner at the time of the survey were over four times more likely to be HIV-positive than other women (South Africa Ministry of Health, *Getting to Success: Improving HIV Prevention Efforts in South Africa, Final Draft Report* [Pretoria, May 31, 2011]). Among sexually active fifteen- to twenty-four-year-olds in the hard-hit South African Province of KwaZulu-Natal, 38 percent of men reported currently having more than one regular sexual partner (Harrison et al., "Young People's Sexual Partnerships in KwaZulu-Natal, South Africa"). In a nationally representative survey of all South Africans of the same ages, 25 percent of men (and 5 percent of women)

reported engaging in concurrent relationships in the past year. Annie E. Steffenson, Audrey E. Pettifor, George R. Seage III, Helen V. Rees, and Paul D. Cleary, "Concurrent Sexual Partnerships and Human Immunodeficiency Virus Risk Among South African Youth." *Sex Transm Dis* 38 (2011): 459–66. Research in Botswana found that 24 percent of men and women reported multiple partners in the past year, and 19 percent having had a concurrent partnership during the same period. Marion W. Carter, Joan M. Kraft, Todd Koppenhaver, et al., " 'A Bull Cannot Be Contained in a Single Kraal': Concurrent Sexual Partnerships in Botswana." *AIDS Behav* 11 (2007): 822–30. Other research findings on the prevalence of concurrency are cited in various review papers, including: Timothy L. Mah and James D. Shelton, "Concurrency Revisited: Increasing and Compelling Epidemiological Evidence." *J Int AIDS Soc* 14 (2011): 33–41; Timothy L. Mah and Daniel T. Halperin, "Concurrent Sexual Partnerships and the HIV Epidemics in Africa: Evidence to Move Forward." *AIDS Behav* 14 (2010): 11–16; Helen Epstein and Martina Morris, "Concurrent Partnerships and HIV: An Inconvenient Truth." *J Int AIDS Soc* 14 (2011): 13; Timothy L. Mah and Daniel T. Halperin, "The Evidence for the Role of Concurrent Partnerships in Africa's HIV Epidemics." *AIDS Behav* 14 (2010): 25–28.

21. Interview with Timberg.

22. Carael later told Timberg that while the sexual behavior in Africa did appear to be well-suited to spreading sexually transmitted infections, the same appeared to be true for some other parts of the world. Carael also said that while the numbers of concurrent relationships appeared to be higher in Africa than in some parts of the world, other measures of sexual behavior were more similar among various regions. In an e-mail to Timberg in September 2011, Carael wrote that he still believes that "broad generalisations about 'Africa' or 'Asia,' or 'South America' would have been totally misleading." Carael also lamented the later use of data from the survey to bolster arguments about the importance of concurrent behavior, which he said were overblown. "This argument, that is making 'concurrency' the major driver of the HIV epidemics in Africa, is simply not supported by the survey data collected in the early nineties. There is no such magic bullet specific to Africa."

23. The concurrency concept emerged a few years before Morris began publishing her work. Christopher P. Hudson, "Concurrent Partnerships Could Cause AIDS Epidemics." *Int J STD AIDS* 4 (1993): 249–53; Charlotte H. Watts and Robert M. May, "The Influence of Concurrent Partnerships on the Dynamics of HIV/AIDS." *Math Biosci* 108 (1992): 89–104.

24. Martina Morris, "Concurrent Partnerships and HIV Transmission," Presentation at the Ninth International Conference on AIDS and STD in Africa, Kampala, Uganda, December 14, 1995. Subsequent modeling and other studies have reached similar conclusions regarding the likely crucial role of concurrency in helping to explain the explosive and continuing severe epidemics in parts of east and southern Africa. Jeffrey Eaton, Timothy Hallett, and Geoffrey Garnett, "Concurrent Sexual Partnerships and Primary HIV Infection: A Critical Interaction." *AIDS Behav* 4 (2011): 687–92; Steven Goodreau, Susan Cassels, Danuta Kasprzyk, Daniel Montaño, April Greek, and Martina Morris, "Concurrent Partnerships, Acute Infection and HIV Epidemic Dynamics Among Young Adults in Zimbabwe." *AIDS Behav* (2010), doi: 10.1007/s10461-010-9858-x ; Martina Morris, Helen Epstein, and Maria Wawer, "Timing Is Everything: International Variations in Historical Sexual Partnership Concurrency and HIV Prevalence." *PLoS One* 5 (2010): e14092; John J. Potterat, Helen

Rogers-Zimmerman, Stephen Q. Muth, et al., "Chlamydia Transmission: Concurrency, Reproduction Number, and the Epidemic Trajectory." *Am J Epidemiol* 150 (1999): 1331–39; Sevgi Aral, "Partner Concurrency and the STD/HIV Epidemic." *Curr Inf Dis Rep* 422 (2010): 134–39; Wim Delva, Carel Pretorius, Stijn Vansteelandt, Marleen Temmerman, and Brian Williams, "Serial Monogamy and the Spread of HIV: How Explosive Can It Get?" Presentation at the XVIII International AIDS Conference, Vienna, Austria; August 2010; Epstein, *The Invisible Cure*; Daniel T. Halperin and Helen Epstein. "Why Is HIV Prevalence So Severe in Southern Africa? The Role of Multiple Concurrent Partnerships and Lack of Male Circumcision: Implications for AIDS Prevention." *South Afr J HIV Med* 16 (2007): 19–25, www.sajhivmed.org.za/index.php/sajhivmed/article/viewFile/54/412.

Mah and Shelton ("Concurrency Revisited: Increasing and Compelling Epidemiological Evidence," and see chapter 4, endnote 19) report that in cohort studies of heterosexual couples a "surprisingly high" proportion of new infections occurred even when neither partner was previously infected. In a Tanzanian study, two-thirds of new infections occurred in such couples, which the researchers concluded was "presumably as a result of extramarital exposure" (Stéphane Hugonnet, Frank Mosha, James Todd, et al., "Incidence of HIV Infection in Stable Sexual Partnerships: A Retrospective Cohort Study of 1802 Couples in Mwanza Region, Tanzania." *J Acq Imm Def Synd* 30 (2002): 73–80). In a Ugandan study, 42 percent of new infections occurred in adults with HIV-negative spouses, while only about 25 percent occurred in adults with HIV-positive spouses. In a national survey in the same country, among seventy-four married people with recent HIV infection, half had spouses who were not infected, and in an additional 13 percent of cases both partners had recently become infected (Lucy M. Carpenter, Anatolia Kamali, Anthony Ruberantwari, Samuel S. Malamba, and James A. G. Whitworth, "Rates of HIV-1 Transmission Within Marriage in Rural Uganda in Relation to the HIV Sero-status of the Partners." *AIDS* 13 [1999]: 1083–89; Jonathan Mermin, Joshua Musinguzi, Alex Opio, et al., "Risk Factors for Recent HIV Infection in Uganda." *J Amer Med Assoc* 300 [2008]: 540–49). Mah and Shelton conclude that such studies "implicate a very active role of concurrency in propagating the HIV epidemic." However, some writers have questioned the importance of concurrency in the African epidemics (see chapter 30).

25. There are glaring discrepancies in rates of sexually transmitted infections across different parts of Africa, as also noted in the appendix. For example, there was a roughly 50 percent chlamydia prevalence in one survey of pregnant women in Conakry, Guinea, and very high rates of such diseases in some other parts of Central and West Africa. Meanwhile, the prevalence of such bacterial infections has for years been vastly lower in countries including Botswana, Zimbabwe, and South Africa, even though HIV rates are much higher in these places than anywhere in West Africa.

26. The Caldwells evidently also drew upon a map prepared by Stephen Moses and colleagues that similarly compared concentrations of male circumcision with HIV rates across regions of sub-Saharan Africa. Stephen Moses, Janet E. Bradley, Nico J. D. Nagelkerke, Allan R. Ronald, Jeckoniah O. Ndinya-Achola, and Francis Plummer, "Patterns of Male Circumcision Practices in Africa: Association with HIV Seroprevalence." *International Journal of Epidemiology* 19 (1990): 693–97.

27. Caldwell and Caldwell "The African AIDS Epidemic." p. 65.

28. Elizabeth Pisani, *The Wisdom of Whores: Bureaucrats, Brothels, and the Business of AIDS* (New York: W. W. Norton, 2008). Although UNAIDS oversaw and ultimately was responsible for the HIV estimates issued in their Global Reports and other documents, they also relied upon the technical contributions of an "external reference group" based at Imperial College in London.

29. Piot said, in an interview with Timberg, that he barely knew Pisani, who was a consultant to UNAIDS rather than a staffer during the years in question. He dismissed her criticisms as off target and unfair.

30. See Lalit Dandona, Vemu Lakshmi, Talasila Sudha, G. Anil Kumar, and Rakhi Dandona, "A Population-Based Study of Human Immunodeficiency Virus in South India Reveals Major Differences from Sentinel Surveillance-Based Estimates." *BMC Med* 4 (2006): 31; Daniel T. Halperin and Glenn L. Post, "Global HIV Prevalence: The Good News Might Be Even Better." *Lancet* 364 (2004): 1035–36. Fears of a severe, African-style epidemic in India turned out to be greatly overblown, but in a country with more than 1 billion people, even a very low HIV rate means that several million people are infected.

31. Pisani, p. 31.

32. Helen Epstein (*The Invisible Cure*, pp. 177–78) reported that she had read through many years worth of UNAIDS documents dating to the organization's founding in 1996 but found few references to partner reduction and none to Zero Grazing. "The agency's 'Best Practice' collection of briefing documents contains issues on condom programs, voluntary testing and counseling, sexually transmitted disease treatment services, and many other things, but as of this writing, there was no Best Practice document about encouraging partner reduction or fidelity," she wrote. Doug Kirby reports searching extensively for youth curricula focusing on multiple sexual partnerships and found only a very small number of them anywhere. He noted that "it appears that relatively few [such curricula] exist and clearly few are widely known or distributed." Kirby, D., Dayton, R., Prickett, A. et al., "Promoting Partner Reduction: Helping Young People Understand and Avoid HIV Risks from Multiple Partnerships. Research Triangle Park, NC: Family Health International, 2012.

Chapter 17: Things Just Fell Apart

1. Collins will be called by his first name in most cases to distinguish him from other members of the same family. The narrative elements of this chapter are based predominantly on his recollections and impressions, as conveyed in interviews with Timberg.

2. It is uncommon today for Luos to remove the lower teeth of adolescents.

3. This description of Luo marriage and some of the rules, rituals, and customs surrounding sexuality represents the cultural ideals, but as with other African patrilineal societies it is difficult to know how strictly they were followed, even in precolonial times. Within this system, there was still room for extramarital sex, according to anthropologist Robert Bailey (personal communication), who has studied Luo culture for many years. For example, both men and women would often visit relatives for extended periods of time, creating opportunities to engage in sexual relations. Similarly, wives would go back to their original home and stay with their family and clan, especially when they were mistreated by their husbands, and this also provided opportunities for liaisons. Dances, weddings, and funerals in neighboring villages allowed for some infidelity as well. However, Bailey believes, in these tradi-

tional settings there were not nearly as many opportunities for longer-term extramarital relationships, and most liaisons would have been brief.

4. E. E. Evans-Pritchard, "Marriage Customs of the Luo of Kenya." *Africa* 20 (1950): 132–42.

5. In some places the hymen itself was carried back to the home of the bride's parents.

6. Betty Potash, "Some Aspects of Marital Stability in a Rural Luo Community," *Africa* 48, no. 4 (1978): 384.

7. After World War I, Cameroon was taken under French control.

8. "Outline of Mission Field," Mission Board of Seventh-day Adventists, 1927, p. 81.

9. Ibid., p. 85.

10. Potash, p. 381.

11. The peak of HIV prevalence can only be roughly estimated because the 2003 Demographic and Health Survey was the first population-based national measure of HIV rates conducted in Kenya, and rates had already been declining substantially by that time (chapter 24, endnote 24). Also, any national estimate includes the Luo populations, which have much higher prevalence, meaning that HIV levels among other Kenyans are considerably lower than the overall national estimate.

12. Erick Otieno Nyambedha and Jens Aagaard-Hansen, "Practices of Relatedness and the Reinvention of *Duol* as a Network of Care for Orphans and Widows in Western Kenya." *Africa* 77 (2007): 522.

Chapter 18: X Factor

1. Studies conducted among African American populations have found relatively high levels of concurrency. In a national survey, black women were two to four times more likely to report having a concurrent partner than women from other ethnic groups (Adaora A. Adimora, Victor J. Schoenbach, Dana M. Bonas, Francis E. A. Martinson, Kathryn H. Donaldson, and Tonya R. Stancil, "Concurrent Sexual Partnerships among Women in the United States." *Epidemiology* 13 [2002]: 320–27). In a study conducted with African Americans in the southern United States, where some heterosexual outbreaks have occurred among this community, 53 percent of men and 31 percent of women reported practicing concurrency during the previous five years, and 61 percent of respondents believed that at least one of their recent partners had engaged in a concurrent relationship (Adaora A. Adimora, Victor J. Schoenbach, Francis Martinson, Kathryn H. Donaldson, Tonya R. Stancil, and Robert E. Fullilove, "Concurrent Sexual Partnerships among African Americans in the Rural South." *Ann Epidemiol* 14 [2004]: 155–60). Research over the years has suggested that concurrency is also one of the likely explanations for the higher rates of other sexually transmitted infections, such as syphilis, in the African American population. Emilia Koumans, Thomas Farley, James Gibson, et al., "Characteristics of Persons with Syphilis in Areas of Persisting Syphilis in the United States: Sustained Transmission Associated with Concurrent Partnerships." *Sex Transm Dis* 28 (2001): 497–503; Martina Morris, Ann E. Kurth, Deven T. Hamilton, James Moody, and Steve Wakefield, "Concurrent Partnerships and HIV Prevalence Disparities by Race: Linking Science and Public Health Practice." *Am J Public Health* 99 (2009): 1023–31; Potterat et al., "Chlamydia Transmission"; Martina Morris, "Concurrent Partnerships and Syphilis Persistence: New Thoughts on an Old Puzzle." *Sex Transm Dis* 28 (2001): 504–7. Furthermore, in some parts of the country—especially in rural areas of the South—African

Americans have somewhat lower rates of male circumcision. (And in eighteen states in the United States the procedure is no longer available, without cost, to poor parents opting to have their newborn sons circumcised; Brian J. Morris, Stefan A. Bailis, Jake H. Waskett, Thomas E. Wiswell, and Daniel T. Halperin, "Medicaid Coverage of Newborn Circumcision: A Health Parity Right of the Poor." *Am J Public Health* 99 [2009]: 969–71; Aaron A. Tobian and Ronald H. Gray, "The Medical Benefits of Male Circumcision." *J Amer Med Assoc* 306 [2011]: 1479–80.) Some researchers speculate that there could also be a possible genetic reason for why Africans and people of African descent may be more susceptible to HIV infection, though investigators involved in this research area believe that such a co-factor would account for at most some 10 percent of the increased risk in these populations. Rupert Kaul, Craig R. Cohen, Duncan Chege, et al., "Biological Factors That May Contribute to Regional and Racial Disparities in HIV Prevalence." *Am J Reprod Immunol* 65 (2011): 317–24; Nicholas Wade, "Gene Variation May Raise Risk of H.I.V., Study Finds," *The New York Times,* July 17, 2008.

2. The estimated number of HIV infections among white American women remains extremely low, on the order of less than three new cases per hundred thousand women annually. The estimated rate among black women is fifteen times higher, and is about three times higher than among Hispanic women. Joseph Prejean, Ruiguang Song, Angela Hernandez, et al., "Estimated HIV Incidence in the United States, 2006–2009," *PLoS One* 6 (2001): 1–13. HIV rates may be higher among Hispanics than whites in part because most Latino men are uncircumcised. One reason that HIV never reached a critical mass of infections among most heterosexual populations globally is because while some bisexual or injecting-drug using men may infect their female sex partners, these women rarely go on to infect other people—especially if they are not part of a risky sexual network or if their partners are circumcised.

3. Reports suggesting that most heterosexuals in the United States were unlikely to contract HIV often prompted backlash. In 1998 Halperin sought to publish an article about one well-publicized case in which a young man from New York City, Nushawn Williams, had spread HIV to fourteen women in a rural upstate community. Some news coverage, such as prominent stories in *Time* and *Newsweek,* implied that his actions threatened to cause many other new infections, and his face appeared on warning posters throughout the region. But when Halperin and a University of California–Berkeley colleague, Ina Roy, wrote an article calling these accounts overblown, editors were reluctant to publish it. Their piece explained why Williams was the "exception that proves the rule" and that widespread heterosexual outbreaks of HIV remained uncommon in the United States. Their examination of this case suggested that a combination of unusual factors—including the timing of the infections, most of which apparently occurred during the early, acute infection phase when contagiousness is unusually high—facilitated the elevated risk of transmission. A CDC investigation subsequently concluded there had not been a single instance of further HIV transmission beyond the fourteen young women who were initially infected, except for two children borne by them who also ended up becoming infected. But at the time many low-risk people feared they were in peril. In an unpublished study conducted in 1996 by Halperin and David Campt, another Berkeley colleague, two hundred university graduate students were asked about the ease of transmitting HIV through vaginal intercourse. A large majority thought that the risk

of getting the virus from one act of unprotected vaginal sex with an infected person was between 50 and 100 percent.

4. The pace of new HIV infections peaked in the early to mid-1990s in East Africa as a whole, in part because these mainly older epidemics were entering the inevitable downturn phase of any epidemic. Yet the shift also resulted from behavior change in places such as Uganda and the Kagera region of Tanzania (chapter 15 endnote 3; Shelton et al., "Has Global HIV Incidence Peaked?").

5. See Lurie et al., "Who Infects Whom?"; Lurie et al., "Circular Migration and Sexual Networking in Rural KwaZulu-Natal"; Epstein, *The Invisible Cure*; Epstein, "The Fidelity Fix."

6. It is widely believed that virtually all Xhosa men in South Africa's Eastern Cape province are circumcised, but a rigorous study conducted in the heart of the Xhosa region found that only 55 percent of men no longer had a foreskin. HIV prevalence, meanwhile, was 60 percent lower in those men who were actually circumcised. (Rachel Jewkes, Kristin Dunkle, Mzikazi Nduna, et al., "Factors Associated with HIV Sero-positivity in Young, Rural South African Men." *Int J Epidimol* 35 [2006]: 1455–60.) In a 2011 study in another part of South Africa, Orange Farm, 45 percent of the men who reported being circumcised still had intact foreskins, and these men had the same HIV rate as those who reported being uncircumcised. Pascale Lissouba, Dirk Taljaard, Dino Rech, et al., "Adult Male Circumcision as an Intervention Against HIV: An Operational Study of Uptake in a South African Community (ANRS 12126)." *BMC Medicine* (2011), in press.

7. A focus group study of sexual behavior among black South African university students in Durban included a question about the value of male circumcision for HIV prevention. Many of the students reacted with confusion or were even upset by the question, insisting that the focus of prevention should be on behavior change. However, when this was changed in subsequent interviews to the more open-ended question of simply "What do you think about male circumcision?" nearly everyone responded favorably, citing reasons such as improved hygiene and sexual pleasure for both genders (Stephanie Psaki, Nono Ayivi-Guedehoussou, and Daniel Halperin, paper in review).

8. Daniel Halperin, "Field Research Findings from Focus Groups and Interviews in South Africa and Botswana on the Acceptability of Male Circumcision for HIV Prevention and Reproductive Health." Presentation at the American Anthropological Association Annual Meeting, San Francisco, November 2000. Eventually many other studies in South Africa and a number of other sub-Saharan African countries where male circumcision is also no longer routinely practiced would document similar findings about the acceptability of the procedure (see appendix).

9. Several decades after HIV began circulating in Mozambique, the four provinces which have high (above 80 percent) male circumcision prevalence have continued to experience substantially lower HIV rates than the rest of the country, despite generally higher levels of risky sexual behavior in the mainly circumcising areas (Mozambique Demographic and Health Survey, 2003, Calverton, MD: ORC MACRO International). Similar patterns are observed in various other countries, including Tanzania, Kenya, and Zambia. In some countries such as Malawi and Lesotho, however, comparable or even slightly higher HIV rates are found among men who report being circumcised. In the case of Lesotho, it has been documented that many rural men who have attended the traditional "circumcision schools"—and would

therefore tell an interviewer, in the Sesotho language, that they have been "circumcised"— still retain at least part of their foreskin. (Anne Thomas, Marcus Cranston, Bonnie Tran, Malerato Brown, Matsotetsi Tlelai, and Rajiv Kumar, "Voluntary Medical Male Circumcision: A Cross-Sectional Study Comparing Circumcision Self-Report and Physical Examination Findings in Lesotho." *PLoS ONE* 6 (2011): e27561. In a Tanzanian study only 62 percent of Muslim males, who are universally assumed to be circumcised, were actually verified upon clinical examination to be without a foreskin. Helen A. Weiss, Mary L. Plummer, John Changalucha, et al., "Circumcision Among Adolescent Boys in Rural Northwestern Tanzania." *Trop Med Int Health* 13 (2008): 1054–61. (Also see chapter 18, endnote 6, and Michelle Poulin and Adamson S. Muula, "An Inquiry into the Uneven Distribution of Women's HIV Infection in Rural Malawi." *Demog Research* 25 (2011): 869–902; www.demographic-research.org/Volumes/Vol25/28.)

10. Although India's HIV epidemic turned out to be considerably smaller than the official estimates had indicated (epilogue, endnote 1), the country still has a much higher rate than the extremely small epidemics in neighboring Pakistan and Bangladesh. Epidemiological research in India has found a similar preventive effect as in the African studies, and despite religious tensions occasionally related to circumcision status, preliminary studies suggest the procedure may be acceptable in some Hindu communities. Stephen J. Reynolds, Mary E. Shepherd, Arun R. Risbud, et al., "Male Circumcision and Risk of HIV-1 and Other Sexually Transmitted Infections in India." *Lancet* 363 (2004): 1039–40; Purnima Madhivanan, Karl Krupp, Varalakshmi Chandrasekaran, Samuel C. Karat, Arthur L. Reingold, and Jeffrey D. Klausner, "Acceptability of Male Circumcision Among Mothers with Male Children in Mysore, India." *AIDS* 22 (2008): 983–88; Purnima Madhivanan, Karl Krupp, Vinay Kulkarni, Sanjeevani Kulkarni, and Jeffrey D. Klausner, "Acceptability of Male Circumcision for HIV Prevention Among High-Risk Men in Pune, India." *Sex Transm Dis* 38 (2011): 571.

11. Gilbert et al., "The Emergence of HIV/AIDS in the Americas and Beyond."

12. Circumcision rituals, which involved sharp utensils that potentially could also have been used as weapons by slaves against their masters, were not preserved in Haiti, but many other aspects of African culture, such as the voodoo religion, have persisted.

13. At this time Halperin, along with most HIV experts, believed that treating other sexually transmitted infections would have a significant impact on the HIV epidemic. Subsequent research findings, as discussed in chapter 27, have called this assumption into question.

14. Cheryl Mattson, Robert C. Bailey, Richard Muga, and Rudi Poulussen, "Acceptability of Male Circumcision and Predictors of Circumcision Preference in Men and Women in Nyanza Province, Kenya." *AIDS Care* 17 (2005): 182–94; Nelly Westercamp and Robert C. Bailey, "Acceptability of Male Circumcision for Prevention of HIV/AIDS in Sub-Saharan Africa: A Review." *AIDS Behav* 11 (2007): 341–55; Weiss et al., "Male Circumcision"; Weiss et al., "Male Circumcision for HIV Prevention"; USAID, "Male Circumcision."

15. Circumcision was not the main focus of these studies. The circumcision associations were in most cases incidental findings whose importance was recognized only in later review by other researchers such as: Stephen Moses, Robert C. Bailey, and Allan R. Ronald, "Male Circumcision: Assessment of Health Benefits and Risks." *Sex Transm Infect* 74 (1998): 368–73; Helen A. Weiss, Maria A. Quigley, and Richard J. Hayes, "Male Circumcision and Risk of HIV Infection in Sub-Saharan Africa: A Systematic Review and Meta-analysis." *AIDS* 14

(2000): 2361–70; Robert C. Bailey, Francis Plummer, and Stephen Moses, "Male Circumcision and HIV Prevention: Current Knowledge and Future Research Directions." *Lancet Infect Dis* 1 (2001): 223–31 (and see chapter 12, endnote 3).

16. Moses's report was never released by WHO, but it was published in a journal (Stephen Moses, Francis A. Plummer, Janet E. Bradley, Jeckoniah O. Ndinya-Achola, Nico J. D. Nagelkerke, and Allan R. Ronald, "The Association Between Lack of Male Circumcision and Risk for HIV Infection: A Review of the Epidemiological Data." *Sex Transm Dis* 21 [1994]: 201–10). At a half-day consultation on the subject organized by the WHO following the 2000 International AIDS Conference in Durban, and attended by two dozen UN officials and other experts, a proposal by Halperin and Robert Bailey to consider developing guidelines for performing safe adult circumcision was rejected. This type of surgical manual eventually was issued by the WHO, after the clinical trial data was published.

17. See Daniel T. Halperin, Helen Weiss, Richard Hayes, et al., "Comments on Male Circumcision and HIV Acquisition and Transmission in Rakai, Uganda." *AIDS* 16 (2002): 810–12; Wamai et al., "Male Circumcision Is an Efficacious, Lasting and Cost-Effective Strategy for Combating HIV in High-Prevalence AIDS Epidemics."

18. Even in places where circumcision is not routinely practiced, including continental Europe, some men occasionally undergo the procedure to address medical issues such as phimosis. Circumcision used to be relatively common in Britain but declined in popularity starting in the 1950s. The main reason appears to be the decision by the National Health Service to stop paying for circumcision after some deaths due to the use of general anesthesia. Stefan A. Bailis and Daniel T. Halperin, "Male Circumcision: Time to Re-examine the Evidence." *Student BMJ* 14 (2006): 179–80. Local anesthesia is normally used for adult and infant circumcisions today. There has also been some confusion caused by mistaken comparisons with "female genital mutilation," which is a very different type of procedure and can have serious negative medical consequences. Further confusing the issue of male circumcision are the protests of a small but vocal community of activists who often call themselves "intactivists" because of their belief that the male genitalia should remain entirely intact. This constituency has launched aggressive campaigns, including one that resulted in getting an initiative on the ballot in San Francisco to ban the performance of any circumcisions on minors in the city. California officials later ruled that cities had no authority over medical produceress. Adam Cohen, "San Francisco's Circumcision Ban: An Attack on Religious Freedom?" *Time*, June 13, 2011, www.time.com/time/nation/article/0,8599,2077240,00 .html#ixzz1ZUKtpIEL; Jessica Pauline Ogilvie, "The Debate over Circumcising Baby Boys: A Safe and Hygienic Option? Or a Medically Unnecessary and Brutal Procedure?" *Los Angeles Times,* July 11, 2011; Brian J. Morris, Jake H. Waskett, Ronald H. Gray, et al., "Exposé of Misleading. Claims that Male Circumcision Will Increase HIV Infections in Africa," *J Public Health Africa* 2 (2011): 117–22; Joya Banerjee, Jeffrey Klausner, Daniel Halperin, et al., "Circumcision Denialism Unfounded and Unscientific." *Amer J Prev Med* 40 (2011): e11–e12; Charles Hirshberg, "Should All Males Be Circumcised? Some U.S. Doctors Are Reconsidering Their Position." *Men's Health*, January 28, 2009, www.menshealth.com/ health/debate-over-circumcision; Brian J. Morris, Robert C. Bailey, Jeffrey D. Klausner, et al., "Review: A Critical Evaluation of Arguments Opposing Male Circumcision for

HIV Prevention in Developed Countries." AIDS Care, 2012, http://dx.doi.org/10.1080 /09540121.2012.661836; Richard G. Wamai, Brian J. Morris, Jake H. Waskett, et al., "Criticisms of African Trials Fail to Withstand Scrutiny: Male Circumcision *Does* Prevent HIV Infection." *J Law Med* (2012): in press.

19. Caldwell and Caldwell, "The African AIDS Epidemic"; Moses et al., "The Association Between Lack of Male Circumcision and Risk for HIV Infection"; Marc Urassa, James Todd, J. Ties Boerma, Richard Hayes, and Raphael Isingo, "Male Circumcision and Susceptibility to HIV Infection Among Men in Tanzania." *AIDS* 11 (1997): 73–80; Edward C. Green, Bongi Zokwe, and John D. Dupree, "Indigenous African Healers Promote Male Circumcision for Prevention of Sexually Transmitted Diseases." *Trop Doct* 23 (1993): 182–83; Daniel T. Halperin and Robert C. Bailey, "Male Circumcision and HIV Infection: 10 Years and Counting." *Lancet* 354 (1999): 1813–15.

20. Helen A Weiss, Sarah L. Thomas, Susan K. Munabi, and Richard J. Hayes, "Male Circumcision and Risk of Syphilis, Chancroid, and Genital Herpes: A Systematic Review and Meta-Analysis." *Sex Transm Infect* 82 (2006): 101–9; Tobian et al., "Male Circumcision for the Prevention of Acquisition and Transmission of Sexually Transmitted Infections"; Brian J. Morris, Ronald H. Gray, Xavier Castellsagué, F. Xavier Bosch, Daniel T. Halperin, Jake H. Waskett, and Catherine A. Hankins, "The Strong Protection Afforded by Circumcision Against Cancer of the Penis." *Adv Urol* (2011): 1–21; Xavier Castellsagué, F. Xavier Bosch, Nubia Munoz, et al., "Male Circumcision, Penile Human Papillomavirus Infection, and Cervical Cancer in Female Partners." *N Engl J Med* 346 (2002): 1105–12; F. Xavier Bosch, Ginesa Albero, and Xavier Castellsagué, "Male Circumcision, Human Papillomavirus and Cervical Cancer: From Evidence to Intervention." *J Fam Plan Reprod H* 35 (2009): 5–7; Paul K. Drain, Daniel T. Halperin, James P. Hughes, Jeffrey D. Klausner, and Robert C. Bailey. "Male Circumcision, Religion, and Infectious Diseases: An Ecologic Analysis of 118 Developing Countries." *BMC Infect Dis* 6 (2006): 172–82; Jonathan L.Wright, Daniel W. Lin, Janet L. Stanford, "Circumcision and the Risk of Prostate Cancer," *Cancer* (2012), doi:10.1002/cncr.26653; Weiss et al., "Male Circumcision for HIV Prevention"; and see appendix. Some researchers have postulated that the main reason for the HIV prevention effect of circumcision might be because it lowers the risk of some other sexually transmitted infections, which are believed to increase the risk of getting HIV. However, careful analyses of such an indirect effect concluded that this probably plays a negligible role: Kamal Desai, Marie-Claude Boily, Geoff P. Garnett, Benoît R. Mâsse, Stephen Moses, and Robert C. Bailey, "The Role of Sexually Transmitted Infections in Male Circumcision Effectiveness Against HIV: Insights from Clinical Trial Simulation." *Emerg Themes Epidemiol* 3 (2006): 19; Ronald H. Gray, David Serwadda, Aaron A. R. Tobian, et al., "Effects of Genital Ulcer Disease and Herpes Simplex Virus Type 2 on the Efficacy of Male Circumcision for HIV Prevention: Analyses from the Rakai Trials." *PLoS Med* 6 (2009): e1000187. Some other researchers have theorized that the improved hygiene resulting from circumcision may be the main reason for its efficacy in preventing HIV infection. But research among African men to test the theory, through promoting improved genital hygiene, has failed to find an impact on transmission; in one trial postcoital washing actually increased HIV risk. Fredrick E. Makumbi, Ronald H. Gray, and Maria Wawer, "Male Post-coital Penile Cleansing and

the Risk of HIV Acquisition in Rural Rakai District, Uganda." Presentation at the International AIDS Society Conference, Sidney, 2007, www.ias2007.org/pag/Abstracts .aspx?SID=55&AID=5536.

21. One exception to the pattern of high prevalence of male circumcision associated with low HIV rates was Kenya. But because the second largest ethnic group, the Luo, is the only major one that does not traditionally practice circumcision, this has resulted in much higher concentrations of HIV in the relatively few places where Luos predominate. Consequently Kenya's overall HIV rate does not meaningfully reflect the much lower levels in most other parts of the country. Also, it later turned out that the nation's rate was significantly overestimated in the official statistics (chapter 26).

22. Halperin and Bailey, "Male Circumcision and HIV Infection." The article also cited a clearinghouse maintained by Johns Hopkins University that contained more than thirty thousand health communications materials from around the world but not a single poster, flyer, or pamphlet that mentioned the possible benefits of male circumcision.

23. UNAIDS, *Report on the Global HIV/AIDS Epidemic.* June 2000, Geneva: UNAIDS, p. 70.

24. In 2004 and 2005, Piot gave speeches at the Woodrow Wilson Center in Washington, D.C. (www.wilsoncenter.org/article/top-leaders-set-the-next-agenda-for-global-aids). Neither time did he mention circumcision, but in the question-and-answer periods Jim Shelton, a senior technical adviser at USAID, asked him about the issue. (On the second occasion, Shelton brought up the South African trial that had recently found a 60 to 75 percent protective effect against HIV.) Both times, Piot responded by deflecting the question. In a recent interview Piot could not recall the specific incident with Bailey but agreed that he felt substantial frustration with those promoting circumcision in this era; he believed they were overstating the evidence for the procedure's importance in HIV's spread. "It's certainly possible that I reacted this way," he said.

Chapter 19: The Interests of the ANC

1. Jon Jeter, "Free of Apartheid, Divided by Disease." *The Washington Post,* July 6, 2000.

2. Several of the facts used for this account of Mbeki's handling of Virodene come from meticulous work by James Myburgh and published on www.politicsweb.co.za under the title "The Virodene Affair," www.politicsweb.co.za/politicsweb/view/politicsweb/en/page71619? oid=83156&sn=Detail.

3. Pride Chigwedere, George R. Seage III, Sofia Gruskin, Tun-Hou Lee, and Max Essex, "Estimating the Lost Benefits of Antiretroviral Drug Use in South Africa." *J Acquir Immune Defic Syndr* 49 (2008): 410–15, www.hsph.harvard.edu/news/hphr/spring-2009/spr09aids.html.

4. The best history of Mbeki's life and legacy is Mark Gevisser's *Thabo Mbeki: The Dream Deferred* (Johannesburg: Jonathan Ball Publishers, 2007).

5. Peter Slevin, "Mbeki Says Diplomacy Needed for Zimbabwe." *The Washington Post,* September 25, 2003. Despite stereotypes about Europeans commonly having extramarital affairs, national survey data does not indicate this behavior is more regularly practiced than in the United States. Even when people in Europe or North America do practice infidelity, they tend to have less sex within their marriages while having affairs (Druckerman, *Lust in Translation*). This limits the development of large interconnected networks of concurrent partnerships such as occur in parts of Africa.

6. Myburgh, "The Virodene Affair."
7. Data on the prevalence of concurrency in South Africa is cited in endnote 20 to chapter 16. Further indications for the presence of concurrency and resulting extensive sexual networks include the extremely high rates of other sexually transmitted infections, notably syphilis, found historically in many parts of the country. For a broader discussion on the experience with syphilis and other sexually transmitted diseases in southern Africa, see: Megan Vaughan, *Curing Their Ills: Colonial Power and African Illness* (Cambridge, UK: Polity Press, 1991); Karen Jochelson, *The Colour of Disease: Syphilis and Racism in South Africa, 1880–1950* (Hampshire, UK: Palgrave/Macmillan, 2001).
8. UNAIDS, "Report on the Global HIV/AIDS Epidemic," Geneva, July 2002.
9. When he took power Mandela had promised whites that he would not purge the bureaucracy.
10. Antoine Van Gelder craftily made a photocopy of the spy's ID, which he proffered to Timberg in an interview.
11. Interview with Timberg.
12. ANC Treasurer General Mendi Msimang refused an interview with Timberg on the matter.
13. Mbeki later told South African journalist Allister Sparks that one of the documents sent by Visser, a newspaper op-ed written by AZT opponent Anthony Brink, was the first he had heard of the dangers of AZT. Mbeki did not confirm, and Sparks did not ask, how he discovered the article. Tellingly, it appeared in *The Citizen* newspaper in March, but Mbeki made no public utterances about it until the October 28 speech, exactly seven days after Visser sent it to him.
14. Myburgh, "The Virodene Affair."
15. Interview with Timberg.
16. Donald G. McNeil Jr., "Neighbors Kill an H.I.V.-Positive AIDS Activist in South Africa," *The New York Times* December 28, 1998.
17. The Treatment Action Campaign did come to support male circumcision in later years and lobbied the government for better access to these services.

Chapter 20: Poverty Trap

1. Daniel Wakin, "President's Claim that AIDS Drug Is Dangerous Sets Off Debate," Associated Press, November 3, 1999.
2. Rachel Swarns, "Safety of Common AIDS Drug Questioned in South Africa," *The New York Times*, November 25, 1999.
3. For more on AIDS denialism, see Kalichman, *Denying AIDS*. There are also various Web sites on this issue, such as: www.newhumanist.org.uk/2165/how-to-spot-an-aids-denialist. When we use the term "denialists" we refer to people who believe that the HIV virus is not the cause of AIDS. By this definition Mbeki was not a denialist because he never categorically denied a link between HIV and AIDS. However, his publicly expressed doubts about the prevailing wisdom are widely believed to have harmed efforts to address the sexual spread of the epidemic, by calling into question the role of behavior in spreading the disease. See, for example: www.aidsmap.com/South-African-women-with-AIDS-conspiracy-beliefs-half-as-likely-to -use-condoms/page/1797893/.
4. Gevisser, *Thabo Mbeki*, p. 738–39.

5. Barton Gellman, "S. African President Escalates AIDS Feud; Mbeki Challenges Western Remedies," *The Washington Post*, April 19, 2000.
6. David Brown, "Statement Assures Doubters: HIV Causes AIDS," *The Washington Post*, July 2, 2000.
7. Mbeki joined the Communist Party in 1962 but later relinquished his membership, according to Gevisser.
8. Speech text issued by the Office of the Presidency, and delivered on July 9, 2000.
9. Countries such as Botswana and South Africa consistently rank highest in Africa on United Nations measures of education, literacy, gender equality, economic growth, and per-capita income, yet also have among the highest HIV rates in the world. See, for example: World Economic Forum, *Global Gender Gap Report*, 2007 (www.weforum.org/pdf/gendergap/rankings2007.pdf) and endnote 7 to the prologue on how most studies have found a positive association between individuals with higher income and higher HIV rates in Africa.
10. Interview by Timberg with Carl Landauer, one of the two heart surgeons who were early Virodene investors.
11. The South Africa Minister of Health, Manto Tshabalala-Msimang, became infamous in South Africa and beyond for her reluctance to support using ARVs in the public health system and for her advocacy of alternative approaches to fighting AIDS. The most notorious was her unproven assertion that a diet rich in beets, lemons, garlic, and some other ingredients would help prevent the descent into this disease.
12. Timberg was never able to determine if those painkillers included acetaminophen, which is widely dispensed in African medical clinics; the London trials suggested it might cause liver problems when mixed with Virodene.
13. Zigi Visser, the Virodene company administrator, wrote Timberg an e-mail in December 2007 replying to several questions about the trials in Tanzania. Visser expressed sorrow about the reported death of a former trial participant but said that no "serious adverse events or deaths" happened during the trial or for the six months afterward. "You interviewed some of the surviving Study participants six and a half years after completion of the study, indeed a miracle of its own, considering that they only received six dosis [*sic*] of Virodene," Visser wrote. "Obviously one would expect these participants to deteriate [*sic*], become ill and die without continued medication and care."

Chapter 21: A, B, and C

1. UNAIDS, "Financial Resources Required to Achieve Universal Access to HIV Prevention, Treatment, Care and Support," Geneva: UNAIDS, September 26, 2007.
2. Interview with Timberg.
3. Addressing risk among youth is important but in most places they comprise only a relatively small proportion of the total number of people infected. In South Africa, for example, more than three-fourths of people with HIV are over age twenty-five, and the majority of new infections also take place in this age group. The number of infections occurring among people under age twenty, the usual target audience of youth prevention programs, is even smaller. Shisana et al., "South African National HIV Prevalence, HIV Incidence, Behaviour and Communication Survey"; Thomas Rehle et al., "National HIV Incidence Measures—

New Insights into the South African Epidemic." *S Afr Med J* 97 (2007): 194–99 and personal communication with Thomas Rehle.

4. Rena Singer, "Is LoveLife Making Them Love Life?" *Mail & Guardian*, August 24, 2005.

5. Daniel Halperin and Brian Williams, "This Is No Way to Fight AIDS in Africa," *The Washington Post*, August 26, 2001. And see Epstein, *The Invisible Cure*; Helen Epstein, "AIDS in South Africa: The Invisible Cure." *New York Review of Books*, July 17, 2003.

6. HIV prevalence among white people (age two and above) in South Africa is estimated to be 0.3 percent (Olive Shisana, Thomas Rehle, Leickness Simbayi, et al., *South African National Prevalence, Incidence, Behaviour and Communication Survey, 2008.* [Cape Town: HSRC Press, 2009]). The esitmate for white women, of about 0.7 percent (personal communication with Thomas Rehle), is much higher than among women in Europe or North America. But the black populations in South Africa, where rates among women are estimated to be 26.6 percent (Ibid), have been much more severely impacted.

7. Singer, "Is LoveLife Making Them Love Life?"

8. Halperin and Williams, "This Is No Way to Fight AIDS."

9. James D. Shelton and Beverly Johnston, "Condom Gap in Africa: Evidence from Donor Agencies and Key Informants." *BMJ* 323 (2001): 139.

10. Stover did not discount condoms entirely, saying that without their promotion the HIV epidemic in Kenya might have worsened. But he stressed that, in the absence of other changes in sexual behavior, the use of condoms was not enough to cause a broad reversal in a widespread epidemic. USAID, "What Happened in Uganda? Declining HIV Prevalence, Behavior Change and the National Response." Washington, D.C.: USAID Technical Report, 2002, www.usaid.gov/pop_health/aids/Countries/africa/uganda_report.pdf. (Kenya eventually saw its own HIV rates decline, also accompanied by reductions in reported multiple sexual partnerships; see chapter 24, endnote 24.)

11. Nantulya at the time was, along with Green, on a midcareer fellowship at Harvard's School of Public Health. His presentation of the "Be Faithful" aspect also included the broader range of partner reduction measures (Ibid).

12. Although the origins of the ABC approach are not entirely clear, according to some accounts it was an activist Filipino health minister, Juan Flavier, who in 1992 first branded the "ABC" strategy. This formulation spread through public campaigns, international conferences, and news reports, gradually making its way across the globe. Karen Hardee, Jay Gribble, Stephanie Weber, Tim Manchester, and Martha Wood, "Reclaiming the ABCs: The Creation and Evolution of the ABC Approach." Washington, D.C.: Population Action International, 2008.

13. Lynne Muthoni Wanyeki, "Kenya—Population: Church Burns Condoms and AIDS Materials," Inter Press News Service, September 5, 1996.

14. Greater access to family planning services leads to fewer unwanted pregnancies and consequently to fewer abortions. Duff Gillespie, "Making Abortion Rare and Safe." *Lancet* 362 (2004): 72; Charles F. Westoff, *Recent Trends in Abortion and Contraception in 12 Countries: DHS Analytical Studies No. 8* (2005) Calverton, MD: ORC Macro, 2005; Mizanur Rahman, Julie DaVanzo, and Abdur Razzaque, "Fertility Transition, Contraceptive Use, and Abortion in Rural Bangladesh: The Case of Matlab," The POLICY Project, February 2000.

15. Despite the skepticism and occasional opposition that Halperin faced at USAID, it almost

certainly was a venue more hospitable to new ideas than if he had gone to work for another U.S. government agency.

16. In late December of 2002, administrator Andrew Natsios issued a cable to USAID missions worldwide mandating that in countries with widespread HIV epidemics, prevention campaigns would henceforth be built around the concept of ABC. A slogan that had begun the year obscure among policy circles ended it as an official U.S. government edict.

17. Bessinger et al., "Sexual Behavior, HIV and Fertility Trends." Since this study was conducted it became clear that HIV rates have also declined in two more of these countries, Kenya and Zimbabwe.

18. USAID, "The 'ABCs' of HIV Prevention: Report of a USAID Technical Meeting on Behavior Change Approaches to Primary Prevention of HIV/AIDS," Washington, D.C.: USAID Office of HIV-AIDS, 2002; "Male Circumcision: Current Epidemiological and Field Evidence; Program and Policy Implications for HIV Prevention and Reproductive Health," Washington, D.C.: USAID Office of HIV-AIDS/AIDSMARK, 2002, www.path.org/files/MC_USAID_02_Meeting_Report.pdf. Also see www.usaid.gov/press/frontlines/fl_nov11/FL_nov11_50_HIV_AIDS.html.

19. Arthur Allen, "Sex Change: Uganda v. Condoms." *The New Republic*, May 27, 2002. The author was among those who found this headline glib and misleading. Green, an anthropologist then based at Harvard University, quickly became an unlikely favorite of some Christian conservatives because of his focus on how Uganda slowed the spread of HIV by campaigning against extramarital sex. Some veterans of these ideological wars came to view Green as an advocate primarily for abstinence during this period, but he remembers it differently; he says that his true focus was on A and B in roughly equal measure. The problem, he said, was that few people focused on the second letter. Years later Green recalled to Timberg how he pleaded with a key Christian conservative that PEPFAR should emphasize fidelity and partner reduction as much as abstinence. "What I was told is, 'There is a constituency for abstinence. No one has thought about fidelity.'"

20. Rod Dreher, "Death in Africa . . . and Stoppable Death at That: How to Combat AIDS," *The National Review*, Feb. 10, 2003.

Chapter 22: On the Jericho Road

1. PEPFAR also had goals to care for ten million people, including orphans and others impacted by AIDS.

2. In his 2011 book, *We Meant Well: How I Helped Lose the Battle for the Hearts and Minds of the Iraqi People* (New York: Metropolitan Books), Peter van Buren recounts how many American officials and contractors in Iraq were blind to the nuances and complexities of the local cultural and political reality. While the costs of the Iraq and Afghanistan wars dwarfs that of the U.S. government's effort to combat AIDS globally, the tens of billions of dollars for this purpose have not always been prudently spent either. During the more than five years Halperin spent working for USAID, he was frequently dismayed at how funds were allocated. USAID missions, for example, would receive reprimands for not spending their entire annual budgets. And some U.S. government representatives working on AIDS, typically overburdened with bureaucratic responsibilities, often were out of touch with local cultural and other subtleties.

3. Jay Leftkowitz, "AIDS and the President—An Inside Account," *Commentary Magazine*, January 2009.

4. The adult HIV prevalence rate in Guyana is officially 1.2 percent, with an estimated 5,900 infected people, and is 0.4 percent in Vietnam, with 140,000 people estimated to be living with HIV. In contrast, Malawi's rate is 11 percent, with 920,000 people infected, and Zimbabwe's rate is 14.3 percent, with an estimated 1.2 million people infected with HIV (UNAIDS 2010 Global Report, Geneva). However, Zimbabwe and other non-"focus" countries have received some PEPFAR and other global HIV funding. Meanwhile some other donors, such as the Global Fund to Fight AIDS, Tuberculosis and Malaria, have awarded considerable grants to countries with virtually no epidemics, such as Bolivia, the Philippines, and Bulgaria.

5. Shelton et al., "Partner Reduction Is Crucial for Balanced 'ABC' Approach." And see Epstein, "The Fidelty Fix"; James D. Shelton, "Why Multiple Sex Partners?" *Lancet* 374 (2009) 367–69; Vinod Mishra, Serene Thaddeus, Jessica Kafuko, et al., *Fewer Lifetime Sexual Partners and Partner Faithfulness Reduce Risk of HIV Infection: Evidence from a National Sero-Survey in Uganda*. In *Uganda Ministry of Health Working Papers* (Kampala: Ministry of Health, ICF Macro, USAID, and CDC, 2009).

6. Asiimwe-Okiro et al., "Change in Sexual Behaviour and Decline in HIV Infection"; Stoneburner and Low-Beer, "Population-Level HIV Declines and Behavioral Risk Avoidance"; Bessinger et al., "Sexual Behavior, HIV and Fertility Trends"; Timothy B. Hallett, Simon Gregson, James J. C. Lewis, Ben Lopman, and Geoffrey P. Garnett. "Behaviour Change in Generalised HIV Epidemics: Impact of Reducing Cross-Generational Sex and Delaying Age at Sexual Debut." *Sex Transm Infect* 83 (2007): i50–i54.

7. In an interview with Timberg, Serwadda said he was unable to recall the name of the senator.

8. David Serwadda, "Beyond Abstinence," *The Washington Post*, May 16, 2003. Despite Serwadda's critique of the Bush administration's overemphasis on abstinence, he also was, for example, coauthor of an article documenting the limits of condom use for impacting the kind of widespread epidemic found in Uganda (Ahmed et al., "HIV Incidence and Sexually Transmitted Disease Prevalence Associated with Condom Use").

9. The abstinence earmark infuriated liberals and many other people working on HIV prevention. It also put Halperin, who had worked to raise the prominence of "ABC" and especially partner reduction, in an awkward spot. He disliked the earmark, but when it became clear that some version of it was politically inevitable, he lobbied some conservatives he had recently gotten to know to support expanding the earmark to include both A and B. This might have directed resources in a way that a little more closely resembled Uganda's original approach. Within a couple years, the earmark was expanded to also include "B" interventions.

10. Some of the new AIDS fighters were, quite literally, missionaries. The Bush administration made certain that religious groups—called "faith-based organizations"—were installed in key roles in PEPFAR-funded programs, much to the chagrin of the decidedly secular public health authorities at the United Nations and other international groups.

11. There was undoubtedly also some value in the greater push for accountability under PEPFAR. Bush administration officials were eager for impact and had seen how many other global initiatives became bureaucratized and inefficient. Their hope was that by setting ambitious and clearly stated goals, they could avoid that fate for PEPFAR.

12. Under Tobias, PEPFAR eventually approved many generic ARVs for use in its programs. That greatly lowered the cost of providing medication, allowing more people to be treated. But other medical costs, including for doctors, nurses, and facilities, soon became important factors in the overall cost per patient treated in the program.

13. Deborah Sontag, Sharon LaFraniere, and Michael Wines, "Early Tests for U.S. in Its Global Fight on AIDS," *The New York Times*, July 14, 2004.

14. Craig Timberg, "Uganda's Early Gains Against HIV Eroding; Message of Fear, Fidelity Diluted by Array of Other Remedies," *The Washington Post*, March 29, 2007.

Chapter 23: Gordon and Thandi

1. Some elements of this chapter appeared in an article Timberg wrote for *The Washington Post* on April 12, 2004, headlined, "Most S. Africans Loyal to Ruling Party; As Elections Approach, Discontent Over AIDS Directed Largely at President Mbeki, Not ANC."

2. People in sexual relationships in which both partners are already infected are commonly counseled to practice safer sex, because of the risk of "reinfection." Not all experts agree, however, on how significant this risk is.

Chapter 24: A Marshall Plan for Botswana

1. Surveys in Botswana show that about 80 percent of adults are currently unmarried, though a sizable proportion report being in a steady relationship. Marriage rates in some other southern African nations, including South Africa, Namibia, and Swaziland, have also become very low. See Figure 2 in Daniel T. Halperin, Owen Mugurungi, Timothy B. Hallett, et al., "A Surprising Prevention Success: Why Did the HIV Epidemic Decline in Zimbabwe?" *PLoS Med* 8 (2011): e1000414; Green and Ruark, *AIDS, Behavior, and Culture*, pp. 187–221; John Bongaarts, "Late Marriage and the HIV Epidemic in Sub-Saharan Africa," *Popul Stud* 61 (2007): 73–83; Epstein, "The Fidelity Fix"; Epstein, *The Invisible Cure*.

2. Groups like Population Services International eventually abandoned such sexually charged approaches in Botswana and other parts of Africa. In recent years the organization, together with the Botswana government, conducted a major campaign directed at concurrency. However, controversy brewed in 2007 over an initiative called "Ma-14," which translates loosely to "Sweet 14," that urged the young female partners of "sugar daddies" to use condoms when having this type of intergenerational sex. The campaign was pulled shortly after some public officials and others complained (www.mmegi.bw/index .php?sid=1&aid=34&dir=2007/May/Thursday10). In 2009, when Halperin was teaching a graduate course on HIV at Harvard's School of Public Health, an insightful but usually shy medical doctor from Botswana blurted out, during a discussion on Museveni's earlier approach to condoms, "You know, we should also stop doing condoms in my country!" (On such occasions Halperin would then need to defend the importance of condom distribution in Africa.)

3. Justin Gillis, "Bill Gates' Hands-on Charity," *The Washington Post*, October 1, 2003.

4. Eventually the Gates Foundation also helped fund male circumcision services in Africa, www .malecircumcision.org.

5. Ian Wilhelm, "Gates Program in Botswana Offers Lessons on Fighting AIDS," *The Chronicle of Philanthropy*, November 11, 2004.

6. Huntly Collins, "Botswana to Benefit from Public-Private Effort to Fight AIDS," *Philadelphia Inquirer*, July 10, 2000.

7. Claire Keeton, "Botswana; A Nation Healing Itself: Botswana Goes to War on HIV/Aids," *Sunday Times* (Johannesburg), April 4, 2004.

8. Condom use rates in Botswana are mentioned in chapter 1, endnote 5. President Mogae, at the end of his tenure and since leaving office, has talked much more openly about the problem of sexual culture, including concurrency.

9. Keeton, "Botswana, a Nation Healing Itself."

10. Epstein, *The Invisible Cure*. The head of WHO's Swaziland office while Halperin was based there was a Ugandan physician. Once while they were meeting in his office, the man pointed out his window toward the main mall in Mbabane, to a very large and sexually suggestive condom ad and said: "We turned HIV around with Zero Grazing. Here they are trying to do so with those billboards. It won't work."

11. The nursing school is especially important in a nation that has no medical school of its own. All doctors must get trained in other countries.

12. Interview with Timberg. In a presentation Selelo-Mogwe made in 2002 at the International AIDS Conference in Barcelona, she recounted with disdain how donor-funded HIV prevention groups would go out to remote villages and—finding "only some grandmas sitting around" available for a condom demonstration—would proceed to instruct them on how to place the condoms on a large wooden model of a penis. For a thought-provoking contrast between the approaches used in Botswana and Uganda, see Timothy Allen and Suzanne Heald, "HIV/AIDS Policy in Africa: What Has Worked in Uganda and What Has Failed in Botswana?" *Journal of International Development* 16 (2004): 1141–54; and Ann Swidler, "Responding to AIDS in Sub-Saharan Africa," in Peter Hall and Michèle Lamont, eds., *Successful Societies: Institutions, Cultural Repertoires and Population Health* (New York: Cambridge University Press, 2009).

13. This effort in Botswana actually relied on South African medical students who were brought in on buses and used a medically safe circumcision technique. The idea to revive traditional initiation rites and combine them with a modern clinical procedure was developed by the chief of the Mochudi area (chapter 1, endnote 9), as part of some "re-Africanization" efforts begun by him and some other traditional leaders in the 1970s. At their peak, in the early 1980s, the Mochudi rituals—which involved staying out in the bush for a month or more—attracted several thousand men each year.

14. Weiss et al., "Male Circumcision and Risk of HIV Infection in Sub-Saharan Africa." Two other important research findings were announced in 2000 and also influenced the decision to begin funding the randomized trials in Africa. One was a Johns Hopkins University–led study in Rakai, Uganda, that followed a number of couples in which only one partner was infected initially over a thirty-month period (Thomas C. Quinn, Maria J. Wawer, Nelson Sewankambo, et al., "Viral Load and Heterosexual Transmission of Human Immunodeficiency Virus Type 1." *N Engl J Med* 342 [2000]: 921–29). The most widely publicized result of the study was that people with higher viral loads were, as expected, more likely to transmit HIV to their partners. But the study also produced an unexpectedly strong finding. Among the couples in which the woman was infected at the beginning of the study, there were 137 couples in which the men were uncircumcised and 50 couples containing circumcised men.

Among the group of uncircumcised men, forty became infected over the duration of the study, whereas none of the circumcised men got the virus. The other crucial study, funded by some European donors, involved analyzing an extensive amount of data collected from four African cities, two with high HIV rates and two with much lower rates, to tease out what might be causing the different levels of infection. Out of more than one hundred biological, behavioral, socioeconomic and other potential explanatory factors that were examined, the two main ones which emerged were male circumcision and genital ulcers caused by certain sexually transmitted diseases such as genital herpes. Bertran Auvert, Ann Buvé, Emmanuel Lagarde, et al., "Male Circumcision and HIV Infection in Four Cities in Sub-Saharan Africa." *AIDS* 15, suppl. 4 (2001): S31–40. In a subsequent, more rigorous analysis conducted by the chief statistician on the study, Bertran Auvert, and a colleague, Brian Ferry, the role of herpes no longer appeared to be as substantial. Instead, the two factors overwhelmingly associated with variations in HIV rates across the four African cities they studied were male circumcision and people's lifetime number of sexual partners (Auvert and Ferry, Presentation at the International AIDS Conference, Barcelona, 2002; USAID, "The 'ABCs' of HIV Prevention").

15. Some Zulu men, including South Africa's President Jacob Zuma, were circumcised, usually for medical reasons. Zuma has not publicly discussed why he was circumcised.

16. The principal investigator of the circumcision trial in Uganda, Ronald Gray, has said that so many men wanted to be circumcised that some who were randomized to the non-circumcision arm of the study came back, trying to enroll under a different name (personal communication with Halperin).

17. See Malcolm Potts, "Circumcision and HIV." *Lancet* 355 (2000): 926–27.

18. Bertran Auvert, Dirk Taljaard, Emmanuel Lagarde, Joëlle Sobngwi-Tambekou, Rémi Sitta, and Adrian Puren, "Randomized, Controlled Intervention Trial of Male Circumcision for Reduction of HIV Infection Risk: The ANRS 1265 Trial." *PLoS Med* 2 (2005): 1112–22. (And see Jon Cohen, "Male Circumcision Thwarts HIV infection." *Science* 309 [2005]: 860.)

19. In the three circumcision trials, some men who were assigned by the researchers to the control (uncircumcised) groups decided to get circumcised on their own, outside the confines of the studies. This had the effect of weakening the apparent impact of the intervention, because those men still received the lower risk conferred by the procedure. Also, a small number of men turned out to have active but initially undetected HIV infections when the trials began. When the Orange Farm researchers took such additional factors into account, the official estimate of 60 percent protection grew to 76 percent (see James Shelton, "Estimated Protection Too Conservative," *Plos Med* 3 [2006]: e65). Similarly more comprehensive analyses of the Ugandan and Kenyan trials indicated the protective effect to be about 55 percent and 60 percent, respectively. (Ronald H. Gray, Godfrey Kigozi, David Serwadda, et al., "Male Circumcision for HIV Prevention in Men in Rakai, Uganda: A Randomised Trial." *Lancet* 369 [2007]: 657–66; Robert C. Bailey, Stephen Moses, Corette B. Parker, et al., "Male Circumcision for HIV Prevention in Young Men in Kisumu, Kenya: A Randomised Controlled Trial." *Lancet* 369 [2007]: 643–56; Donald G. McNeil, "Circumcision's Anti-AIDS Effect Found Greater Than First Thought," *The New York Times,* February 23, 2007). Moreover, the study participants included men at low risk of infection, including some who reported no sexual activity during the trials. This would have further underestimated the

actual biological effect of circumcision because whether such men got circumcised or not presumably would not have affected their risk of HIV infection. In the Uganda trial, for example, when the analysis was restricted to men more likely to be exposed to HIV, such as those reporting multiple or nonmarried partners, the protective effect was about 70 percent, even in the more conservative type of analysis (Gray et al, p. 662). This was consistent with the findings of the most rigorous observational studies—those that followed groups of circumcised and uncircumcised men over time, such as the Pumwani study in Nairobi—which usually have found an effect of 75 percent or greater. (Weiss et al., "Male Circumcision and Risk of HIV Infection in Sub-Saharan Africa"; Halperin and Bailey, "Male Circumcision and HIV Infection"; Bailey et al., "Male Circumcision and HIV Prevention"; Quinn, et al., "Viral Load and Heterosexual Transmission of Human Immunodeficiency Virus Type 1"; Reynolds et al., "Male Circumcision and Risk of HIV-1 and Other Sexually Transmitted Infections in India." This level of protective effect was later confirmed by the long-term follow-up data from the three trial sites, which was announced in 2011 to be in the roughly 70 percent range overall. Bertran Auvert, Dirk Taljaard, Dino Rech, et al., "Effect of the Roll-Out of Male Circumcision in Orange Farm (South Africa) on the Spread of HIV (ANRS-12126)." Presented at the International AIDS Society Meetings, Rome, 2011, Abstract #WELBC02; Xiangrang Kong, Godfrey Kigozi, Victor Ssempija, et al., "Longer-Term Effects of Male Circumcision on HIV Incidence and Risk Behaviors During Post-Trial Surveillance in Rakai, Uganda." Presented at the 18th Conference on Retroviruses and Opportunistic Infections, Boston, Feb 27–Mar 2, 2011. Abstract 36, www.hivandhepatitis.com/2011conference/croi2011/docs/0311_2010c.html; Robert C. Bailey, Stephen Moses, et al., "The Protective Effect of Adult Male Circumcision Against HIV Acquisition Is Sustained for at Least 54 Months: Results from the Kisumu, Kenya Trial." Presented at the International AIDS Conference, Vienna, 2010, Abstract FRLBC101.

20. PEPFAR also issued a brief, cautious statement at the International AIDS Society meeting in Rio de Janeiro. Soon afterward PEPFAR announced that while it could fund limited preliminary research on the acceptability and other pre-implementation aspects of circumcision, it would not—while awaiting guidance from "normative bodies" such as the WHO—directly pay for providing the service.

21. It is ethical to test a potentially valuable intervention only if there is legitimate uncertainty about whether it will provide health benefits. Researchers could not, for example, withhold access to a proven vaccine from someone in danger of contracting a potentially fatal disease it likely would prevent. The same principle now applies to circumcision. Its ability to significantly lessen the risk of contracting HIV is sufficiently clear that it would be unethical to withhold it from men selected to participate in a medical trial.

22. In this assertion from an interview with Timberg (and also stated by Bailey in a speech at the Intertional AIDS Society Meeting in Sydney, Australia, in 2007), he was referring not only to the 1999 *Lancet* article, but also to the Weiss et al. meta-analysis review and other important research findings announced in 2000 (chapter 24, endnote 14).

23. Michael Grunwald, "All-out Effort Fails to Halt AIDS Spread," *The Washington Post*, December 2, 2002.

24. For more information on behavior change in countries where HIV has declined see: Potts et al., "Reassessing HIV Prevention"; Shelton et al., "Partner Reduction Is Crucial for Balanced

'ABC' Approach to HIV Prevention"; Bessinger et al., "Sexual Behavior, HIV and Fertility Trends"; Halperin and Epstein, "Why Is HIV Prevalence So Severe"; Hallett et al., "Declines in HIV Prevalence"; Yared Mekonnen, Eduard Sanders, Mathias Aklilu, et al., "Evidence of Changes in Sexual Behaviours Among Male Factory Workers in Ethiopia." *AIDS* 17 (2003): 223–31; George Bello, Bertha Simwaka, Tchaka Ndhlovu, Felix Salaniponi, and Timothy B. Hallett, "Evidence for Changes in Behaviour Leading to Reductions in HIV Prevalence in Urban Malawi." *Sex Transm Infect* 2011, doi:10/1136/sti.2010.043786; Southern African Development Community, *Expert Think Tank Meeting on HIV Prevention in High-Prevalence Countries in Southern Africa: Report*, May 10–12, 2006, www.sadc.int/downloads/news/SADCPrevReport.pdf137. In Kenya, for example, men of all ages reported large reductions in multiple sexual partnerships between 1993 and 2003, which corresponds roughly to the main period of declining HIV infections in the country (see chart on page 312 of appendix, and Shelton, "Confessions of a Condom Lover").

25. Though HIV rates eventually did decline modestly in Botswana, at least among youth, the sizable number of infections in some recent prevention trials conducted there suggests the pace of new infections has still not slowed greatly.

Chapter 25: What Shall We Do?

1. Scott Kraft, "Africa's Death Sentence," *Los Angeles Times*, March 1, 1992.
2. "Anti-AIDS Campaign Launched in Zimbabwe," Xinhua News Agency, November 24, 1997. President Mugabe spoke about AIDS more than other African leaders, most of whom rarely uttered the word until recent years, but he did not approach the focus or openness displayed by Museveni. And sometimes years passed during which Mugabe barely mentioned the disease. In addition, Mugabe's conversation on the subject often included hateful comments about gay people. (Museveni and some other politicians in Uganda have also made homophobic statements over the years.) Another country whose leader publicly raised the issue of AIDS was Zambia (where HIV probably declined significantly in the 1990s, at least in urban areas; see Bessinger et al., "Sexual Behavior, HIV and Fertility Trends"). In the late 1980s Kenneth Kaunda's own son died from the affliction, and in recent years the former president has once again been speaking out about HIV in the region.
3. Interview with Timberg. See Wilson, "Partner Reduction and the Prevention of HIV/AIDS."
4. "'Kill AIDS Sufferers,' Says Zimbabwe MP," Associated Press, March 8, 1994.
5. "Official Breaks with Tradition in Announcing AIDS Death, Blames West," Associated Press, April 6, 1996.
6. Interview with Timberg. The video for "Todii" was produced with funding from the United Nations (UNFPA).
7. Craig Timberg, "Number of People with HIV Doubled in Past Decade, U.N. Finds," *The Washington Post*, November 22, 2005. Even in this era, officials from UNAIDS and other UN organizations continued to warn that HIV was still a rising threat in much of the world and might soon reach dangerous new levels in regions where rates had until then remained modest. In a talk Piot gave on November 20, 2004, at the Woodrow Wilson Center, he said, "The situation we face in China, India, and Russia bears alarming similarities to the situation we faced twenty years ago in Africa. The virus in these populous countries is perilously close to a tipping point. If it reaches that point, it could transition from a series of concentrated outbreaks

and hot spots into a generalized explosion across the entire population—spreading like a wildfire from there . . . When the very act essential to furthering the human race also threatens it, we are in a very precarious place" (www.wilsoncenter.org/article/top-leaders-set-the-next -agenda-for-global-aids). Many other prominent experts—as well as politicians, Hollywood actors, rock stars, and other influential people—were similarly warning of the likelihood of an explosive global epidemic.

8. Halperin et al., "A Surprising Prevention Success." Other kinds of behavior change were also occurring in Zimbabwe around that time. The number of men who reported paying for sex, for example, declined by almost 50 percent between 1999 and 2005. And in the Manicaland study, the number of men reporting a casual partner in the past month fell by half, during roughly the same years. Simon Gregson, Geoffrey P. Garnett, Constance A. Nyamukapa, et al., "HIV Decline Associated with Behavior Change in Eastern Zimbabwe." *Science* 311 (2006): 664–66.

9. In focus groups and key informant interviews conducted with several hundred people throughout the country, various ways in which sexual behavior had changed were frequently mentioned. For example, years ago men who got a sexually transmitted disease were often seen as "heroes" by their comrades (as in Uganda previously; chapter 14). But everyone who was interviewed agreed that, by around the late 1990s, this was no longer the case, and that if someone had such an infection it would have become deeply embarrassing for this to be discovered. Backson Muchini, Clemens Benedikt, Exnevia Gomo, et al., "Local Perceptions of the Forms, Timing and Causes of Behavior Change in Response to the AIDS Epidemic in Zimbabwe." *AIDS Behav* 2010, doi 10.1007/s10461-010-9783-z.

10. Elements of this chapter appeared in a story Timberg wrote for *The Washington Post* on July 13, 2007, headlined "In Zimbabwe, Fewer Affairs and Less HIV."

11. The authors have a rare area of disagreement over whether economic deterioration or other kinds of social changes were more important in causing HIV to decline in Zimbabwe. Timberg believes that the political repression, financial chaos, and the related exodus of young, middle-class Zimbabweans was crucial in the drop in new HIV infections experienced during this time. Halperin's research suggests that the intense experience with AIDS deaths in a society with high rates of education and social cohesion was ultimately decisive. However, we agree that both sets of factors played important roles in the historic reversal of this epidemic.

12. Halperin et al., "A Surprising Prevention Success."

13. There were small amounts of ARVs available through private doctors and, eventually, through programs offered by organizations such as Doctors Without Borders. But Zimbabwe for many years was an uncommonly difficult place to procure these medicines.

14. If Botswana's diamonds had been discovered in the 1860s rather than the 1960s, it surely would have been claimed by either the British Cape Colony in South Africa or Cecil Rhodes, who was then developing what later became Zimbabwe and Zambia.

15. Brazil's HIV epidemic is driven mainly by men having sex with other men, though there are also substantial contributions from sex work and injecting drug users in some areas. An increasingly significant proportion of new infections also occur in the female sex partners of bisexual males, drug users, and other high-risk men. An analysis of the role of anal sex in the Brazilian epidemic suggests that half or more of all sexually transmitted HIV infections

among women may result from this practice, not from vaginal sex (Halperin et al., "High Level of HIV-1 Infection"; Daniel T. Halperin, Stephen C. Shiboski, Joel M. Palefsky, and Nancy S. Padian, "High Risk of HIV-1 Infection from Anal Intercourse: A Substantial, Overlooked Co-factor in Heterosexual HIV Transmission," paper in review).

16. In some regions of Brazil, as many as 80 percent of heterosexuals report practicing anal intercourse at least occasionally. Yet the topic is largely taboo for discussion, including in most HIV prevention messages aimed at the wider population. Daniel T. Halperin, "Heterosexual Anal Intercourse: Prevalence, Cultural Factors, and HIV Infection and Other Health Risks." *AIDS Patient Care* 13 (1999): 717–30; Daniel Halperin, "HIV, STDs, Anal Sex and AIDS Prevention Policy in a Northeastern Brazilian City." *Int J STD AIDS* 9 (1998): 294–98.

17. Daniel T. Halperin, Markus Steiner, Michael Cassell, et al., "The Time Has Come for Common Ground on Preventing Sexual Transmission of HIV." *Lancet* 364 (2004): 1913–15.

18. When James Dobson, whose Focus on the Family group thought some PEPFAR policies had strayed too far from the conservative line, asked Peterson point-blank, "Where do you stand on condoms?" she replied that they had a role to play, along with programs promoting abstinence and faithfulness. Her answer was not good enough. Focus on the Family and some other very conservative groups singled her out for attack. Peterson resigned in frustration a few months later. Michael Kranish, "Religious Right Wields Clout," *Boston Globe*, October 9, 2006.

19. Hill, in a later conversation with Halperin, expressed how he and several other Bush appointees felt at the time that they were often caught between some conservatives who opposed condoms and some liberals who felt pushing abstinence and be faithful messages was morally presumptuous. Some people criticized the administration, for example, for reaching out to Muslims, traditional healers, and other non-Christian groups. Hill said that although he could not remember asking Halperin about whether the potential candidate went to church, he supposes that the comment resulted from his desire to learn whether the candidate was open to working with Christian and other religious groups. He added, "If I asked the question that way, it was stupid, and it was wrong of me." Ken Yamashita (who in 2011 was the USAID mission director in Afghanistan) ended up being chosen by Hill for the HIV-AIDS Office position. Yamashita said, in a conversation with Halperin, that he is not a Christian, let alone a church-goer, suggesting that Hill "didn't have any kind of litmus test." He added that "Hill always tried to include a big tent—to include people from many different perspectives on issues, including HIV prevention. In the two years I worked with him, he never raised religious views as any kind of filter."

Chapter 26: Raymond the Great

1. As word started getting out that male circumcision was available at the pilot project in Swaziland, more and more men were coming in for the service. However, in October 2006 PEPFAR decided against renewing the grant underwriting the program. PEPFAR had recently decreed that it could not support any research into circumcision that involved providing actual services (chapter 24, endnote 20), and officials decided that renewing the Swazi grant would have violated the new rule. The Swazis, who had built the program from scratch using a modest USAID investment, were frustrated by the decision. Dudu Simelane, deputy executive director for the Family Life Association of Swaziland, which had been performing

the surgeries, told Timberg for an article he wrote at the time, "It's best we try by all means to continue, but funding is the determinant. We wouldn't like to stop, really." ("U.S. to End Funding of Anti-AIDS Program in Swaziland," *The Washington Post*, October 13, 2006.)

2. Myron S. Cohen, George M. Shaw, Andrew J. McMichael, and Barton F. Haynes, "Acute HIV-1 Infection." *N Engl J Med* 364 (2011): 1943–54; Steven D. Pinkerton, "Probability of HIV Transmission During Acute Infection in Rakai, Uganda." *AIDS Behav* 12 (2008): 677–84; T. Déirdre Hollingsworth, Roy M. Anderson, and Cristoff Fraser, "HIV-1 Transmission, by Stage of Infection." *J Infect Dis* 198 (2008): 687–93. Helen Epstein ("Why Is AIDS Worse in Africa?") was one of the first writers to emphasize the importance of how concurrency and acute infection reinforce one another; this has been subsequently explored by several modelers and other researchers, such as Goodreau et al., "Concurrent Partnerships, Acute Infection and HIV Epidemic Dynamics Among Young Adults in Zimbabwe" and Eaton et al., "Concurrent Sexual Partnerships and Primary HIV Infection." Jim Shelton, who has been scrutinizing the African epidemic for many years, aptly characterizes acute infection as "the underlying driver, like gasoline poured on the fire of the epidemic" ("A Tale of Two-Component Generalised HIV Epidemics").

3. The DHS-type surveys may underestimate the actual prevalence level if, for example, many higher-risk individuals are not at home when surveyors are conducting their interviews. However, statistical analyses of this possibility in countries including Kenya, the Dominican Republic, and India suggest that only a relatively minor underestimate is likely to occur (Demographic and Health Survey Reports, Calverton, MD: IFC MACRO International; Dandona et al., "A Population-based Study of HIV in South India").

4. Chin, *The AIDS Pandemic*; Shelton et al., "Has Global HIV Incidence Peaked?"

5. Asiimwe-Okiror et al., "Change in Sexual Behaviour and Decline in HIV Infection." Stoneburner kept working on the Uganda data while taking a series of temporary jobs and consulting contracts, but for years he struggled to get his analysis published in a major journal. His best hope appeared to be in 1997, when the editor of *AIDS* expressed interest in his research, according to Stoneburner. But as that article was under review, he was surprised to see the same journal publish a similar piece of research—coauthored by Caraël (by Asiimwe-Okiror et al.; see chapter 14). In it, Caraël wrote that casual sex among Ugandan men had dropped only modestly, from 22.6 percent to 18.1 percent during the crucial years when HIV's spread declined. Casual sex among young women, meanwhile, had gone in the wrong direction, rising from 6 percent to 8 percent. The average number of sex partners had stayed steady at 1.8 per urban Ugandan. The data appeared much more impressive for changes in condom use, for which they used the same number Caraël had cited at the Kampala conference, showing that the percentage of men who had ever used condoms increased from 15 to 55 percent. Overall, Uganda's reversal in HIV, Caraël and his coauthors wrote, resulted from a "slight decrease" in casual sex and a "sharp increase" in condom use. The article listed several Ugandan coauthors but did not mention Stoneburner.

6. In late 2002 Halperin was working with a consultant from the University of North Carolina, Ruth Bessinger, on the USAID-funded "ABC Study" (chapter 21, endnote 17). To be able to conduct the analysis she needed access to the 1989 data set on Ugandan sexual behavior. After she tried unsuccessfully to locate it for some time, Halperin sought assistance from Paul Delay, then head of the HIV-AIDS Office at USAID and who had close ties with some of-

ficials at UNAIDS. Delay eventually succeeded in acquiring the documents, which were shipped in both paper and electronic form. Halperin later shared them with Stoneburner, who was working as a consultant to the study.

7. Stoneburner and Low-Beer, "Population-Level HIV Declines and Behavioral Risk Avoidance in Uganda." Relatively similar conclusions about what had led to the HIV decline in Uganda appeared in several other articles the same year, including by Green et al., "Uganda's HIV Prevention Success," Epstein, "Why Is AIDS Worse in Africa?" and Slutkin et al., "How Uganda Reversed Its HIV Epidemic." The latter paper was coauthored by several prominent experts, including Michel Carael and Paul Delay, both of whom worked for UNAIDS.

8. Rand Stoneburner and Daniel Low-Beer, "Analyses of HIV Trend and Behavioral Data in Uganda, Kenya, and Zambia: Prevalence Declines in Uganda Relate More to Reductions in Sex Partners than Condom Use." Presentation at XIII International AIDS Conference, Durban, South Africa, July 7–14, 2000. Abstract #ThOrC735.

9. Based mainly on statistics from pregnant women in urban areas, the earlier methodologies for estimating HIV rates were less reliable for measuring actual prevalence levels, but they were useful for tracking changes in epidemiological dynamics over time. And household-based surveys, while having major limitations for measuring the true extent of risky sexual behavior in a community (see chapter 30), are helpful for documenting historical trends in behavior. In fact, most of the data upon which the prevention successes in places such as Uganda and Zimbabwe are based comes from this type of behavioral survey and earlier HIV monitoring evidence.

10. For references on "transactional sex" see appendix and chapter 16, endnote 19.

11. Halperin et al., "Understanding the HIV Epidemic in the Dominican Republic."

12. Chin makes a case in his book *The AIDS Pandemic* that the opposite was true (also see Halperin and Post, "Global HIV Prevalence").

13. The 2007 Demographic and Health Survey in Liberia found a rate of 1.5 percent, compared to the previous UNAIDS estimate of 6.5 percent. In Burkina Faso the official rate had been 6.5 percent, but the DHS estimated it at 1.8 percent. Such population-based surveys have been conducted in several other West African countries, for example in Ghana and Sierra Leone. In each case they have found considerably lower HIV rates than the previous UNAIDS and country estimates (see Halperin and Post, "Global HIV Prevalence").

14. In an interview with Timberg, Piot vigorously disputed the suggestion that he had allowed political considerations to affect his scientific judgments. He also said that Chin's work underestimated HIV rates and that the statistics improved after the establishment of UNAIDS. When it became clear that some numbers were too high, Piot said, the agency moved swiftly to make adjustments.

15. Interview with Timberg.

16. Much of this account appeared in *The Washington Post* in an article by Timberg on March 29, 2007, headlined "Uganda's Early Gains Against HIV Eroding."

17. Interview with Timberg.

18. Following Museveni's speech in Bangkok, some prominent experts wrote newspaper editorials criticizing his comments on the role of condoms in Uganda's HIV decline.

19. In 1995, 10 percent of men and 1 percent of women reported having multiple sex partners in Uganda. In the 2000/2001 survey, this had risen to 24 percent and 2 percent, respectively,

and to 28 percent and 4 percent in the 2004/2005 survey. (The recall period for the 1995 survey was for the previous six months; for the other surveys it was for the previous twelve months. As a result, the increases beginning in 1995 may not have been quite as large as these comparisons suggest.) Sources: Uganda Demographic and Health Surveys, Uganda AIDS Information Survey, Kampala/Calverton, MD: ORC MACRO International.

20. When Halperin and Doug Kirby conducted research in Uganda in 2003 about what had happened during the Zero Grazing period (chapter 13, endnote 12), they interviewed a security guard at the same Makerere parking lot, who reported that—during those (few) key years in the late 1980s—the lot was "almost always empty."

Chapter 27: *Makhwapheni Uyabulala*

1. Suzanne J. Maman, Jessie K. Mbwambo, Nora M. Hogan, Ellen Weiss, Gad P. Kilonzo, and Michael D. Sweat, "High Rates and Positive Outcomes of HIV-Serostatus Disclosure to Sexual Partners: Reasons for Cautious Optimism from a Voluntary Counseling and Testing Clinic in Dar es Salaam, Tanzania." *AIDS Behav* 7 (2003): 373–82.

2. The rules of PEPFAR, for example, prohibited direct salary payments to Africans working in public sector jobs but allowed fixed per-diem expenses, such as for meals and travel to conferences, to be paid in cash. One outcome was a surge in training workshops, typically involving pep talks interspersed with exercises intended to stimulate conversation. At the end of these meetings participants walked away with envelopes full of bills. Meanwhile, more urgent work back in offices and clinics often went undone. (See Ann Swidler and Susan C. Watkins, "'Teach a Man to Fish': The Sustainability Doctrine and its Social Consequences." *World Dev* [2009] 37: 1182–96.)

3. See Potts et al., "Reassessing HIV Prevention."

4. In September 2002, shortly after Hearst was commissioned by UNAIDS to conduct a review of condom effectiveness, he attended a day-long USAID meeting on behavior change (USAID, "The 'ABCs' of HIV Prevention: Report of a USAID Technical Meeting on Behavior Change Approaches to Primary Prevention of HIV/AIDS"). On the flight back to San Francisco he had an epiphany that, as he later told Halperin, "changed my thinking about HIV prevention forever," and prompted him to rethink a trial on condom promotion in Uganda that his research group had recently completed. Upon exiting the plane he immediately called his biostatistician, Sammy Chen, and ask her to rerun the data. They soon learned that their seemingly successful condom promotion study appeared to have caused more harm than good. People in the intensive condom promotion arm did report higher condom use, but also a larger proportion of multiple sex partners, which subsequent analysis suggested had probably increased their overall risk of infection. Phoebe Kajubi, Moses R. Kamya, Sarah Kamya, et al., "Increasing Condom Use Without Reducing HIV Risk: Results of a Controlled Community Trial in Uganda." *J Acq Immun Def Syn* 40 (2005): 77–82. And see John Richens, John Imrie, and Andrew Copas, "Condoms and Seat Belts: The Parallels and the Lessons." *Lancet* 355 (2000): 400–3; Michael M. Cassell, Daniel T. Halperin, James D. Shelton, and David Stanton, "Risk Compensation: The Achilles' Heel of Innovations in HIV Prevention?" *BMJ* 332 (2006): 605–7.

5. One experiment in Tanzania in the early 1990s to test such treatment efforts reported a modest effect. But in eight other subsequent trials, none have shown any measurable impact

on HIV transmission from treating these infections (see appendix; Ronald H. Gray and Maria J. Wawer, "Reassessing the Hypothesis on STI Control for HIV Prevention," *Lancet* 371 [2008]: 2064–65; Potts et al., "Reassessing HIV Prevention"; Shelton, "Ten Myths."). There are very good public health reasons to treat syphilis, chlamydia, and herpes—such as preventing congenital diseases in newborn babies—but increasingly it has become clear that preventing HIV is perhaps not one of them. At a conference on the issue held in Geneva in July 2006, attended by more than one hundred of the world's leading scientists in this area, Halperin found himself badly outnumbered by supporters of treating other sexually transmitted infections as an HIV prevention tool. During a coffee break in the meeting, Kevin de Cock, then head of the World Health Organization's HIV-AIDS Division, commiserated with Halperin. "You have to understand," de Cock said in an aside, "these views are not evidence-based. They are faith-based."

The Swaziland Ministry of Health's Sexually Transmitted Diseases department examined its syndromic management program against sexually transmitted infections in 2006. About four out of every five cases turned out to be for vaginal discharge, which generally is caused not by sexually transmitted infections but by other conditions such as yeast infections. The routine use of antibiotics in these cases offered no relief to the women, and actually could have harmed them by increasing the likelihood of developing resistant strains of bacteria. In addition, the improper diagnosis of a sexually transmitted infection can impart an unwarranted stigma (see appendix, endnote 27).

6. See for example Elizabeth L. Corbett, Beauty Makamure, Yin B. Cheung, et al., "HIV Incidence During a Cluster-Randomized Trial of Two Strategies Providing Voluntary Counselling and Testing at the Workplace, Zimbabwe." *AIDS* 21 (2008): 483–89; Daniel T. Halperin, "AIDS Prevention: What Works?" *The Washington Post*, October 22, 2007, p. A23; Cassell et al., "Risk Compensation"; and appendix.

7. See chapter 26, endnote 2; and Kimberly A. Powers, Azra C. Ghani, William C. Miller, et al., "The Role of Acute and Early HIV Infection in the Spread of HIV and Implications for Transmission Prevention Strategies in Lilongwe, Malawi: A Modelling Study." *Lancet* 378 (2011): 256–68.

8. Male circumcision is not a major HIV cofactor in Europe or other areas where heterosexual transmission accounts for only a minority of overall infections. But it is notable that the proportion of heterosexual HIV infections occurring in men is much greater there than in North America, where instances of men acquiring the virus sexually from women typically are rare. In a discordant couple study in California (Padian et al., "Heterosexual Transmission of HIV in Northern California"), the female partners were about eight times more likely to contract the virus from an infected male partner than vice versa, while in several European discordant couple studies women were only about twice as likely to become infected as were men. This difference likely results from a greater vulnerability to heterosexual HIV infection among uncircumcised European men. (See Robert C. Bailey and Daniel T. Halperin, "Circumcision and HIV." *Lancet* 355 (2000): 927; Quinn et al., "Viral Load and Heterosexual Transmission of Human Immunodeficiency Virus Type 1" [response letter].)

9. The Swazis who developed the Makhwapheni campaign were eager to employ a hard-hitting message that would break through the denial and complacency surrounding HIV and the behaviors that were spreading it. But with ARVs increasingly available, a message such as secret

lovers "will kill you" no longer seemed reasonable or accurate. After some discussion a compromise message—saying that secret lovers "can" kill you—was agreed upon.

10. Much of this account appeared in an article by Timberg in *The Washington Post* on October 29, 2006, headlined "In Swaziland, 'Secret Lovers' Confronted in Fight Against AIDS."

11. Aldo Spina, "Secret Lovers Kill: A National Mass Media Campaign to Address Multiple and Concurrent Partnerships." AIDSTAR-One: Case Studies Series 2009; www.comminit.com/en/node/305575/cchangepicks/.

Chapter 28: The Flood

1. Much of this account appeared in an article written by Timberg in *The Washington Post* on July 23, 2007, headlined "In Botswana, Step to Cut AIDS Proves a Formula for Disaster."

2. In the absence of ARVs taken by the mother or infant or other medical interventions, between 5 to 10 percent of babies born to mothers with HIV will end up infected in utero, and another 10 to 15 percent will become infected at delivery. As noted the risk from breast-feeding can approach 1 percent a month.

3. The rough estimate of nearly 1 percent risk of infection per month is based mainly on studies of babies whose mothers practiced mixed feeding. With those who practice exclusive breast-feeding, the risk is substantially lower. (See appendix and, for example: Peter J. Iliff, Ellen G. Piwoz, and Naume V. Tavengwa. "Early Exclusive Breastfeeding Reduces the Risk of Postnatal HIV-1 Transmission and Increases HIV-Free Survival." *AIDS* 19 [2005]: 699–708.)

4. At a 2002 meeting at USAID headquarters in Washington, D.C., discussion grew heated about how best to feed babies whose mothers have HIV. Nearly all the HIV experts in attendance became upset when Miriam Labbok, a breast-feeding specialist in the Health and Nutrition Office, and Halperin, who had previously conducted research on infant nutrition in Central America, suggested that HIV-positive mothers might be able to keep nursing their children. Their argument was inspired by a recent intriguing study from South Africa: Anne Coutsoudis, Kubendran Pillay, Elizabeth Spooner, Louise Kuhn, and Hoosen M. Coovadia, "Influence of Infant-Feeding Patterns on Early Mother-to-Child Transmission of HIV-1 in Durban, South Africa: A Prospective Cohort Study." *Lancet* (1999) 354: 471–76 (and see Daniel T. Halperin, "Breastfeeding and HIV Transmission: The Realities May Be Much More Complex Than 'Breast vs. Bottle.'" *Global AIDSLink* [1999] 58: 8–10.)

5. Arjan de Wagt and David Clark, "A Review of UNICEF's Experience with the Distribution of Free Infant Formula for Infants of HIV-Infected Mothers in Africa." Presentation for the LINKAGES Art and Science of Breastfeeding Series, April 14, 2004, Academy for Educational Development, Washington, D.C.

6. Ibou Thior, Shahin Lockman, Laura M. Smeaton, et al., "Breastfeeding Plus Infant Zidovudine Prophylaxis for 6 Months vs Formula Feeding Plus Infant Zidovudine for 1 Month to reduce Mother-to-Child HIV Transmission in Botswana: A Randomized Trial: The Mashi Study." *JAMA* 296 (2006): 794–805.

7. Iliff et al, "Early Exclusive Breastfeeding"; Thior et al., "Breastfeeding Plus Infant Zidovudine Prophylaxis"; Coutsoudis et al., "Influence of Infant-Feeding Patterns; Halperin, "Breastfeeding and HIV Transmission."

8. A reliable HIV test generally was not possible until babies reached their first birthdays.

Chapter 29: Mother and Son

1. We have chosen to not disclose the last names of Yvonne, Sarah, and Sifiso, all of whom knew Timberg personally, in order to protect their privacy.

2. While conducting qualitative research in southern Africa since the 1990s, Halperin heard similar notions on many occasions. In a town in Botswana, for example, a group of young men explained to him, "If a girl accepts a beer or two, she's yours for the evening" (Halperin 2000, "Old Ways and New Spread AIDS in Africa").

Chapter 30: What Shall We Do? Part II

1. http://go.worldbank.org/OPC5GJ5RD0.

2. Larry Sawers and Eileen Stillwaggon, "Concurrent Sexual Partnerships Do Not Explain the HIV Epidemics in Africa: A Systematic Review of the Evidence." *J Int AIDS Soc* 13 (2010): 1–24. This paper criticized much of the research on concurrency for supposed methodological flaws or inaccuracies. Some of the specific critiques have some validity. For example, Sawers and Stillwaggon rejected the merits of a study cited in a review paper by Halperin and his former Harvard colleague Tim Mah ("Concurrent Sexual Partnerships and the HIV Epidemics in Africa") because they had reported that concurrency among married men in Bangladesh was 5 percent when the original article said it was 6 percent. In fact, the cited article had stated 5 percent in the text but an accompanying chart said 6 percent, and Halperin and Mah failed to note this discrepancy. Such flaws are regrettable. But identifying such minor errors offerred little useful insight into the totality of the evidence, which strongly suggests that concurrency plays a crucial role in HIV's spread in the severely affected parts of Africa. After dismissing the role of sexual behavior in these epidemics, Sawers and Stillwaggon offered an alternative explanation, arguing that other conditions prominent in Africa because of its poverty—such as malaria, malnutrition, intestinal worms, and unsafe medical care—were the real factors driving the spread of the virus. However, this view overlooks that within Africa HIV is now worst in the relatively developed countries of southern Africa, where health care standards are higher than elsewhere on the continent and such maladies are comparatively rare. Conversely, these other diseases are much more common in places such as West Africa with comparatively little HIV. See Mah and Shelton, "Concurrency Revisited" and Epstein and Morris, "Concurrent Partnerships and HIV." The issue was also debated as part of the series "Emerging Issues in Today's HIV Response: A Debate Co-hosted by the World Bank and USAID." Washington, D.C., http://go.worldbank.org/OPC5GJ5RD0.

3. Some researchers have drawn upon international comparisons of standard survey data to conclude there is insufficient regional variation in reported sexual behavior to explain the much higher HIV rates in sub-Saharan Africa. An often-cited 2006 paper pointed to such data as indicating that Africans, on average, do not report having higher numbers of sexual partners than people elsewhere. (Kaye Wellings, Martine Collumbien, Emma Slaymaker, et al., "Sexual Behaviour in Context: A Global Perspective." *Lancet* 368 [2006]: 1706–28. Similarly: Audrey E. Pettifor, Brooke A. Levandowski, Catherine Macphail, et al., "A Tale of Two Countries: Rethinking Sexual Risk for HIV among Young People in South Africa and the United States," *Journal of Adolescent Health* 49 [2011]: 237–43.) The authors of that *Lancet* paper concluded that factors such as poverty should be investigated instead of sexual behavior. But they also acknowledged they had examined data only on multiple partnerships

and not specifically on concurrency: "These data do not capture whether partnerships exist concurrently or serially. Concurrent sexual partnerships (those that overlap in time) allow more rapid spread of sexually transmitted infections than do the same rate of new sequential partnerships." (See Helen Epstein and Daniel Halperin, "Global Sexual Behaviour." *Lancet* 369 [2007]: 557.) Some research that did attempt to examine concurrency has not found an association with HIV, such as: Emmanuel Lagarde, Bertran Auvert, Michel Carael, et al., "Concurrent Sexual Partnerships and HIV Prevalence in Five Urban Communities of Sub-Saharan Africa." *AIDS* 15 (2001): 877–84; Vinod Mishra and Simona Bignami-Van Assche, "Concurrent Sexual Partnerships and HIV Infection: Evidence from National Population-Based Surveys." Calverton, Maryland: Macro International Inc, 2009. However, these studies overlooked the fact that the elevated risk for concurrency results mainly from the behavior of a person's sex partners and consequently from the broader connections to the sexual networks in a society. That makes comparing individuals' reports of their own concurrency with HIV risk much less useful than ones that attempt to measure concurrency in their partners. See UNAIDS Reference Group on Estimates, Modelling, and Projections: Working Group on Measuring Concurrent Sexual Partnerships, "HIV: Consensus Indicators Are Needed for Concurrency," *Lancet* 375 (2009): 621–22; Mah and Halperin, "Concurrent Sexual Partnerships and the HIV Epidemics in Africa"; Epstein and Morris, "Concurrent Partnerships and HIV"; Mah and Shelton, "Concurrency Revisited"; Goodreau et al., "Concurrent Partnerships, Acute Infection and HIV Epidemic Dynamics; Epstein, *The Invisible Cure*. Recent studies that have employed genetic techniques to trace the source of new infections in couples (chapter 4, endnote 2) are shedding light on these issues. In a 2011 multi-country trial among discordant couples, in which only 5 percent of participants initially reported having multiple partners, based on genetic analysis of the HIV strains 28 percent of all new infections appeared to result from sex outside the relationship. In addition, the proportion of infections from outside the couple was much higher in Africa than in other regions; in the Soweto site, for example, all the infections were determined to stem from sex outside the primary relationship. Cohen et al., "Prevalence of HIV-1 Infection," www .nejm.org/doi/suppl/10.1056/NEJMoa1105243/suppl_file/nejmoa1105243_appendix.pdf.

4. In the Kenyan study, men also substantially underreported their level of concurrency in face-to-face interviews compared to responses voted into a ballot box. Hesban Ooko, Mary Ann Seday, Edit Akom, and Paul Kuria, "Variations in Self-Reported Concurrent Sexual Partnerships Using Two Methodologies." Population Services International (2010), www.psi .org/iac2010. The limits of self-reported data on intimate behavior that is collected in such standard surveys have long been recognized. (Kevin A. Fenton, Anne M. Johnson, Sally McManus, and Bob Erens, "Measuring Sexual Behaviour: Methodological Challenges in Survey Research." *Sex Transm Infect* 77 [2001]: 84–92; Dianne Morrison-Beedy, Michael P. Carey, and Xin Tu, "Accuracy of Audio Computer-Assisted Self-Interviewing [ACASI] and Self-Administered Questionnaires for the Assessment of Sexual Behavior." *AIDS Behav* 10 [2006]: 541–52; Nancy Luke, Shelley Clark, and Eliya Zulu, "Using the New Relationship History Calendar Method to Improve Sexual Behavior Data." Fifth African Population Conference. Arusha, Tanzania, 2007.) In the 2007 Swaziland DHS only 2.2 percent of women reported having more than one sex partner during the previous year, but in a 2006 study conducted in a rural area of the country and relying on an interviewing technique that en-

sured greater discretion and confidentiality, 62 percent of females reported having two or more partners in just the past three months; the corresponding rate for men was 70 percent. Victoria James and Richard Matikanya, "Protective Factors: A Case Study for Ngudzeni ADP (Swaziland)" (Australia/Swaziland: World Vision, 2006).

Research utilizing biological techniques capable of detecting recent sexual encounters has also confirmed that people often are less than forthcoming when interviewers inquire about such private matters in standard household surveys. In Zimbabwe, a 2009 study conducted by CDC and other researchers used vaginal swabs to test for the presence of prostate-specific antigen (PSA), which is only produced by men (chapter 4, endnote 5). Among the women who tested positive for PSA contact within the previous two days, about half had told interviewers they had used condoms for any recent sex act, or hadn't had recent sex at all. (See Alexandra M. Minnis, Markus J. Steiner, Maria F. Gallo, et al., "Biomarker Validation of Reports of Recent Sexual Activity: Results of a Randomized Controlled Study in Zimbabwe." *Am J Epidemiol* 170 [2009]: 918–24, including on the methodological reasons why the researchers concluded this finding had probably underestimated the actual level of misreporting in the face-to-face interviews.) Attempting to overcome such limitations, researchers in the Lake Malawi study utilized a sophisticated network analysis methodology through which many people who reported not having multiple partners were later identified by someone else in the community as having been a sexual partner. Stéphane Helleringer, Hans-Peter Kohler, Linda Kalilani-Phiri, James Mkandawire, and Benjamin Armbruster, "The Reliability of Sexual Partnership Histories: Implications for the Measurement of Partnership Concurrency During Surveys." *AIDS* 25 (2011): 503–11.

5. Prior to the publication of the randomized trial data from Kenya and Uganda in 2007, skepticism surrounding the potential role of male circumcision for HIV prevention remained pervasive. In 2000, for example, Halperin submitted a proposal to the National Institutes of Health for a small pilot project on how South African men viewed the practice. An NIH program officer curtly rejected the idea, suggesting that even studying the issue would be "unethical." See Arthur Allen, "Cultural Baggage Stymies AIDS Prevention in Africa," *Washington Independent*, January 18, 2008, http://washingtonindependent.com/2592/cultural-baggage-stymies-aids-prevention-in-africa; Ann Swidler, "Who Is Afraid of Male Circumcision? Donor Resistance and HIV Prevention in Sub-Saharan Africa," Presentation to Sociology Department Colloquium, University of Pennsylvania, March 3, 2010.

6. Some global health organizations and donors, including WHO and UNAIDS, became much more supportive of male circumcision after the publication of the trial results. They organized a number of meetings and clinical training sessions, established an active Web site (www.malecircumcision.org), and facilitated modeling studies to help governments understand the potential impact of scaling up services.

7. One country that has experienced a dramatic shift in the practice of male circumcision is South Korea. The procedure once was rarely perfomed in the country, but now over 90 percent of males are estimated to be circumcised. Part of the reason may be that hundreds of thousands of U.S. servicemen were stationed in the nation during the Korean War in the early 1950s. Myung-Geol Pang and DaiSik Kim, "Extraordinary High Rates of Male Circumcision in South Korea: History and Underlying Causes." *BJU Int* 89 (2002): 48–54;

Seung-June Oh, Tae Kim, Dae J. Lim, and Hwang Choi, "Knowledge of and Attitude Towards Circumcision of Adult Korean Males by Age." *Acta Paediatr* 93 (2004): 1530–34.

8. A defining moment in Kenya came when Prime Minister Raila Odinga, himself a Luo, publicly announced that he had undergone the procedure, and he urged other Luo men to also become circumcised.

9. See www.ft.com/intl/cms/s/0/6c5726a8-19c1-11e1-ba5d-00144feabdc.0.html#axzz1fE1D RumY. For more information on male circumcision and sexual pleasure see appendix, endnote 18.

10. Guidelines in most programs call for men to fully abstain from any kind of sexual activity, including masturbation, for six weeks following circumcision. Many clinicians believe that nearly all men are fully healed after just a few weeks, and some have advocated a four-week abstinence period instead. But most authorities argue that it's safer to have a longer period, because some men will not follow the guidelines strictly. There also is a concern, however, that some men will not undertake the procedure because they do not want to forego sexual activity for that long a period of time. Anecdotes from countries including Swaziland and a recent Kenyan study suggest that some men are also worried that their partners would have sex with other men during the period of mandated abstinence. Amy Herman-Roloff, Nixon Otieno, Kawango Agot, Jeckoniah O. Ndinya-Achola, and Robert C. Bailey, "Acceptability of Medical Male Circumcision Among Uncircumcised Men in Kenya One Year After the Launch of the National Male Circumcision Program." *PLoS One* 6 (2011): e19814. One study suggested that men who have sex before being fully healed from the procedure may have an elevated risk for contracting HIV, or possibly of transmitting it, but the study numbers were not statistically significant. (Maria J. Wawer, Frederick Makumbi, Godfrey Kigozi, et al., "Randomized Trial of Male Circumcision in HIV-Infected Men: Effects on HIV Transmission to Female Partners, Rakai, Uganda." *Lancet* 374 [2009]: 229–37.) Although a subsequent review of data from all the trials found no evidence of this risk, programs should continue to counsel men and couples to abstain from sex until full wound healing. Supriya D. Mehta, Stephen Moses, Ronald Gray, et al., "Does Sex in the Early Period After Circumcision Increase HIV-Seroconversion Risk? Pooled Analysis of Adult Male Circumcision Clinical Trials." *AIDS* 23 (2009): 1557–64.

11. Another country that has achieved a measure of success in expanding circumcision services is Tanzania where, as of October 2011, more than one hundred thousand males had received the procedure during the previous year. Modeling based on the probably conservative estimate of a 60 percent protective effect (chapter 24, endnote 19) suggested that for every five circumcisions performed in the Iringa region of southern Tanzania, which has the country's worst HIV epidemic, one new infection would be prevented. One important reason that Kenya and Tanzania have so far been more successful in providing services is that in East African settings lower-level clinicians—such as nurses and clinical officers—perform most minor surgeries like male circumcision. (See Amy Herman-Roloff, Emma Llewellyn, Walter Obiero, et al., "Implementing Voluntary Medical Male Circumcision for HIV Prevention in Nyanza Province, Kenya: Lessons Learned During the First Year." *PLoS One* 6, no. 4 [2011]: e18299; and Nathan Ford, Kathryn Chu, Edward J. Mills, "Safety of Task-Shifting for Male Medical Circumcision: A Systematic Review and Meta-Analysis." *AIDS* (2012) 26: 559–66.) In south-

ern Africa, usually only physicians are allowed to conduct these procedures, making it harder to offer circumcision services on a large scale.

12. For further examples of the ongoing controversy around concurrency, see Mark N. Lurie and Samantha Rosenthal, "Concurrent Partnerships as a Driver of the HIV Epidemic in Sub-Saharan Africa: The Evidence Is Limited." *AIDS Behav* 14 (2010): 17–24; H. Epstein, "The Mathematics of Concurrent Partnerships and HIV: A Commentary on Lurie and Rosenthal, 2009." *AIDS Behav* 14 (2010): 29–30; M. Morris, "Barking Up the Wrong Evidence Tree: Comment on Lurie & Rosenthal, 2009," *AIDS Behav* 14 (2010): 31–33; Mah and Halperin, "The Evidence for the Role of Concurrent Partnerships in Africa's HIV Epidemics"; http://go.worldbank.org/OPC5GJ5RD0. A South African study that relied on indirect ecological evidence reported a strong correlation between reported number of lifetime sex partners and HIV risk but evidently no association between concurrency and HIV. This finding has been cited by some skeptics as proof of the need to eliminate behavior change programs. Yet the authors had actually concluded that, "Our findings suggest that in similar hyperendemic sub-Saharan African settings, there is a need for clear messages aimed at the reduction of multiple partnerships, irrespective of whether those partnerships overlap in time." (Frank Tanser, Till Barnighausen, Lauren Hund, Geoffrey P. Garnett, Nuala McGrath, and Marie-Louise Newell, "Effect of Concurrent Sexual Partnerships on the Rate of New HIV Infections in a High-Prevalence, Rural South African Population: A Cohort Study." *Lancet* 378 [2011]: 247–55; and see the *Lancet* response letters by Martina Morris/Helen Epstein and by James D. Shelton and Stephane Helleringer, questioning the study methodology.) The dichotomy between concurrent and other multiple partnerships may be a moot point; most studies in Africa have found that nearly all multiple partnerships are also concurrent in nature. In a large Mozambican survey, for example, in two provinces with severe HIV epidemics, Gaza and Sofala, 49 and 37 percent of men, respectively, reported having more than one sex partner during the previous year, and 43 and 31 percent of men having at least one concurrent partner during the same period (Population Services International, "HIV Behavioral (TRaC) Study Among Men and Women Aged 15–35," Maputo, September 2008). Out of 12,515 people surveyed in Rakai, Uganda, between 1999 and 2005, 28 percent reported having multiple partners during the previous year, and every one of those relationships turned out to be concurrent ones as well. Tom Lutalo, "Association of HIV Incidence with Concurrent Partnerships in the Rakai study." Presented at the UNAIDS Reference Group on Estimates, Modelling and Projections, London, February 29, 2008, www.epidem.org/publications/london2008.pdf.

13. John Bongaarts and Mead Over, "Global HIV/AIDS Policy in Transition." *Science* 328 (2010): 1359–60; Ward Cates, "HPTN 052 and the Future of HIV Treatment And Prevention," *Lancet* 378 (2011): 224–25; Bernard Schwartländer, John Stover, Timothy Hallett, et al., "Towards an Improved Investment Approach for an Effective Response to HIV/AIDS." *Lancet* 377 (2011):2031–41.

14. Some published research from low-resource settings has reported ARV adherence rates that are encouragingly high, but a rigorous review of published and unpublished studies from across Africa concluded that overall levels of adherence were, in the aggregate, relatively disappointing. Christopher J. Gill, Davidson H. Hamer, Jonathan L. Simon, Donald M. Thea, and Lora L. Sabin, "No Room for Complacency About Adherence to Antiretroviral

Therapy in Sub-Saharan Africa." *AIDS* 19 (2005): 1243–49. A subsequent review, also based on many African studies, found that two years after commencing ARV treatment about half of all patients had either stopped taking the medications, died, or ceased coming to the clinics where they had begun receiving the drugs. Sidney S. Rosen, Matthew P. Fox, and Christopher J. Gill, "Patient Retention in Antiretroviral Therapy Programs in Sub-Saharan Africa: A Systematic Review." *PLoS Med* (2007): e298, doi:10.1371/journal.pmed.0040298. Even Brazil, which had achieved widespread access to locally produced generic ARVs in a large, middle-income country, is beginning to face financial constraints as increasing treatment failure is causing the government to import much more expensive second- and third-line drugs.

15. In recent years, a tenet of leading agencies such as PEPFAR and UNAIDS is to foster "local ownership" of AIDS and other health and development initiatives, yet it remains to be seen how successful this will be.

16. Patients on ARV regimens face the risk of various types of long-term toxicity from these medicines. Brendan A. I. Payne, Ian J. Wilson, Charlotte A. Hateley, et al., "Mitochondrial Aging Is Accelerated by Anti-Retroviral Therapy Through the Clonal Expansion of mtDNA Mutations." *Nat Genet* 43 (2011): 806–10; James D. Shelton, "ARVs as HIV Prevention: A Tough Road to Wide Impact." *Science* (2011) 334: 1645–46; www.sciencemag.org/content/334/6063/1645.summary?sid=3ae49239-e0de-4e55-ac7e-5b20509474e7.

17. Jon Cohen, "Complexity Surrounds HIV Prevention Advances." *Science* 333 (2011): 393. And see Myron S. Cohen and Cynthia L. Gay, "Treatment to Prevent Transmission of HIV-1." *Clinical Infectious Diseases* 50, no. S3 (2010): S85–S95; Kumi Smith, Kimberly A. Powers, Kathryn E. Mucssig, William C. Miller, Myron S. Cohen, "HIV Treatment as Prevention: The Utility and Limitations of Ecological Observation." *PLoS Med* 9 (2012): e1001260. doi:10.1371/journal.pmed.1001260; Shelton, "ARVs as HIV Prevention"; Till Barnighausen, David Bloom, Salal Humair, "Is Treatment as Prevention the New Game-Changer? Costs and Effectiveness." Oral presentation at the International AIDS Conference, Washington D.C., July 23, 2012, Abstract #MOAE0202; http://pag.aids2012.org/abstracts.aspx?aid=19531. And see discussion on Treatment-as-Prevention in the appendix.

18. In some places where ARVs have been widely available for many years, such as Australia, behavioral risk compensation appears to have outweighed the potential benefit of broadly reducing new HIV infections from the use of these medications. See appendix; Andrew E. Grulich and David P. Wilson, "Is Antiretroviral Therapy Modifying the HIV Epidemic?" *Lancet* 376 (2010): 1824; James D. Shelton, Myron Cohen, Matthew Barnhart, and Timothy Hallett, "Is Antiretroviral Therapy Modifying the HIV Epidemic?" *Lancet* 376 (2010): 1824–25; Cassell et al., "Risk Compensation."

19. Even in the United States, after many years of widespread ARV availability, the number of people who are not reached or who drop out at some stage of the treatment process is sobering. Less than 30 percent of those infected with HIV are currently being treated successfully enough to have suppressed viral loads. This is unlikely to be a high enough percentage to make a significant impact, at the population level, on the number of new infections. Edward M. Gardner, Margaret P. McLees, John F. Steiner, Carlos Del Rio, and William J. Burman, "The Spectrum of Engagement in HIV Care and Its Relevance to Test-and-Treat Strategies for Prevention of HIV Infection." *Clin Infect Dis* 52 (2011): 793–800.

20. The Lancet, "HIV Treatment as Prevention—It Works." *Lancet* 377 (2011): 1719. (And see response letters by Helen Epstein and Martina Morris and by Willard Cates, *Lancet* 378 (2011): 224–25.)

21. Rechristened as "medical male circumcision," this is now classified by international donors and health organizations as a biomedical intervention.

22. Some approaches that are more amenable to being tested in a clinical trial setting, such as those relying on counseling sessions, may be able to produce some results regarding the impact of concurrency-related interventions. The first randomized trial to focus on addressing the risk of concurrency, which was conducted among young African American women, found a significant reduction in reported concurrency and also had a biological impact. Women who participated in intensive group counseling sessions were 65 percent less likely to get a sexually transmitted infection, such as the virus that causes cervical cancer. Gina Wingood, Ralph DiClemente, LaShun Robinson-Simpson, et al., "Efficacy of an HIV Intervention in Reducing High-Risk HPV, Non-viral STIs and Concurrency Among African American Women: A Randomized Controlled Trial," *J Acq Immun Def Synd*, 2012.

23. A satiric article in 2003 purported to be a literature review of the use of randomized controlled trials to determine whether people jumping out of airplanes should use parachutes (Gordon Smith and Jill Pell, "Parachute Use to Prevent Death and Major Trauma Related to Gravitational Challenge: Systematic Review of Randomised Controlled Trials." *BMJ* 327 [2003]: 1459). Of course, there had been no such trials because doing so would have required half of any study group to fall thousands of feet to the earth unprotected. Certainly, the evidence provided by clinical trials has an important role in determining how best to fight HIV. And alternative methods to rigorously measure the effectiveness of behavior change and other prevention approaches are also urgently needed. Yet the quest for perfect evidence can delay the adoption of lifesaving strategies. A later paper also employed the parachute metaphor to argue a similar point: Malcolm Potts, Ndola Prata, Julia Walsh, Amy Grossman, "Parachute Approach to Evidence-Based Medicine," *BMJ* 333 (2006):701–3.

24. A preliminary calculation in 2000 attempted to capture the effect of male circumcision on HIV rates in fifteen African and Asian countries where the practice was common (Potts, "Circumcision and HIV"). The analysis compared infection rates in these nations to those in neighboring, otherwise similar ones where circumcision was rare and HIV, as a consequence, was more prevalent. It estimated that, even if variations in circumcision rates accounted for only half of the differences in HIV rates, the procedure had prevented about eight million infections in the fifteen countries. And see Kate Orroth, Richard G. White, Esther E. Freeman, et al., "Attempting to Explain Heterogeneous HIV Epidemics in Sub-Saharan Africa: Potential Role of Histroical Changes in Risk Behaviour and Male Circumcision." *Sex Transm Infect* (2011), doi:10.1136/sextrans-2011-050174.

25. Gay men may be as likely to reduce their numbers of sex partners as to use condoms consistently, according to some research (see Rotello, *Sexual Ecology*). See prologue, endnote 9 on the potential benefits of circumcision for some men who have sex with other men.

26. Approaches such as pre-exposure prophylaxis might benefit high-risk populations including sex workers, but this could also be problematic. For example, if some people with such very high-risk behaviors were to use condoms less often because they now felt safe, they might actually become more likely to contract or spread HIV (see appendix).

27. See chapter 24, endnote 24; Paul Arora, Rajesh Kumar, Madhulekha Bhattacharya, Nico J. D. Nagelkerke, and Prabhat Jha, "Trends in HIV Incidence in India from 2000 to 2007." *Lancet* 372 (2008): 289–90; Halperin et al., "Understanding the HIV Epidemic in the Dominican Republic"; appendix.

28. Conversation with Halperin in 2007.

29. Epstein, *The Invisible Cure*; Stoneburner and Low-Beer, "Population-Level HIV Declines and Behavioral Risk Avoidance in Uganda"; Green, *Rethinking AIDS Prevention*; Green et al., "Uganda's HIV Prevention Success"; Allen and Heald, "HIV/AIDS Policy in Africa"; appendix.

30. Soul City Institute Regional Programme, *Multiple and Concurrent Sexual Partnerships in Southern Africa: A Ten Country Research Report,* Johannesburg, 2008, www.soulcity.org.za/projects/onelove.

31. There is a history of such "edutainment" soap opera methodologies being used effectively for promoting issues such as family planning in countries including Egypt and Mexico.

32. The Population Services International program was developed with technical input from Helen Epstein. While Halperin initially was very encouraged by this project, on a later visit to Mozambique he observed that the effort appeared to have potentially become more complicated and less focused by also introducing themes of condoms and HIV testing into the methodology. The Mozambican campaign addressing concurrency also involved a number of other local and international partners, with funding primarily from USAID. An evaluation conducted in September 2011 included some encouraging findings of the campaign's effects. Maria Elena Figueroa and Lawrence Kincaid, *The Impact of a Multimedia Campaign on HIV Prevention Behaviors in Mozambique* (Research Brief Series). Baltimore: USAID/Project Search: Research to Prevention, 2012; www.jhsph.edu/r2p/publications.html.

33. A 2011 review article noted that the findings from the circumcision trials were remarkably similar to the observational evidence from more than a decade earlier, and suggested that public health may have consequently been compromised by delaying in the pursuit of "gold standard" evidence: Reidar K. Lie and Franklin G. Miller. "What Counts as Reliable Evidence for Public Health Policy: The Case of Circumcision for Preventing HIV Infection." *BMC Medical Research Methodology* 11 (2011): 34–41. Also see Potts, "Parachute Approach to Evidence-Based Medicine."

34. There is growing concern that the fight against AIDS is receiving a disproportionate level of attention in some poor countries. In Haiti, Rwanda, and Ethiopia, for example, the majority of health spending goes for this disease, even though many more people die from other causes, such as lack of access to clean drinking water. Jeremy Shiffman, "Has Donor Prioritization of HIV/AIDS Displaced Aid for Other Health Issues?" *Health Policy Plann* 23 (2008): 95; Devi Sridhar and Rajaie Batniji, "Misfinancing Global Health: Case for Transparency in Disbursements and Decision Making." *Lancet* 372 (2008): 1185–91; Laurie Garrett, "The Challenge of Global Health." *Foreign Affairs,* January/February 2007, pp. 14–38; www.lauriegarrett.com/blog/media/1/20070112-garrett.pdf; Daniel Halperin, "Putting a Plague in Perspective," *The New York Times*, January 1, 2008; Roger England, "Are We Spending Too Much on HIV?" *BMJ* 334 (2007): 344; David Berlan and Daniel T. Halperin, "Apparent Displacement of Global Health Funding Priorities: Time for Some Difficult Decisions?" *Soc Sci Med* (2011) doi:10.1016/j.socscimed.2011.05.043. In the decade between 1999 and

2009, U.S. foreign assistance spending on HIV-AIDS initiatives grew from $215 million annually to more than $6.5 billion. During the same years, U.S. funding for all international family planning programs rose slowly, from $385 million to just $495 million. (USAID, "Child Survival and Health Programs Fund: Annual Report to Congress." USAID annual reports, 2000–2011; PEPFAR, "PEPFAR Funding: Investments that Save Lives and Promote Security [Updated June 2011]." www.pepfar.gov/press/80064.htm; Kaiser Family Foundation, "U.S. Government Funding for Global HIV/AIDS Through FY 2005," www.kff.org/hivaids/loader.cfm?url=/commonspot/security/getfile.cfm&PageID=38708.)

In some PEPFAR "focus countries," including Zambia, Kenya, and Mozambique, fertility rates rose slightly during the years when PEPFAR and other donors were increasing support dramatically for the fight against HIV. Some experts believe this had the inadvertent effect of pulling professionals away from such basic health fields as family planning to work in the higher-paying, donor-funded AIDS programs. (Malcolm Potts and Martha Campbell, "The Myth of 9 Billion: Why Ignoring Family Planning Overseas Was the Worst Foreign-Policy Mistake of the Century." *Foreign Policy*, May 9 2011, www.foreignpolicy.com/articles/2011/05/09/the_myth_of_9_billion?page¼full; Jeremy Shiffman, David Berlan, and Tamara Hafner, "Has Aid for AIDS Raised All Health Funding Boats?" *J Acq Immun Def Synd* 52 [2009]: S45.) Paul Farmer has cautioned: "The influx of AIDS funding can indeed strangle primary care, distort public health budgets, and contribute to brain drain . . . These untoward or 'perverse' effects are not inevitable; they occur only when programs are poorly designed." Paul Farmer, "Intelligent Design: From 'Marvelous Momentum' to Health Care for All: Success Is Possible With the Right Programs." *Foreign Affairs* (2007); and see David Egilman, Tess Bird, Fernando Mora, and Nicholas Druar, "Get AIDS and Survive? The 'Perverse' Effects of Aid: Addressing the Social and Environmental Determinants of Health, Promoting Sustainable Primary Care, and Rethinking Global Health Aid." *Int J Occup and Envir Health* 17 (2011): 364–82. President Obama's administration has pledged to redress some of this disproportionality in international health assistance through the Global Health Initiative and other endeavors. It remains to be seen whether such efforts will make a significant impact on the problem. John Donnelly, "Healing the World, Part 1: A Slow, Stumbling Start to Obama's Global Health Plan Undercuts Its Ambitious Goals." *Global Post*, May 9, 2011, www.globalpost.com/dispatch/news/health/110428/healing-the-world.

Epilogue

1. Some of the global reduction in mortality was because new HIV infections had begun declining about a decade earlier, so the number of deaths would consequently begin to decline around this time. In addition, the number of new infections was not only recalibrated downward, to an estimated 2.4 million per year (from an estimated 4.3 million in 2006), but this figure was also forecasted to continue falling. Momentum had been building for a major revision of HIV estimates, especially after a household survey estimated in 2007 that India's infection rate was about one-quarter of 1 percent of adults (Dandona et al., "A Population-Based Study of Human Immunodeficiency Virus in South India"; Shelton et al., "Has Global HIV Incidence Peaked?"; Halperin and Post, "Global HIV Prevalence"). This meant that India had about 2.5 million infections, which was less than half what UNAIDS had previously estimated. Halperin's Harvard research office was located at the time in

Cambridge, Massachusetts, where soon after the news broke he ran into Piot taking a coffee break near where UNAIDS was having a meeting that day. When Halperin asked Piot what he thought of the report, he was glum. "This is actually bad news," Piot replied. It would be harder now to show AIDS as a global problem, he said, as opposed to mainly an African one. And it would give fuel to those who thought the world's response had grown overly expensive.

2. Interview with Timberg.

3. Interview with Timberg.

4. Mbeki's economic policies earned him high marks internationally, as did his initiatives bringing better roads, services, and housing to traditionally impoverished parts of South Africa. But his policy toward Zimbabwean president Robert Mugabe was almost universally denounced. Zimbabwe's steep deterioration hurt the region politically and economically, and Mbeki's reluctance to act forcefully was almost as puzzling as were his policies on AIDS.

5. Between the presidencies of Mbeki and Zuma there was an eight-month interim period during which Zuma ally Kgalema Mothlanthe held the position.

6. The announcement by the South African Minister of Health to embrace exclusive breast-feeding immediately stirred controversy in a country which had grown accustomed to use of substitute infant-feeding practices. Plus News, "Breastfeeding Policy Turnaround," www.plusnews.org/report.aspx?ReportID=93600; http://mg.co.za/article/2011-08-24-motsoaledi-argues-breast-is-best.

7. Auvert et al., "Effect of the Roll-out of Male Circumcision in Orange Farm (South Africa)." Subsequent research will further examine the long-term impact of male circumcision on the overall community, including on women. Long-term data from the circumcision trial sites in Uganda and Kenya were also presented in 2010 and 2011, showing an overall protective effect of about 70 percent (Kong et al., "Longer-term effects of Male Circumcision"; Bailey et al., "The Protective effect of Adult Male Circumcision").

8. South Africa still has the highest number of new HIV infections annually of any nation, and the prevalence of male circumcision remains low. So it would potentially stand to benefit more than any other country from making the procedure widely accessible. See Pam Belluck, "Obstacles Slow an Easy Way to Prevent HIV in Men," *The New York Times*, September 27, 2011, page D3. Other nations also have been slow to start expanding circumcision services. Among them is Swaziland, where despite a large boost in funding from PEPFAR and the Gates Foundation, demand has been flagging in recent years. The initiative, largely carried out by U.S.-based contractors, neglected to get the crucial support of traditional Swazi cultural leaders, some of whom began undermining support for the circumcision effort; http://articles.latimes.com/2012/jul/24/opinion/la-oe-halperin-HIV-circumcision-africa-20120724. In July 2011 the Swazi king, however, finally endorsed male circumcision for HIV prevention, http://mg.co.za/article/2011-07-16-swazi-king-endorses-mass-circumcision.

9. By many accounts, family planning has been one of the most effective initiatives in the history of foreign assistance to poor countries. (John Bongaarts and Steven Sinding, "Population Policy in Transition in the Developing World," *Science* 333 [2011]: 574–76; Potts and Campbell, "The Myth of 9 Billion"; Michael Gerson, "Family Planning as Pro-life Cause," *The Washington Post*, Aug. 29, 2011). In addition, several modeling studies have found that increasing access to such services in Africa could have a substantial impact on the numbers of children

born with HIV, and on the numbers who end up as orphans, because infected women will, in many cases, prevent pregnancy by using contraceptives. The cumulative effect of such decisions could go at least as far toward reducing the numbers of infected children as programs that use ARVs to block the spread of the virus between mothers and their babies. Heidi W. Reynolds, Barbara Janowitz, Rose Wilcher, and Ward Cates, "Contraception to Prevent HIV-Positive Births: Current Contribution and Potential Cost Saving in PEPFAR Countries." *Sex Transm Infect* 84, supplement II (2008): ii49–ii53; and Daniel T. Halperin, John Stover, and Heidi Reynolds, "Benefits and Costs of Expanding Access to Family Planning Programs to Women Living with HIV." *AIDS* 23 (2009): S123–S130. In addition to the effect on HIV infections of making such services available to women who know they are HIV positive, increasing the availability of family planning services in severely affected regions more generally would further reduce mother-child transmission, because many women who are unaware of their HIV status would also end up choosing to have fewer children, including fewer HIV-positive ones. (See appendix, including on the possibly increased HIV risk from some hormonal-based methods.)

10. See the beginning of chapter 15 and Shelton et al., "Has Global HIV Incidence Peaked?" regarding the crucial difference between when a decline in new HIV infections begins and when—at least several years later—the overall prevalence rate within a society begins to decline.

11. Museveni has also been outspokenly critical in recent years of the use of male circumcision for HIV prevention.

12. "Global Fund Withdraws Support for LoveLife." *Africa News*, December 19, 2005.

13. Even in countries with severe HIV epidemics such as South Africa, relatively few men of Sifiso's age are infected yet (Shisana et al., "South African National HIV Prevalence"; South Africa Ministry of Health, *Getting to Success*).

Appendix

1. This more technical Appendix is loosely based on a paper published in the journal *Science* by ten experts, mostly from the University of California and Harvard University, (1) and an article in *The Lancet*, whose lead author is David Wilson, head of HIV Programs at the World Bank. (2)

2. UNAIDS had traditionally characterized epidemics in which pregnant women in urban areas had HIV rates of more than than 5 percent as being generalized, while less severe epidemics were considered concentrated. In 2000, the agency changed its categorization so that any country in which pregnant women in urban areas have an estimated infection rate of more than 1 percent is now considered to have a generalized epidemic.

3. Increased access to circumcision services for men who have sex with men also may be useful, especially in some parts of the world, including much of Latin America, Asia, and Africa, where substantial numbers of such men are exclusively the insertive partners in anal sex and also have sex with women. For these men, circumcision appears to be similarly protective as it is for men practicing only vaginal sex, based on findings from recent studies conducted in South Africa, Peru, and Australia (prologue, endnote 9).

4. Although the brothels were then the main driver of the Thai epidemic, a considerable proportion of infections were also due to injecting drug use and men having sex with other men.

Unfortunately prevention efforts for these populations do not appear to have been nearly as successful. Over time, the wives and other female partners of the clients of sex workers have accounted for an increasing proportion of those infected with HIV, but this pattern of secondary transmission was also substantially interrupted by the campaign to halt the epidemic at its source, by promoting consistent condom use in the brothels. There also was a large decline in the numbers of Thai men frequenting the brothels, which is no longer a common practice there. Similar shifts, involving both increased consistent condom use for sex work and fewer clients overall, have occurred in some other countries where HIV rates have declined, such as Cambodia and the Dominican Republic (see chapter 15, endnote 8).

5. In Washington, D.C., for example, when city health officials reported an adult HIV rate of 3 percent this prompted confusing comparisons to African epidemics. But Washington's epidemic is driven mainly by a combination of men having sex with men (including, in some communities, many men who also have sex with women) and users of injecting drugs. In cities such as Washington high rates of incarceration also contribute, primarily because of the common practice of anal sex among men unable to have sex with women while in prison. In addition, the African American population—while not nearly as affected as the hardest-hit parts of Africa—tends to have much higher rates of heterosexually spread HIV than other groups in the United States. Part of the reason, as discussed in chapter 18, is because of a greater prevalence of concurrent sexual partnerships. Another factor in Washington's relatively high HIV rate is, perhaps counterintuitively, the nearly universal availability of effective ARV treatment in the city. This success means that with few people now dying of AIDS, even a relatively modest pace of new infections will likely keep the total number of people living with HIV rising steadily for years to come. See Craig Timberg, "AIDS in the District Is Serious, but Not Critical." *The Washington Post,* March 22, 2009; www.washingtonpost.com/wp-dyn/content/article/2009/03/20/AR2009032001761.html.

6. Some areas, such as parts of the Caribbean, West and Central Africa, and the island of Papua (which includes both the nation of Papua New Guinea and the Papua region of Indonesia) appear to have mixed epidemics, characterized by substantial aspects of both concentrated and generalized epidemic patterns. (2) And in countries with very large populations, even a low prevalence level means that a large number of people are infected. India, for example, with an estimated adult infection rate of about 0.3 percent, has some 2.4 million people living with HIV.

7. Most studies examining rape and HIV infection in Africa have not found a significant association, probably because though rape is a horrific crime, it typically is confined to a single event in a victim's life. (9) However, one study in South Africa showed an increased HIV risk for women who reported that their male partner was violent (10), probably in large part because men who are violent toward their partners are also more likely to have other partners and consequently to become infected, then transmit the virus to their main partner. And women who have other partners are at higher risk of infection as well as of violence, when this is discovered. So infidelity appears to be an important root cause of both higher HIV rates and violence in sexual relationships. (5, 11, 12)

8. This summary is not intended to be exhaustive. In addition, we agree with many experts who call for a "combination prevention" approach, although we believe it would be most effective

to prioritize those strategies that are likely to have the greatest impact within a particular epidemic. (2, 13, 14)

9. Many of these unplanned pregnancies result in unsafe abortions, which contribute to high rates of illness and death among women in many poorer countries (see epilogue, endnote 9).

10. The most rigorous reviews of male condoms have estimated that, when used consistently, they reduce risk of HIV infection by roughly 80 to 90 percent (18, 20), although one older study estimated an effectiveness of 95 percent. (19) Some evidence suggests that female condoms may be slightly less effective, for both pregnancy and HIV prevention, than male condoms. The likely cause is incorrect use, usually involving slippage from the correct position, which can result in semen leaking. CDC researchers in Brazil found numerous cases of such incorrect use, even among experienced and well-trained users (chapter 4, endnote 5). In some instances, such as with inebriated clients of sex workers, these products can provide women with a method that gives them greater personal control—something that continues to be an important priority in the prevention world. However, there have been reports of poor women in places such as Zimbabwe buying the product (whose price is heavily subsidized by donors) only to utilize the outer ring as a bracelet. And many users have complained of irritating sounds during intercourse and other problems. Newer versions of the female condom attempt to address some of these issues.

11. Condoms are less effective against some other infections, including HSV-2 (genital herpes) and HPV (which can cause cervical cancer), that are spread through skin-to-skin contact, including areas of skin that are not protected by a condom. In earlier years the evidence for condoms preventing these infections was not definitive, but more recent studies have confirmed they do provide protection, although less so than for HIV and other infections spread primarily through direct contact with semen or other genital fluids. (19)

12. In one widely cited multisite study published in *The Lancet* (46) some decrease in risky sexual behavior was reported following counseling and testing. However, there was no impact on the rate of new sexually transmitted infections from the intervention, even in this very large randomized trial. And some of the reported change in behavior was, for example, among men who tested negative while their partners tested positive, which in many cases led to the men simply leaving the relationship or ceasing sexual relations with their partners.

13. There were more new HIV infections recorded among the participants randomized to receive intensive access to HIV testing compared to those who were not, but this difference did not reach standard statistical measures for "significance." The threshold for statistical significance, however, was nearly reached, prompting the study authors to conclude there had likely been some negative impact from the intervention. In another study, also conducted in Zimbabwe, the researchers similarly concluded that "Increased sexual risk following receipt of a negative result may be a serious unintended consequence of Voluntary Counseling and Testing." (48)

14. The advocates of the "test-and-treat" approach—a variation of "treatment-as-prevention" that would entail a massive expansion of testing programs and immediately placing all those who test positive on ARVs—believe this would significantly impact on the epidemic. However, a number of experts have questioned the methodology used in the preliminary modeling (61) upon which the strategy is based. (62–64) Because of potential limitations such as inconsistent ARV adherence, development of drug resistance, and behavioral risk compensation,

some researchers have cautioned that this approach could even do more harm than good by dramatically boosting program costs without actually preventing many new infections. (65) For these and other reasons discussed in chapter 30 and the Treatment-as-Prevention section of this Appendix, it remains unclear whether such approaches would contribute a large population-level benefit. An additional concern related to HIV testing regards the potential for partner violence against women who test positive. In one study in Tanzania, 16 percent of women who tested positive for HIV and subsequently disclosed their status to their male partner reported being blamed for the infection; 4 percent reported being victims of violence. (66)

15. Such a population-level effect of the widespread practice of male circumcision can be seen in regions such as West Africa, where HIV has been present for many decades since its birth there, yet prevalence in the general population remains relatively low (see chart on page 179). (68–72, 88, 89)

16. A randomized trial to confirm some encouraging observational data from Uganda was terminated early due to "futility," meaning that an interim analysis did not show evidence of benefit and projections suggested the study would be unable to enroll enough serodiscordant couples to be able to detect a statistically meaningful effect. (99) It would require a much larger trial—also involving couples with an HIV-positive man and an uninfected female partner, with half of the men getting circumcised—to definitively confirm the observational data. For ethical and logistical reasons, it is unlikely that such a trial will be conducted. (100) In HIV epidemiology, generally the next best evidence after randomized trials are studies that follow serodiscordant couples over time, to observe which factors may help predict who is more likely to become infected. (Because it would be considered unethical to conduct a randomized trial on the efficacy of condoms, for example, most estimates for the level of protection they provide comes from such serodiscordant couple studies.) The two serodiscordant couple studies that examined the effect of circumcision on HIV-infected men transmitting the virus to their female partners suggested that women may have been about 40 percent less likely to become infected if their male partners were circumcised. These findings approached but did not reach statistical significance. (100–102) However, the evidence for the protective effect against some other sexually transmitted infections is more clear-cut. In the randomized trial in Uganda, for example, male circumcision reduced the risk of HPV infection among long-term female partners by 77 percent. (96)

17. A 2007 review of thirteen acceptability studies conducted in a number of African countries found that many uncircumcised men—up to 80 percent in high-HIV-prevalence countries such as Botswana and Swaziland—as well as their female partners are interested in and supportive of having the procedure performed, including for their male children. (103) At least seven subsequent acceptability studies have been conducted, including some in the Caribbean and Asia. (68–71, 105–11)

18. In many countries where male circumcision services have recently been established—such as Kenya, Tanzania, and South Africa—a number of ethnic groups still traditionally perform the procedure. Because it is readily apparent that members of these groups can also become infected with HIV, it seems unlikely that the practice would be widely perceived in such places as offering full protection. Meanwhile, some critics have suggested that male circumcision is similar to "female genital mutilation" because it allegedly also reduces sexual func-

tioning and pleasure. Unlike male circumcision, however, these practices—particularly the most extreme forms such as infibulation—can pose significant health risks for women. In the rigorous studies that have investigated male circumcision's effect on sexual pleasure, (115–28) nearly all men and their female partners report that after men become circumcised sexual pleasure is the same or enhanced, for both partners. During the 2005–2006 Swaziland pilot circumcision program mentioned in chapter 26, many women began saying that after getting circumcised their partners could have sex longer before reaching orgasm. Some of the clinic nurses reported that women would use metaphors such as, "He used to go from here [Mbabane] to Manzini [a city half an hour's drive away], now he can go all the way to the border."

19. Another potential challenge in many parts of Africa is the widespread practice of "dry sex," in which powders, herbs, leaves or even newspapers are inserted into the vagina before intercourse to make it drier, and often "tighter" as well. (133–37) Some research indicates that such practices, which can result in greater trauma and lesions during intercourse, may be a risk factor for HIV infection. (133–34) Their popularity—and the related beliefs that women with lubricated vaginas are promiscuous—could also complicate the future popularity of "wet" topical microbicides, though a South African study suggests this may not necessarily be the case. (138)

20. Clinicians have observed that increasing numbers of people are using these post-exposure prophylaxis services in a way similar to a "morning-after pill," often saying that a condom had broken during sex the night before.

21. However, one trial of this pre-exposure prophylaxis approach, conducted among women in three African countries, did not succeed, likely due to low levels of adherence. (170)

22. A recurring issue in regions with high HIV rates regards the possibility that some hormonal-based contraceptives may increase women's risk of HIV infection. Findings from a study published in 2011 suggested that injectable methods of hormonal contraception, and perhaps oral ones as well, can significantly heighten the risk of women contracting HIV. (180) The study found that such methods may also increase the risk of an HIV-infected woman transmitting the virus to a male sexual partner, apparently because their use is associated with higher HIV levels in female genital secretions. Some previous studies that have examined these issues similarly found higher risk, but most did not. (181–83) Yet the 2011 study findings were alarming, especially because the analysis involved a large sample size, employed sophisticated statistical techniques, and may have found a biological explanation in the increased levels of virus in genital secretions. However, most experts urged caution in interpreting the findings, mainly because, as with previous research, these results were based on observational data, which can be biased by various factors. Women who use hormonal contraceptives, for example, may have more frequent sex or use condoms less often than those who do not use these methods. (182, 183) For such reasons, many scientists have called for a randomized trial to provide more definitive evidence on this issue. If such research confirms the risk of hormonal contraceptives, it would present a difficult policy dilemma. In deciding whether to make changes in family planning programs, public health officials would need to weigh this risk against other important considerations, including: 1) the consequences of unintended pregnancy, including increased risk of maternal illness and mortality if effective

contraceptive methods were made less available; 2) the actual risk of acquiring HIV in a given geographical setting; 3) various social and economic benefits of using contraception; and 4) the availability and feasibility of providing alternative, non-hormonal contraceptive methods.

23. When mothers with HIV have access to birth control, they often choose to not have more children, in part because of concerns about caring for them as their own condition may worsen. (184–87)

24. It is possible there may still be a few exceptions, particularly in some of the world's poorest countries where unsterile injection methods remain common. But even in such places, non-sexual transmission is unlikely to account for a large share of total HIV infections. It may, however, help spread other pathogens, such as hepatitis, that are much more contagious.

25. The main treatment approach in the Tanzania trial was syndromic management, meaning that people with symptoms such as genital ulcers or urethral and vaginal discharges were presumptively treated with antibiotics. (203) Some of the other trials employed a mass treatment strategy, or a combination of the two. It is perhaps noteworthy that the declines in other sexually transmitted infections in this Tanzanian study were lower than the reported decline in HIV itself. Meanwhile, some researchers have speculated that this trial may have showed an effect because the HIV epidemic there was at an early and still rapidly accelerating stage, while the subsequent studies were conducted in more mature epidemics. (204) Even if this hypothesis was true, it is probably a moot point, as there are now virtually no places having such a new and rapidly increasing HIV epidemic spread primarily by sex.

26. Several studies—including the extensive "Four Cities" investigation in the 1990s (chapter 24, endnote 14)—have examined the puzzling contrast in parts of West and Central Africa between the persistently low or modest HIV rates alongside often very high levels of bacterial infections such as gonorrheas and chlamydia. As that study concluded, the main reason for this disconnect is the pervasive practice in those regions of male circumcision,(88) which substantially reduces the risk of infection from HIV and some other sexually transmitted infections, but has no effect on ones that are mainly spread in men through contact with the urethral opening in the penis. Meanwhile, in places such as Uganda and Thailand, and among gay men in the United States, rates of both HIV and other sexually transmitted infections do appear to have declined around the same time (5, 31, 151, 207) but probably the main reasons were the profound changes in sexual behavior discussed in this book.

27. These programs may also pose some risk to public health. Aggressive efforts to combat sexually transmitted infections often lead to incorrect diagnoses of vaginal discharge, which is typically the most common symptom diagnosed by clinics that treat such infections. Such discharges are more likely to be caused by yeast infections or bacterial vaginosis than by an infection spread through sex. (211) The continuing routine use of antibiotics in these cases may increase the likelihood of women developing resistant strains of bacteria. The incorrect diagnosis of having a sexually transmitted infection also can create stigma and rancor in relationships when the male partner learns—often when contacted by health authorities and told that he too must be treated for a sexually transmitted infection—that his wife or girlfriend supposedly has such an infection. This issue is exacerbated because the World Health Organization has yet to disseminate clear guidelines on the specific problem of the misdiagnosis of vaginal discharge, or on the larger question of the role in general of treating other

sexually transmitted infections for HIV prevention. In July 2006 the WHO and UNAIDS organized a three-day experts consultation on the matter (chapter 27, endnote 5), yet no public report was ever issued. In addition, bacterial vaginosis, which is very common throughout much of Africa and in some other regions, may be a risk factor for HIV infection in women. (212) This problem is often made worse by the use of traditional and commercial vaginal douching products, (131–34, 213) which are produced and marketed by a large global industry.

28. Treatment-as-prevention might also be useful for other high-risk populations, such as sex workers and users of injection drugs. In fact, some experts are beginning to worry that, particularly during the global financial squeeze, it may become necessary to ration the use of ARVs to those people for whom the medications are most likely to prevent further transmission of HIV. (62, 219, 220) Concerns have been raised, however, regarding the feasibility of ensuring consistent adherence in such populations, as well as the possibility of risk compensation if taking ARVs becomes perceived as a substitute for condom use or other effective prevention measures. Resistance to these medications, especially when adherence is poor, remains a concern. ARV resistance has declined in recent years in Europe and the United States, probably as a result of more effective and better-tolerated drugs, better adherence programs, and more careful clinical monitoring strategies. (221) Yet experts worry that resistance levels could rise in regions where there is less access to effective medical care.

29. Some clinical benefits have also been found for people initiating ARVs while their immune systems are still in the early stages of deterioration. (173, 218) However, it is also important to take into account the potential long-term toxicity of these powerful medicines. (64, 173, 231)

30. The initial analysis of the trial data, which compared rates of infection in that half of study participants who were randomized to receive the vaccine compared with those who were assigned to only receive a placebo, found a 29 percent reduction in HIV risk. However, a subsequent analysis that compared only those who actually received or did not receive all six vaccine injections found a 26 percent reduction, which was not statistically significant. (237, 238)

References to the Appendix

1. Potts, M., D. T. Halperin, D. Kirby, et al. "Reassessing HIV Prevention." *Science* 320 (2008): 749–50 (including the Supporting Online Supplemental Material).

2. Wilson, D., and D. T. Halperin. "'Know Your Epidemic, Know Your Response': A Useful Approach—If We Get It Right." *Lancet* 372 (2008): 423–26.

3. Cohen, J. "Asia and Africa: On Different Trajectories?" *Science* 304 (2004): 1932–38.

4. Shelton. J. D. "Ten Myths and One Truth About Generalised HIV Epidemics." *Lancet* 370 (2007): 1809–11.

5. Epstein, E. *The Invisible Cure: Why We Are Losing the Fight Against AIDS in Africa*. New York: Picador, 2008.

6. Leclerc-Madlala, S. "Transactional Sex and the Pursuit Of Modernity." *Social Dynamics* 29 (2004): 1–21.

7. Hunter, M. "The Materiality of Everyday Sex: Thinking Beyond Prostitution." *African Studies* 61 (2002): 99–119.

8. Swidler, A., and S. C. Watkins. "Ties of Dependence: AIDS and Transactional Sex in Rural Malawi." *Studies in Family Planning* 38 (2007): 147–62.

9. Anema, A., M. R. Joffres, E. Mills, and P. B. Spiegel. "Widespread Rape Does Not Directly Appear to Increase the Overall HIV Prevalence in Conflict-Affected Countries: So Now What?" *Emerging Themes in Epidemiology* 5 (2008): 11, doi:10.1186/1742-7622-5-11.

10. Jewkes, R. K., K. Dunkle, M. Nduna, and N. Shai. "Intimate Partner Violence, Relationship Power Inequity, and Incidence of HIV Infection in Young Women in South Africa: A Cohort Study." *Lancet* 376 (2010): 41–48.

11. Castor, D., S. Cook, S. Leclerc-Madlala, and J. Shelton. "Intimate-Partner Violence and HIV in South African Women." *Lancet* 376 (2010): 1219–20.

12. Epstein H. "Intimate-Partner Violence and HIV in South African Women." *Lancet* 376 (2010): 1219.

13. Halperin, D. T. "Combination HIV Prevention Must Be Based on Evidence." *Lancet* 373 (2009): 544–45.

14. Wilson D. "Partner Reduction and the Prevention of HIV/AIDS." *BMJ* 328 (2004): 848–49.

15. Halperin, D. T., M. Steiner, M. Cassell, et al. "The Time Has Come for Common Ground on Preventing Sexual Transmission of HIV." *Lancet* 364 (2004): 1913–15.

16. Hallett, T. B., S. Gregson, J. J. C. Lewis, B. Lopman, and G. P. Garnett. "Behaviour Change in Generalised HIV Epidemics: Impact of Reducing Cross-Generational Sex and Delaying Age at Sexual Debut." *Sex Transm Infect* 83 (2007): i50–i54.

17. Kirby, D. B. "The Impact of Abstinence and Comprehensive Sex and STD/HIV Education Programs on Adolescent Sexual Behavior." *Sexuality Research & Social Policy* 5 (2008): 18–27.

18. Weller, S., and K. Davis. "Condom Effectiveness in Reducing Heterosexual HIV Transmission." *Cochrane Database Systematic Review* 3 (2001): CD003255.

19. Warner, L., and M. J. Steiner. "Male Condoms." In: *Contraceptive Technology*, 20th revised ed., R. A. Hatcher, J. Trussell, A. L. Nelson, et al., eds. New York: Ardent Media, 2011.

20. Hearst, N., and S. Chen. "Condom Promotion for AIDS Prevention in the Developing World: Is It Working?" *Studies in Family Planning* 35 (2004): 39–47.

21. Shelton, J. D. "Confessions of a Condom Lover." *Lancet* 368 (2006): 1947–49.

22. Cohen, J. "Asia—The Next Frontier for HIV/AIDS: Two Hard-Hit Countries Offer Rare Success Stories: Thailand and Cambodia." *Science* 301 (2003): 1658–62.

23. Rojanapithayakorn, W., and R. Hanenberg. "The 100% Condom Program in Thailand." *AIDS* 10 (1996): 1–7.

24. Kilmarx, P. H., S. Supawitkul, M. Wankrairoj, et al. "Explosive Spread and Effective Control of Human Immunodeficiency Virus in Northernmost Thailand: The Epidemic in Chiang Rai Province, 1988–99." *AIDS* 14 (2000): 2731–40.

25. Kerrigan, D. D., J. M. Ellen, C. L. Moreno, et al. "Environmental-Structural Factors Significantly Associated with Consistent Condom Use Among Female Sex Workers in the Dominican Republic." *AIDS* 17 (2003): 415–23.

26. Halperin, D. T., A. de Moya, E. Perez-Then, G. Pappas, and J. M. Garcia Calleja. "Understanding the HIV Epidemic in the Dominican Republic: A Prevention Success Story in the Caribbean?" *Journal Acquired Immunod Syndrome* 51 (2009): S52–59.

27. Mukerjee, M. "The Prostitutes' Union." *Scientific American*, March 26, 2006.

28. Reza-Paul, S., T. Beattie, H. U. Syed, et al. "Declines in Risk Behaviour and Sexually Trans-mitted Infection Prevalence Following a Community-Led HIV Preventive Intervention Among Female Sex Workers in Mysore, India." *AIDS* 22 Suppl 5 (2008): S91–100.

29. UNAIDS. "Acting Early to Prevent AIDS: The Case of Senegal," June 1999.

30. Ghys, P. D., M. O. Diallo, V. Ettiègne-Traoré, et al. "Increase in Condom Use and Decline in HIV and Sexually Transmitted Diseases Among Female Sex Workers in Abidjan, Côte d'Ivoire, 1991–1998." *AIDS* 16 (2002): 251–58.

31. Luchters, S., M. F. Chersich, and A. Rinyiru. "Impact of Five Years of Peer-Mediated Inter-ventions on Sexual Behavior and Sexually Transmitted Infections Among Female Sex Work-ers in Mombasa, Kenya." *BMC Public Health* 8 (2008): 143, doi: 10.1186/1471-2458-8-143.

32. Hanenberg, R. S., W. Rojanapithayakorn, P. Kunasol, and D. Sokal. "Impact of Thailand's HIV-Control Programme as Indicated by the Decline in Sexually Transmitted Diseases." *Lancet* 344 (1994): 243–45.

33. Gregson, S., E. Gonese, T. B. Hallett, et al. "HIV Decline in Zimbabwe Due to Reductions in Risky Sex? Evidence from a Comprehensive Epidemiological Review." *Int J Epidemiol* 39 (2010): 1311–23.

34. Halperin, D. T., O. Mugurungi, T. B. Hallett, et al. "A Surprising Prevention Success: Why Did the HIV Epidemic Decline in Zimbabwe?" *PLoS Med* 8 (2011): e1000414.

35. Peterman, T. A., L. H. Tian, and L. Warner. "Condom Use in the Year Following a Sexually Transmitted Disease Clinic Visit." *Internat J STD & AIDS* 20 (2009): 9–13.

36. Ahmed, S., T. Lutalo, M. Wawer, D. Serwadda, N. K. Sewankambo, F. Nalugoda, et al. "HIV Incidence and Sexually Transmitted Disease Prevalence Associated with Condom Use: A Population Study in Rakai, Uganda." *AIDS* 15 (2001): 2171–79.

37. Tavory, I. and A. Swidler. "Condom Semiotics: Meaning and Condom Use in Rural Malawi." *American Sociological Review* 74 (2009): 171–89.

38. Westercamp, N., C. Mattson, M. Madonia, et al. "Determinants of Consistent Condom Use Vary by Partner Type Among Young Men in Kisumu, Kenya: A Multi-Level Data Analysis." *AIDS and Behavior* 14 (2010): 949–59.

39. Weinhardt, L. S., M. P. Carey, B. T. Johnson, et al. "Effects of HIV Counseling and Testing on Sexual Risk Behavior: A Meta-Analytic Review of Published Research, 1985–1997." *Am J Public Health* 89, no. 9 (1999): 1397–1405.

40. Glick, P. "Scaling Up HIV Voluntary Counseling and Testing in Africa: What Can Evalua-tion Studies Tell Us About Potential Prevention Impacts?" *Evaluation Review* 29 (2005): 331–57.

41. Cassell, M. M., and A. Surdo. "Testing the Limits of Case Finding for HIV Prevention." *Lancet Infect. Dis* 7 (2007): 491–95.

42. Shelton, J. D. "Counseling and Testing for HIV Prevention." *Lancet* 372 (2008): 273–75.

43. Denison, J. A., K. R. O'Reilly, G. P. Schmid, C. E. Kennedy, and M. D. Sweat. "HIV Counseling and Testing, and Behavioral Risk Reduction in Developing Countries: A Meta-Analysis, 1990–2005." *AIDS Behav* 12 (2008): 363–73.

44. Allen, S., J. Tice, P. Van de Perre, et al. "Effect of Serotesting with Counseling on Condom Use and Seroconversion Among HIV Discordant Couples in Africa." *BMJ* 304 (1992): 1605–9.

45. Allen, S., J. Meinzen-Derr, M. Kautzman, et al. "Sexual Behavior of HIV Discordant Couples after HIV Counseling and Testing." *AIDS* 17, no. 5 (2003): 733–40.

46 The Voluntary HIV-1 Counseling and Testing Efficacy Study Group. "Efficacy of Voluntary HIV-1 Counselling and Testing in Individuals and Couples in Kenya, Tanzania, and Trinidad: A Randomised Trial." *Lancet* 356 (2000): 103–12.

47. Matovu, J. K., et al. "Repeat HIV Counseling and Testing (VCT), Sexual Risk Behavior and HIV Incidence in Rakai, Uganda." *AIDS and Behav* 11 (2007): 71–81.

48. Sherr, L., B. Lopman, M. Kakowa, et al. "Voluntary Counselling and Testing: Uptake, Impact on Sexual Behaviour, and HIV Incidence in a Rural Zimbabwean Cohort." *AIDS* 21 (2007): 851–60.

49. Corbett, E. L., B. Makamure, Y. B. Cheung, et al. "HIV Incidence During a Cluster-Randomized Trial of Two Strategies Providing Voluntary Counselling and Testing at the Workplace, Zimbabwe." *AIDS* 21 (2008): 483–89.

50. Kamb, M., M. Fishbein, J. M. Douglas, et al. "Efficacy of Risk-Reduction Counseling to Prevent Human Immunodeficiency Virus and Sexually Transmitted Diseases: A Randomized Controlled Trial." *JAMA* 280 (1998): 1161–67.

51. Cassell, M. M., D. T. Halperin, J. D. Shelton, and D. Stanton. "Risk Compensation: The Achilles' Heel of Innovations in HIV Prevention?" *BMJ* 332 (2006): 605–7.

52. Richens, J., J. Imrie, and A. Copas. "Condoms and Seat Belts: The Parallels and the Lessons." *Lancet* 355 (2000): 400–3.

53. Kajubi, P., M. R. Kamya, S. Kamya, et al. "Increasing Condom Use Without Reducing HIV Risk: Results of a Controlled Community Trial in Uganda." *J Acq Immunodef Syn* 40 (2005): 77–82.

54. Williamson, L. M., J. P. Dodds, D. E. Mercey, G. J. Hart, and A. M. Johnson. "Sexual Risk Behaviour and Knowledge of HIV Status Among Community Samples of Gay Men in the UK." *AIDS* 22 (2008): 1063–70.

55. Bezemer, D., F. de Wolf, M. C. Boerlijst, et al. "A Resurgent HIV-1 Epidemic Among Men Who Have Sex with Men in the Era of Potent Antiretroviral Therapy." *AIDS* 22 (2008): 1071–77.

56. Coates, T. "What Is to Be Done?" *AIDS* 22 (2008): 1079–80.

57. Metcalf, C. A., J. M. Douglas, C. K. Malotte, et al. "Relative Efficacy of Prevention Counseling with Rapid and Standard HIV Testing: A Randomized, Controlled Trial (RESPECT-2)." *Sex Transm Dis* 32 (2005): 130–38.

58. Pinkerton, S. D. "Probability of HIV Transmission During Acute Infection in Rakai, Uganda." *AIDS & Behav* 12 (2008): 677–84.

59. Hollingsworth, T. D., R. M. Anderson, and C. Fraser. "HIV-1 Transmission, by Stage of Infection." *J Infect Dis* 198 (2008): 687–93.

60. Cohen, M. S., G. M. Shaw, A. J. McMichael, and B. F. Haynes. "Acute HIV-1 Infection." *N Engl J Med* 364 (2011): 1943–54.

61. Granich, R. M., C. F. Gilks, C. Dye, K. M. De Cock, and B. G. Williams. "Universal Voluntary HIV Testing with Immediate Antiretroviral Therapy as a Strategy for Elimination of HIV Transmission: A Mathematical Model." *Lancet* 373 (2009): 48–59.

62. Wagner, B. G., and S. Blower. "Voluntary Universal Testing and Treatment Is Unlikely to

Lead to HIV Elimination: A Modeling Analysis." *Nature Precedings*, October 2009, http://precedings.nature.com/documents/3917/version/1.

63. Herling Ruark, A. H., J. D. Shelton, D. T. Halperin, M. Wawer, and R. Gray. "Universal Voluntary HIV Testing and Immediate Antiretroviral Therapy." *Lancet* 373 (2009): 1078.

64. Cohen, M. S., and C. L. Gay. "Treatment to Prevent Transmission of HIV-1." *Clinical Infectious Diseases* 50, no. S3 (2010): S85–S95.

65. Dodd, P. J., G. P. Garnett, and T. B Hallett. "Examining the Promise of HIV Elimination by 'Test and Treat' in Hyperendemic Settings." *AIDS* 24 (2010): 729–35.

66. Maman, S., J. K. Mbwambo, N. M. Hogan, E. Weiss, G. P. Kilonzo, and M. D. Sweat. "High Rates and Positive Outcomes of HIV-Serostatus Disclosure to Sexual Partners: Reasons for Cautious Optimism from a Voluntary Counseling and Testing Clinic in Dar es Salaam, Tanzania." *AIDS and Behav* 7 (2003): 373–82.

67. Tobian, A. A., R. H. Gray, and T. C. Quinn. "Male Circumcision for the Prevention of Acquisition and Transmission of Sexually Transmitted Infections." *Arch Pediatr Adolesc Med* 16 (2010): 478–84.

68. Weiss, H. A., D. Halperin, R. C. Bailey, R. J. Hayes, G. Schmid, and C. A. Hankins. "Male Circumcision for HIV Prevention: From Evidence to Action?" *AIDS* 22 (2008): 567–74.

69. Wamai, R. G., H. A. Weiss, C. Hankins, et al. "Male Circumcision Is an Efficacious, Lasting and Cost-Effective Strategy for Combating HIV in High-Prevalence AIDS Epidemics: Time to Move Beyond Debating the Science." *Future HIV Ther* 2 (2008): 399–405.

70. Halperin, D. T., and R. C. Bailey. "Male Circumcision and HIV Infection: 10 years and Counting." *Lancet* 354 (1999): 1813–15, www.circumcisioninfo.com/halperin_bailey.html.

71. Weiss, H., J. Polonsky, R. Bailey, C. Hankins, D. Halperin, G. Schmid. *Male Circumcision: Global Trends and Determinants of Prevalence, Safety and Acceptability*. Geneva: WHO/UNAIDS (2007), www.malecircumcision.org/media/documents/MC_Global_Trends_Determinants.pdf.

72. Drain, P. K., D. T. Halperin, J. P. Hughes, J. D. Klausner, and R. C. Bailey. "Male Circumcision, Religion, and Infectious Diseases: An Ecologic Analysis of 118 Developing Countries." *BMC Infect Dis* 6 (2006): 172–82.

73. Lie, R. K, and F. G. Miller. "What Counts as Reliable Evidence for Public Health Policy: The Case of Circumcision for Preventing HIV Infection." *BMC Medical Research Methodology* 11 (2011): 34–41.

74. Cohen, J. "Male Circumcision Thwarts HIV Infection." *Science* 309 (2005): 860.

75. Auvert, B., D. Taljaard, E. Lagarde, J. Sobngwi-Tambekou, R. Sitta, and A. Puren. "Randomized, Controlled Intervention Trial of Male Circumcision for Reduction of HIV Infection Risk: The ANRS 1265 Trial." *PLoS Med* 2 (2005): 1112–22.

76. Bailey, R. C., S. Moses, C. B. Parker, et al. "Male Circumcision for HIV Prevention in Young Men in Kisumu, Kenya: A Randomised Controlled Trial." *Lancet* 369 (2007): 643–56.

77. Gray, R. H., G. Kigozi, and D. Serwadda. "Male Circumcision for HIV Prevention in Men in Rakai, Uganda: A Randomised Trial." *Lancet* 369 (2007): 657–66.

78. Shelton, J., "Estimated Protection Too Conservative." *Plos Med* 3 (2006): e65.

79. Auvert, B., D. Taljaard, D. Rech, et al. "Effect of the Roll-Out of Male Circumcision in

Orange Farm (South Africa) on the Spread of HIV (ANRS-12126)." Presented at the International AIDS Society Meetings, Rome, 2011, Abstract #WELBC02.

80. Kong, X., G. Kigozi, V. Ssempija, et al. "Longer-Term Effects of Male Circumcision on HIV Incidence and Risk Behaviors During Post-trial Surveillance in Rakai, Uganda." Presented at the 18th Conference on Retroviruses and Opportunistic Infections, Boston, Feb. 27–Mar. 2, 2011. Abstract 36, www.hivandhepatitis.com/2011 conference/croi2011docs/0311_2010c .html.

81. Hallett, T. B., R. A. Alsallaq, J. M. Baeten, et al. "Will Circumcision Provide Even More Protection from HIV to Women and Men? New Estimates of the Population Impact of Circumcision." *Sex Transm Infect* 87 (2010): 88–93.

82. Williams, B. G., J. L. Lloyd-Smith, E. Gouws, et al, "The Potential Impact of Male Circumcision on HIV in Sub-Saharan Africa." *PLoS Med* 3 (2006): e262.

83. Nagelkerke, N. J., S. Moses, S. J. de Vlas, and R. C. Bailey. "Modelling the Public Health Impact of Male Circumcision for HIV Prevention in High Prevalence Areas in Africa." *BMC Infect Dis* 7 (2007): doi: 10.1186/1471-2334-7-16.

84. Gray, R. H., X. Li, G. Kigozi, et al. "The Impact of Male Circumcision on HIV Incidence and Cost per Infection Prevented: A Stochastic Simulation Model from Rakai, Uganda." *AIDS* 21 (2007): 845–50.

85. Kahn, J. G., E. Marseille, and B. Auvert. "Cost-Effectiveness of Male Circumcision for HIV Prevention in a South African Setting." *PLoS Med* 3 (2006): e517.

86. Auvert, B., E. Marseille, E. L. Korenromp, et al. "Estimating the Resources Needed and Savings Anticipated from Roll-out of Adult Male Circumcision in Sub-Saharan Africa." *PLoS One* 3 (2008): e2679.

87. UNAIDS/WHO/SACEMA Expert Group on Modelling the Impact and Cost of Male Circumcision for HIV Prevention. "Male Circumcision for HIV Prevention in High HIV Prevalence Settings: What Can Mathematical Modelling Contribute to Informed Decision Making?" *PLoS Med* 6 (2009): e1000109.

88. Orroth, K. K., R. G. White, E. E. Freeman, et al. "Attempting to Explain Heterogeneous HIV Epidemics in Sub-Saharan Africa: Potential Role of Historical Changes in Risk Behaviour and Male Circumcision." *Sex Transm Infect* (2011): doi: 10.1136/sextrans-2011-050174.

89. Caldwell, J. C., and P. Caldwell. "The African AIDS Epidemic." *Scientific American*, March 1996.

90. Weiss, H. A., S. L. Thomas, S. K. Munabi, and R. J. Hayes. "Male Circumcision and Risk of Syphilis, Chancroid, and Genital Herpes: A Systematic Review and Meta-Analysis." *Sex Transm Infect* 82 (2006): 101–9.

91. Tobian, A. A., D. Serwadda, T. C. Quinn, et al. "Male Circumcision for the Prevention of HSV-2 and HPV Infections and Syphilis." *N Engl J Med* 360 (2009): 1298–1309.

92. Auvert, B., J. Sobngwi-Tambekou, E. Cutler, et al. "Effect of Male Circumcision on the Prevalence of High-Risk Human Papillomavirus in Young Men: Results of a Randomized Controlled Trial Conducted in Orange Farm, South Africa." *J Infect Dis* 19 (2009): 914–19.

93. Gray, R. H., D. Serwadda, X. Kong, et al. "Male Circumcision Decreases Acquisition and Increases Clearance of High-Risk Human Papillomavirus in HIV-Negative Men: A Randomized Trial in Rakai, Uganda." *J Infect Dis* 201 (2010): 1455–62.

94. Gray, R. H., G. Kigozi, D. Serwadda, et al. "The Effects of Male Circumcision on Female Partners' Genital Tract Symptoms and Vaginal Infections in a Randomized Trial in Rakai, Uganda." *Am J Obstet Gynecol* 200 (2009): e1–e7.

95. Backes, D. M., M. C. Bleeker, C. J. Meijer, et al. "Male Circumcision Is Associated with a Lower Prevalence of Human Papillomavirus-Associated Penile Lesions Among Kenyan Men." *Int J Cancer* (2011) doi: 10.1002/ijc.26196.

96. Wawer, M. J., A. A. R. Tobian, G. Kigozi, et al. "Effect of Circumcision of HIV-Negative Men on Transmission of Human Papillomavirus to HIV-Negative Women: A Randomised Trial in Rakai, Uganda." *Lancet* 377 (2011): 209–18.

97. Castellsagué, X., F. X. Bosch, N. Munoz, et al. "Male Circumcision, Penile Human Papillomavirus Infection, and Cervical Cancer in Female Partners." *N Engl J Med* 346 (2002): 1105–12.

98. Bosch, F. X., G. Albero, X. Castellsagué. "Male Circumcision, Human Papillomavirus and Cervical Cancer: From Evidence to Intervention." *J Fam Plann Reprod Health Care* 35 (2009): 5–7.

99. Wawer, M. J., F. Makumbi, G. Kigozi, et al. "Randomized Trial of Male Circumcision in HIV-Infected Men: Effects on HIV Transmission to Female Partners, Rakai, Uganda." *Lancet* 374 (2009): 229–37.

99. Weiss, H. A., C. A. Hankins, and K. Dickson. "Male Circumcision and Risk of HIV Infection in Women: A Systematic Review and Meta-Analysis." *Lancet Infect Dis* 9 (2009): 669–77.

101. Gray, R. H., N. Kiwanuka, T. C. Quinn, et al. "Male Circumcision and HIV Acquisition and Transmission: Cohort Studies in Rakai, Uganda. Rakai Project Team." *AIDS* 14 (2000): 2371–81.

102. Baeten, J. M., D. Donnell, S. H. Kapiga, et al., for the Partners in Prevention HSV/HIV Transmission Study Team. "Male Circumcision and Risk of Male-to-Female HIV-1 Transmission: A Multinational Prospective Study in African HIV-1-Serodiscordant Couples." *AIDS* 24 (2010): 737–44.

103. Westercamp, N., and R. C. Bailey. "Acceptability of Male Circumcision for Prevention of HIV/AIDS in Sub-Saharan Africa: A Review." *AIDS Behav* 11 (2007): 341–55.

104. Halperin, D. T., K. Fritz, W. McFarland, and G. Woelk. "Acceptability of Adult Male Circumcision for Sexually Transmitted Disease and HIV Prevention in Zimbabwe." *Sexually Transmitted Diseases* 32 (2005): 238–39.

105. Brito, M. O., L. M. Caso, H. Balbuena, and R. C. Bailey. "Acceptability of Male Circumcision for the Prevention of HIV/AIDS in the Dominican Republic." *PLoS One* 4 (2009): e7687.

106. Plank, R. M., J. Makhema, P. Kebaabetswe, et al. "Acceptability of Infant Male Circumcision as Part of HIV Prevention and Male Reproductive Health Efforts in Gaborone, Botswana, and Surrounding Areas." *AIDS Behav* 10 (2010): 1007/s10461-009-9632-0.

107. Westercamp, M., K. E. Agot, J. Ndinya-Achola, and R. C. Bailey. "Circumcision Preference Among Women and Uncircumcised Men Prior to Scale-up of Male Circumcision for HIV Prevention in Kisumu, Kenya." *AIDS Care*, Aug. 22, 2011. [Epub ahead of print, www.ncbi.nlm.nih.gov/pubmed/21854351].

108. Madhivanan P., K. Krupp, V. Chandrasekaran, S. C. Karat, A. L. Reingold, and J. D. Klaus-

ner. "Acceptability of Male Circumcision Among Mothers with Male Children in Mysore, India." *AIDS* 22 (2008): 983–88.

109. Madhivanan, P., K. Krupp, V. Kulkarni, S. Kulkarni, and J. D. Klausner. "Acceptability of Male Circumcision for HIV Prevention Among High-Risk Men in Pune, India." *Sex Transm Dis* 38 (2011): 571.

110. Sullivan, S. G., W. Ma, S. Duan, F. Li, Z. Wu, and R. Detels. "Attitudes Towards Circumcision Among Chinese Men." *J Acquir Immune Defic Syndr* 50 (2009): 238–40.

111. PSI-Haiti/CRESHM. "Pilot Circumcision Project: Follow-up and Final Evaluation Report." April 2005.

112. Mattson, C. L., R. T. Campbell, R. C. Bailey, K. Agot, J. O. Ndinya-Achola, and S. Moses. "Risk Compensation Is Not Associated with Male Circumcision in Kisumu, Kenya: A Multi-Faceted Assessment of Men Enrolled in a Randomized Controlled Trial." *PLoS One* 3 (2008): e2443.

113. Agot, K. E., J. N. Kiarie, H. Q. Nguyen, J. O. Odhiambo, T. M. Onyango, and N. S. Weiss. "Male Circumcision in Siaya and Bondo Districts, Kenya: Prospective Cohort Study to Assess Behavioral Disinhibition Following Circumcision." *J Acquir Immune Defic Syndr* 44 (2007): 66–70.

114. Riess, T. H., M. M. Achieng', S. Otieno, J. O. Ndinya-Achola, and R. C. Bailey. "'When I Was Circumcised I Was Taught Certain Things': Risk Compensation and Protective Sexual Behavior Among Circumcised Men in Kisumu, Kenya." *PLoS One* 5 (2010): e12366, doi: 10.1371/journal.pone.0012366.

115. Krieger, J. N., S. D. Mehta, R. C. Bailey, et al. "Adult Male Circumcision: Effects on Sexual Function and Sexual Satisfaction in Kisumu, Kenya." *J Sex Med* 5 (2008): 2610–22.

116. Kigozi, G., S. Watya, C. B. Polis, et al. "The Effect of Male Circumcision on Sexual Satisfaction and Function: Results from a Randomized Trial of Male Circumcision for Human Immunodeficiency Virus Prevention, Rakai, Uganda." *BJU Int* 10 (2008): 165–70.

117. Kigozi G., I. Lukabwe, J. Kagaayi, et al. "Sexual Satisfaction of Women Partners of Circumcised Men in a Randomized Trial of Male Circumcision in Rakai, Uganda." *BJU Int* 104 (2009): 1698–1701.

118. Collins, S., J. Upshaw, S. Rutchik, C. Ohannessian, J. Ortenberg, and P. Ibertsen. "Effects of Circumcision on Male Sexual Function: Debunking a Myth?" *J Urol* 167 (2002): 2111–12.

119. Masood, S., H. R. H. Patel, R. C. Himpson, J. H. Palmer, G. R. Mufti, and M. K. M Sheriff. "Penile Sensitivity and Sexual Satisfaction After Circumcision: Are We Informing Men Correctly?" *Urol Int* 75 (2005): 62–66.

120. Senkul, T., C. Iseri, B. Sen, K. Karademir, F. Saracoglu, and D. Erden. "Circumcision in Adults: Effect on Sexual Function." *Urology* 63 (2004): 155–58.

121. Bleustein, C. B., J. D. Fogarty, H. Eckholdt, J. C. Arezzo, and A. Melman. "Effect of Neonatal Circumcision on Penile Neurological Sensation." *Urology* 65 (2005): 773–77.

122. Schober, J. M., H. F. Meyer-Bahlburg, and C. Dolezal. "Self-Ratings of Genital Anatomy, Sexual Sensitivity and Function in Men Using the 'Self-Assessment of Genital Anatomy and Sexual Function, Male' Questionnaire." *BJU Int* 103 (2009): 1096–1103.

123. Payne, K., L. Thaler, T. Kukkonen, S. Carrier, and Y. Binik. "Sensation and Sexual Arousal in Circumcised and Uncircumcised Men." *J Sex Med* 4 (2007): 667–74.

124. Fink, K. S., C. C. Carson, and R. F. deVellis. "Adult Circumcision Outcomes Study: Effect on Erectile Function, Penile Sensitivity, Sexual Activity and Satisfaction." *J Urol* 167 (2002): 2113–16.

125. Ferris, J. A., J. Richters, M. K. Pitts, et al. "Circumcision in Australia: Further Evidence on Its Effects on Sexual Health and Wellbeing." *Aust NZ J Publ Hlth* 341 (2010): 60–64.

126. Senol, M. G., B. Sen, K. Karademir, H. Sen, and M. Saraçoğlu. "The Effect of Male Circumcision on Pudendal Evoked Potentials and Sexual Satisfaction." *Acta Neurol Belg* 10 (2008): 890–93.

127. Laumann, E. O., C. M. Maal, and E. W. Zuckerman. "Circumcision in the United States: Prevalence, Prophyactic Effects, and Sexual Practice." *J Am Med Assoc* 277 (1997): 1052–57.

128. Williamson, M. L., and P. S. Williamson. "Women's Preferences for Penile Circumcision in Sexual Partners." *J Sex Educ Ther* 14 (1988): 8–12.

129. Abdool-Karim, Q., S. Abdool-Karim, J. A. Frohlich, et al. "Effectiveness and Safety of Tenofovir Gel, an Antiretroviral Microbicide, for the Prevention of HIV Infection in Women." *Science* 329 (2010): 1168–74.

130. Centers for Disease Control and Prevention. "CDC Statement on Study Results of Products Containing Nonoxynol-9." *Morbidity and Mortality Weekly Report* 49 (2000): 717–18.

131. McMahon, J. M., K. M. Morrow, M. Weeks, D. Morrison-Beedy, and A. Coyle. "Potential Impact of Vaginal Microbicides on HIV Risk Among Women with Primary Heterosexual Partners." *J Assoc Nurses AIDS Care* 22 (2011): 9–16.

132. Hearst, N. "Condoms in the Context of Microbicides." Presentation at the Microbicides 2004 Conference, March 28–31, 2004, London, United Kingdom.

133. Brown, J. E., and R. C. Brown. "Traditional Intravaginal Practices and the Heterosexual Transmission of Disease: A Review." *Sex Trans Dis* 27 (2000): 183–88.

134. Van de Wijgert, J. H., P. R. Mason, L. Gwanzura, et al. "Intravaginal Practices, Vaginal Flora Disturbances and Acquisition of Sexually Transmitted Diseases in Zimbabwean Women." *Journal Infectious Diseases* 18 (2000): 587–94.

135. Dallabetta, G., P. Miotti, J. Chiphangwi, et al. "Traditional Vaginal Agents: Use and Association with HIV Infection in Malawian Women." *AIDS* 9 (1995): 293–97.

136. Beksinska, M. E., H. V. Rees, I. Kleinschmidt, and J. McIntyre. "The Practice and Prevalence of Dry Sex Among Men and Women in South Africa: A Risk Factor for Sexually Transmitted Infections?" *Sex Transm Inf* 5 (1999): 178–80.

137. Halperin, D. T. "Dry Sex Practices and HIV Infection in the Dominican Republic and Haiti." *Sex Trans Infect* 75 (1999): 445–46.

138. Gafos, M., M. Mzimela, S. Sukazi, R. Pool, C. Montgomery, and J. Elford. "Intravaginal Insertion in KwaZulu-Natal: Sexual Practices and Preferences in the Context of Microbicide Gel Use." *Cult Health Sex* 12 (2010): 929–42.

139. Padian, N. S., A. van der Straten, G. Ramjee, et al. "Diaphragm and Lubricant Gel for Prevention of HIV Acquisition in Southern African Women: A Randomized Controlled Trial." *Lancet* 370 (2007): 251–61.

140. Shelton, J. D., D. T. Halperin, V. Nantulya, M. Potts, H. D. Gayle, and K. K. Holmes. "Partner Reduction Is Crucial for Balanced 'ABC' Approach to HIV Prevention." *BMJ* 328 (2004): 891–94.

141. Halperin, D. T., and H. Epstein. "Why Is HIV Prevalence So Severe in Southern Africa?

The Role of Multiple Concurrent Partnerships and Lack of Male Circumcision: Implications for AIDS Prevention." *South African J HIV Med* 16 (2007): 19–25, www.sajhivmed.org.za/index.php/sajhivmed/article/viewFile/54/412.

142. Southern African Development Community. "Expert Think Tank Meeting on HIV Prevention in High-Prevalence Countries in Southern Africa: Report." May 10–12, 2006, www.sadc.int/downloads/news/SADCPrevReport.pdf137.

143. Mah, T., and D. T. Halperin. "The Evidence for the Role of Concurrent Partnerships in Africa's HIV Epidemics." *AIDS and Behavior* 2010, doi 10.1007/s10461-009-9617-z.

144. Shelton, J. D. "Why Multiple Sex Partners?" *Lancet* 374 (2009): 367–69.

145. Green, E. C., T. L. Mah, A. Ruark, and N.Hearst. "A Framework of Sexual Partnerships: Risks and Implications for HIV Prevention in Africa." *Studies in Family Planning* 40 (2009): 63–70.

146. Mah T. L., and J. D. Shelton. "Concurrency Revisited: Increasing and Compelling Epidemiological Evidence." *J Int AIDS Society* 14 (2011): 3341.

147. Schwartländer, B., J. Stover, T. Hallett, et al. "Towards an Improved Investment Approach for an Effective Response to HIV/AIDS." *Lancet* 377 (2011): 2031–41.

148. Eaton, J., T. Hallett, G. Garnett. "Concurrent Sexual Partnerships and Primary HIV Infection: A Critical Interaction." *AIDS Behav* 4 (2011): 687–92.

149. Ghys, P. D., E. Gouws, R. Lyerla, et al. "Trends in HIV Prevalence and Sexual Behavior Among Young People Aged 15–24 Years in Countries Most Affected by HIV." *Sexually Transm Infect* 86 (2010): ii72–ii83.

150. Hallett, T. B., J. Aberle-Grasse, E. K. Alexandre, et al. "Declines in HIV Prevalence Can Be Associated with Changing Sexual Behaviour in Uganda, Urban Kenya, Zimbabwe and Urban Haiti." *Sex Transm Infect* 82 (2006): i1–i8. doi: 10.1136/sti.2005.0160.

151. Stoneburner, R. L., and D. Low-Beer. "Population-Level HIV Declines and Behavioral Risk Avoidance in Uganda." *Science* 304 (2004): 714–18.

152. Green, E. C., D. T. Halperin, V. Nantulya, and J. A. Hogle. "Uganda's HIV Prevention Success: The Role of Sexual Behavior Change and the National Response." *AIDS Behav* 10 (2006): 335–46.

153. Slutkin, G., S. Okware, W. Naamara, et al. "How Uganda Reversed Its HIV Epidemic." *AIDS Behav* 10 (2006): 351–60.

154. Epstein, H. "The Fidelity Fix." *New York Times Magazine*, June 13, 2004, pp. 54–59.

155. Bessinger, R., P. Akwara, and D. T. Halperin. "Sexual Behavior, HIV and Fertility Trends: A Comparative Analysis of Six Countries; Phase I of the ABC Study." Washington, D.C.: Measure Evaluation/USAID, 2003, www.cpc.unc.edu/measure/publications/pdf/sr-03-21b.pdf.

156. Cheluget, B., G. Baltazar, P. Orege, M. Ibrahim, L. H. Marum, and J. Stover. "Evidence for Population Level Declines in Adult HIV Prevalence in Kenya. *Sex Transm Infect* 82, Suppl 1 (2006): i21–26.

157. Gregson, S., G. P. Garnett, C. A. Nyamukapa, et al. "HIV Decline Associated with Behavior Change in Eastern Zimbabwe." *Science* 311 (2006): 664–66.

158. Muchini, B., C. Benedikt, E. Gomo, et al. "Local Perceptions of the Forms, Timing and Causes of Behavior Change in Response to the AIDS Epidemic in Zimbabwe." *AIDS Behav* 2010, doi 10.1007/s10461-010-9783-z.

159. Mekonnen, Y., E. Sanders, M. Aklilu, et al. "Evidence of Changes in Sexual Behaviours Among Male Factory Workers in Ethiopia." *AIDS* 17 (2003): 223–31.

160. Bello, G., B. Simwaka, T. Ndhlovu, F. Salaniponi, and T. B. Hallett. "Evidence for Changes in Behaviour Leading to Reductions in HIV Prevalence in Urban Malawi." *Sex Transm Infect* 2011, doi: 10.1136/sti.2010.043786.

161. Michelo, C., I. Sandøy, and K. Fylkesnes. "Antenatal Clinic HIV Data Found to Underestimate Actual Prevalence Declines: Evidence from Zambia." *Trop Med Internat Health* 13 (2008): 171–79.

162. Morris, M., S. Goodreau, J. Moody. "Sexual Networks, Concurrency, and STD/HIV." In K. K. Holmes, P. F. Sparling, W. E. Stamm, P. Piot, J. N. Wasserheit, L. Corey, et al., eds., *Sexually Transmitted Diseases*. 4th ed. (New York: McGraw-Hill, 2007), pp. 109–25.

163. Goodreau, S. M., S. Cassels, D. Kasprzyk, D. E. Montaño, A. Greek, and M. Morris. "Concurrent Partnerships, Acute Infection and Epidemic Dynamics in Zimbabwe." *AIDS Behav* 2010, doi: 10.1007/s10461-010-9858-x.

164. Spina, A. "Secret Lovers Kill: A National Mass Media Campaign to Address Multiple and Concurrent Partnerships" (Concise Report Summary). AIDSTAR-One: Case Studies Series, www.comminit.com/en/node/305575/cchangepicks.

165. Soul City Institute Regional Programme. "Multiple and Concurrent Sexual Partnerships in Southern Africa: A Ten Country Research Report." 2008, www.soulcity.org/za.

166. Halperin, D. T. "The Controversy over Fear Arousal in AIDS Prevention and Lessons from Uganda." *J Health Commun* 11 (2006): 266–67.

167. Grant, R. M., J. R. Lama, P. L. Anderson, et al. "Preexposure Chemoprophylaxis for HIV Prevention in Men Who Have Sex with Men." *N Engl J Med* 363 (2010): 2587–99.

168. Baeten, J., D. Donnell, and C. Celum. "Pre-exposure Prophylaxis with Antiretroviral Drugs in Men and Women in Africa." Presentation at the 6th IAS Conference on HIV Pathogenesis, Treatment and Prevention, Rome, Italy, July 18, 2011.

169. Thigpen, M., D. Smith, and L. Paxton. "Oral Tenofovir/Emtricitabine as Pre-exposure Prophylaxis in Heterosexual Couples in Botswana." Presentation at the 6th IAS Conference on HIV Pathogenesis, Treatment and Prevention, Rome, Italy, July 18, 2011.

170. Family Health International. "FHI Statement on the FEM-PrEP HIV Prevention Study." April 18, 2011, http://www.fhi.org/en/AboutFHI/Media/Releases/FEM-PrEP_statement 041811.htm.

171. Cohen J. "Complexity Surrounds HIV Prevention Advances." *Science* 333 (2011): 393.

172. Smith K., K. A. Powers, A. D. M. Kashuba, M. S. Cohen. "HIV-1 Treatment as Prevention: The Good, the Bad, and the Challenges." *Curr Opin HIV AIDS* 6 (2011): 315–25.

173. Shelton, James D. "ARVs as HIV Prevention: A Tough Road to Wide Impact." *Science* 334 (2011): 1645–46.

174. Shapiro, R. L., M. D. Hughes, A. Ogwu, et al. "Antiretroviral Regimens in Pregnancy and Breastfeeding in Botswana." *N Engl J Med* 362 (2010): 2282–94.

175. Kourtis, A. P., D. J. Jamieson, and I. de Vincenzi. "Prevention of Human Immunodeficiency Virus-1 Transmission to the Infant Through Breastfeeding: New Developments." *Am J Obstet Gynecol* 197 (2007): S113–22.

176. Thior, I., S. Lockman, et al. "Breastfeeding Plus Infant Zidovudine Prophylaxis for 6 Months

vs. Formula Feeding Plus Infant Zidovudine for 1 Month to Reduce Mother-to-Child HIV Transmission in Botswana: A Randomized Trial: The Mashi Study." *JAMA* 296 (2006): 794–805.

177. Iliff, P. J., E. G. Piwoz, and N. V. Tavengwa. "Early Exclusive Breastfeeding Reduces the Risk of Postnatal HIV-1 Transmission and Increases HIV-Free Survival." *AIDS* 19 (2005): 699–708.

178. Halperin, D. T. "Breastfeeding and HIV Transmission: The Realities May Be Much More Complex Than 'Breast vs. Bottle.' " *Global AIDSLink* 58 (1999): 8–10.

179. WHO/UNICEF/UNFPA/UNAIDS. "Guidelines on HIV and Infant Feeding: Principles and Recommendations for Infant Feeding in the Context of HIV and a Summary of Evidence." 2010.

180. Heffron, R., D. Donnell, H. Rees, et al. "Use of Hormonal Contraceptives and Risk of HIV-1 Transmission: A Prospective Cohort Study." *Lancet Infectious Diseases* (2011); doi: 10.1016/S1473-3099(11)70247-X.

181. Morrison, C. S., A. N. Turner, and L. B. Jones. "Highly Effective Contraception and Acquisition of HIV and Other Sexually Transmitted Infections." *Best Practice & Research Clinical Obstetrics and Gynaecology* 23 (2009): 263–84.

182. USAID. "USAID Response to Publication of Findings on Hormonal Contraception and HIV Acquisition in Uninfected Women and HIV Transmission from Infected Women to Male Partners, Washington: USAID, October 7, 2011, www.usaid.gov/our_work/global_health/pop/techareas/docs/usaid_hc_hiv.pdf.

183. Shelton, J. D. "Scientists and the Media Must Give a Balanced View." *Nature* 79 (2011): 7.

184. Reynolds, H. W., B. Janowitz, R. Homan, and L. Johnson. "The Value of Contraception to Prevent Perinatal HIV Transmission." *Sexually Transmitted Diseases* 33 (2006): 350–56.

185. Duerr, A., S. Hurst, A. P. Kourtis, N. Rutenberg, and D. J. Jamieson. "Integrating Family Planning and Prevention of Mother-to-Child HIV Transmission in Resource-Limited Settings." *Lancet* 336 (2005): 261–63.

186. Halperin, D. T., J. Stover, and H. Reynolds. "Benefits and Costs of Expanding Access to Family Planning Programs to Women Living with HIV." *AIDS* 23 (2009): S123–30.

187. Peck R., D. W. Fitzgerald, B. Liautaud, et al. "The Feasibility, Demand, and Effect of Integrating Primary Care Services with HIV Voluntary Counseling and Testing: Evaluation of a 15-Year Experience in Haiti, 1985–2000." *J Acquir Immune Defic Syndr* 33 (2003): 470–75.

188. The World Bank. "Malawi and Tanzania Research Shows Promise in Preventing HIV and Sexually-Transmitted Infections." July 18, 2010, http://web.worldbank.org/WBSITE/EXTERNAL/NEWS/0,contentMDK:22649337~pagePK:34370~piPK:34424~theSitePK:4607,00.html.

189. Baird, S., E. Chirwa, C. McIntosh, and B. Özler. "The Short-Term Impacts of a Schooling Conditional Cash Transfer Program on the Sexual Behavior of Young Women." *Health Economics* 19, S1 (2010): 55–68.

190. Baird, S., C. McIntosh, and B. Özler. "Schooling, Income, and HIV Risk: Experimental Evidence from a Cash Transfer Program." Presentation at the XVIII International AIDS Conference; Vienna, Austria; July 18–23, 2010.

191. Dunbar, M. S., M. C. Maternowska, M. S. Kang, I. Mudekunye-Mahaka, and

N. S. Padian. "Findings from SHAZ!: A Feasibility Study of a Microcredit and Life-Skills HIV Prevention Intervention to Reduce Risk Among Adolescent Female Orphans in Zimbabwe." *J Prev Interv Community* 38 (2010): 147–61.

192. Luke, N. "Confronting the 'Sugar Daddy' Stereotype in Urban Kenya." *Int Fam Plan Perspect* 31 (2005): 6–14.

193. Chesson, H. W., P. Harrison, and R. Stall. "Changes in Alcohol Consumption and in Sexually Transmitted Disease Incidence Rates in the United States: 1983–1998." *J Stud Alcohol* 64 (2003): 623–30.

194. Chesson, H., P. Harrison, and W. J. Kassler. "Sex Under the Influence: The Effect of Alcohol Policy on Sexually Transmitted Disease Rates in the United States." *J. Law Econ* 43 (2000): 215–38.

195. Gisselquist, D., J. J. Potterat, S. Brody, F. Vachon. "Let It Be Sexual: How Health Care Transmission of AIDS in Africa Was Ignored." *Int J STD AIDS* 14 (2003): 148–61.

196. Gisselquist, D., J. J. Potterat, J. S. St. Lawrence, et al. "How to Contain Generalized HIV Epidemics? A Plea for Better Evidence to Displace Speculation." *Int J STD AIDS* 20 (2009): 443–46.

197. Brewer, D. D., J. J. Potterat, M. Okinyi. "Data Trump Speculation and Distortion of HIV Transmission Routes in Sub-Saharan Africa." *Int J STD AIDS* 22 (2011): 118–20.

198. Sawers, L., and E. Stillwaggon. "Concurrent Sexual Partnerships Do Not Explain the HIV Epidemics in Africa: A Systematic Review of the Evidence." *Journal of the International AIDS Society* (2010) 13: 1–24.

199. Boily, M. C., R. G. White, M. Alary, C. M. Lowndes, and K. Orroth. "Transmission of HIV via Unsafe Injection or Unsafe Sex? Anomalies or Misunderstanding?" *Int J STD AIDS* 15 (2004): 61–63.

200. Garnett, G. P., and C. Fraser. "Let It Be Sexual—Selection, Aggregation and Distortion Used to Construct a Case Against Sexual Transmission." *Int J STD & AIDS* 14 (2003): 782–84.

201. Lopman, B. A., G. P. Garnett, P. R. Mason, S. Gregson. "Individual Level Injection History: A Lack of Association with HIV Incidence in Rural Zimbabwe." *PLoS Med* 2 (2005): e37.

202. Schmid, G. P., A. Buve, P. Mugyeny, et al. "Transmission of HIV-1 Infection in sub-Saharan Africa and Effect of Elimination of Unsafe Injections." *Lancet* 363 (2004): 482–88.

203. Grosskurth, H., F. Mosha, J. Todd, et al. "Impact of Improved Treatment of Sexually Transmitted Diseases on HIV Infection in Rural Tanzania: Randomised Controlled Trial." *Lancet* 346 (1995): 530–36.

204. Korenromp, E. L., R. G. White, K. K. Orroth, et al. "Determinants of the Impact of Sexually Transmitted Infection Treatment on Prevention of HIV Infection: A Synthesis of Evidence from Mwanza, Rakai and Masaka Intervention Trials." *J Infect Dis* 191, suppl. 1 (2005): 168–78.

205. Gray, R. H., and M. Wawer. "Reassessing the Hypothesis on STI Control for HIV Prevention." *Lancet* 371 (2008): 2064–65.

206. Celum, C., A. Wald, J. R. Lingappa, et al. "Acyclovir and Transmission of HIV-1 from Persons Infected with HIV-1 and HSV-2." *New England Journal of Medicine* 362 (2010): 427–39.

207. Korenromp, E. L., R. Bakker, S. J. de Vlas, et al. "HIV Dynamics and Behaviour Change

as Determinants of the Impact of Sexually Transmitted Disease Treatment on HIV Transmission in the Context of the Rakai Trial," *AIDS* 16 (2002): 2209–18.

208. CDC-Botswana (courtesy of F. J. Ndowa). "HIV and Syphilis Prevalence Among Pregnant Women in Botswana, 1990–2002." Presented by D. Halperin at WHO/UNAIDS Meeting on STI Control for HIV Prevention, July 2006, Geneva, Switzerland.

209. Botswana National AIDS Control Program (courtesy of F. J. Ndowa). "Changes in the Aetiology of GUD in Botswana from 1993 to 2002." Presented by D. Halperin at WHO/UNAIDS Meeting on STI Control for HIV Prevention, July 2006, Geneva.

210. Johannesburg Metropolitan Council (courtesy of F. J. Ndowa). "Prevalence of Reactive Syphilis Serological Tests (RPR) among STD Patients in Johannesburg (1993–2001)." Presented by D. Halperin at WHO/UNAIDS Meeting on STI Control for HIV Prevention, July 2006, Geneva.

211. Sloan, N. L., B. Winikoff, N. Haberland, C. Coggins, and C. Elias. "Screening and Syndromic Approaches to Identify Gonorrhea and Chlamydial Infection Among Women." *Stud Fam Plann* 31 (2000): 55–68.

212. Bukusi, E. A., C. R. Cohen, A. S. Meier, et al. "Bacterial Vaginosis: Risk Factors Among Kenyan Women and Their Male Partners." *Sex Transm Dis* 33 (2006): 361–67.

213. Hassan, W. M., L. Lavreys, V. Chohan, et al. "Associations Between Intravaginal Practices and Bacterial Vaginosis in Kenyan Female Sex Workers Without Symptoms of Vaginal Infections. *Sex Transm Dis* 34 (2007): 384–88.

214. Anglemyer, A., G. W. Rutherford, M. Egger, and N. Siegfried. "Antiretroviral Therapy for Prevention of HIV Transmission in HIV-Discordant Couples." *Cochrane Database of Systematic Review* 5 (2011).

215. Attia, S., M. Egger, M. Muller, et al. "Sexual Transmission of HIV According to Viral Load and Antiretroviral Therapy: Systematic Review and Meta-Analysis." *AIDS* 23 (2009): 1397–1404.

216. Del Romero, J., J. Castilla, V. Hernando, et al. "Combined Antiretroviral Treatment and Heterosexual Transmission of HIV-1: Cross-Sectional and Prospective Cohort Study." *BMJ* 340 (2010): 2205.

217. Donnell, D., J. M. Baeten, et al. "Heterosexual HIV-1 Transmission After Initiation of Antiretroviral Therapy: A Prospective Cohort Analysis." *Lancet* 375 (2010): 2092–98.

218. Cohen, M. S., Y. Chen, M. McCauley, et. al. "Prevention of HIV-1 Infection with Early Antiretroviral Therapy." *New Engl J Med* 365 (2011): 493–505.

219. Cates, W. "Who Gets the HAART? Policy Implications for a Limited Resource." In review.

220. Cates, W. "HPTN 052 and the Future of HIV Treatment and Prevention." *Lancet* 378 (2011): 224–25.

221. Paquet, A.C., et al. "Significant Reductions in the Prevalence of Protease Inhibitor and 3-Class Resistance: Recent Trends in a Large HIV-1 Protease/Reverse Transcriptase Database," Presentation at the 51st Interscience Conference on Antimicrobial Agents and Chemotherapy, Abstract #H2-800, Chicago, September 2011, www.aidsmap.com/page/2080135.

222. Epstein, H., and M. Morris. "HPTN 052 and the Future of HIV Treatment and Prevention." *Lancet* 378 (2011): 225.

223. Grulich, A. E., D. P. Wilson. "Is Antiretroviral Therapy Modifying the HIV Epidemic?" *Lancet* 376 (2010): 1824.

224. Shelton, J. D., M. Cohen, M. Barnhart, T. Hallett. "Is Antiretroviral Therapy Modifying the HIV Epidemic?" *Lancet* 376 (2010): 1824–25.

225. McFarland, W. "HIV/AIDS Epidemiological Update, San Francisco." Slide presentation for HIVInsite, University of California, San Francisco, June 10, 2009, http://hivinsite.ucsf.edu/InSite?page=cfphp-mcfarland2009-sl.

226. Montaner, J.S.G., V. D. Lima, R. Barrios, et al. "Association of Highly Active Antiretroviral Therapy Coverage, Population Viral Load, and Yearly New HIV Diagnoses in British Columbia, Canada: A Population-Based Study." *Lancet* 376 (2010): 532–39.

227. Das, M., P. L. Chu, G. M. Santos, et al. "Decreases in Community Viral Load Are Accompanied by Reductions in New HIV Infections in San Francisco." *PLoS One* 5 (2010): e11068.

228. Walque, D., H. Kazianga, and M. Over. "Policy Research Working Paper 5486: Antiretroviral Therapy Awareness and Risky Sexual Behaviors: Evidence from Mozambique." The World Bank Development Research Group Human Development and Public Services Team, November 2010.

229. Powers, K. A., A. C. Ghani, W. C. Miller, et al. "The Role of Acute and Early HIV Infection in the Spread of HIV and Implications for Transmission Prevention Strategies in Lilongwe, Malawi: A Modelling Study." *Lancet* 378 (2011): 256–68.

230. Shelton, J. D. "A Tale of Two-Component Generalised HIV Epidemics." *Lancet* 375 (2010): 964–66.

231. Payne, B. A. I., I. J. Wilson, C. A. Hateley, et al. "Mitochondrial Aging Is Accelerated by Anti-retroviral Therapy Through the Clonal Expansion of mtDNA Mutations." *Nature Genetics* 43 (2011): 806–10.

232. Bongaarts, J., and M. Over. "Global HIV/AIDS Policy in Transition." *Science* 328 (2010): 1359–60.

233. Cohen, J. "Promising AIDS Vaccine's Failure Leaves Field Reeling." *Science* 318 (2007): 28–29.

234. Cohen, J. "Did Merck's Failed HIV Vaccine Cause Harm?" *Science* 318 (2007): 1048–49.

235. Horton, R. "AIDS: The Elusive Vaccine." *The New York Review of Books*, September 23, 2004.

236. Rerks-Ngarm, S., P. Pitisuttithum, S. Nitayaphan, et al. "Vaccination with ALVAC and AIDSVAX to Prevent HIV-1 Infection in Thailand." *N Engl J Med* 361 (2009): 2209–20.

237. Butler, D. "Jury Still Out on HIV Vaccine Results." *Nature* 461 (2009): 1187.

238. Cohen, J. "Unrevealed Analysis Weakens Claim of AIDS Vaccine 'Success.'" *Science Insider*, Oct. 5, 2009, http://news.sciencemag.org/scienceinsider/2009/10/unrevealed-anal.html.

ADDITIONAL SUGGESTED READINGS

Barnett, Tony, and Alan Whiteside. *AIDS in the Twenty-First Century: Disease and Globalization*, second edition. Basingstoke, UK: Palgrave, 2006.

Behrman, Greg. *The Invisible People: How the U.S. Has Slept Through the Global AIDS Pandemic, the Greatest Humanitarian Catastrophe of Our Time*. New York: Free Press, 2004.

Carter, David. *Stonewall: The Riots That Sparked the Gay Revolution*. New York: St. Martin's Press, 2004.

Chin, James. *The AIDS Pandemic: The Collision of Epidemiology with Political Correctness*. Milton Keynes, UK: Radcliffe Publishing, 2007.

Cohen, Jon. *Shots in the Dark: The Wayward Search for an AIDS Vaccine*. New York: W. W. Norton, 2001.

Druckerman, Pamela. *Lust in Translation: The Rules of Infidelity from Tokyo to Tennessee*. New York: The Penguin Press, 2007.

Easterly, William. *The White Man's Burden: Why the West's Efforts to Aid the Rest Have Done So Much Ill and So Little Good*. New York: Penguin Books, 2006.

Epstein, Helen. *The Invisible Cure: Why We Are Losing the Fight Against AIDS in Africa*. New York: Picador, 2008.

Ewens, Graeme. *Congo Colossus: The Life and Legacy of Franco & OK Jazz*. North Walsham, UK: Buku Press, 1994.

Garrett, Laurie. *The Coming Plague: Newly Emerging Disease in a World Out of Balance*. New York: Penguin Books, 1994.

Geffen, Nathan. *Debunking Delusions: The Inside Story of the Treatment Action Campaign*. Johannesburg: Jacana Media, 2010.

Gevisser, Mark. *Thabo Mbeki: The Dream Deferred*. Cape Town: Jonathan Ball Publishers, 2007.

Geschiere, Peter. *Village Communities and the State : Changing Relations Among the Maka of South-Eastern Cameroon Since the Colonial Conquest*, James J. Ravell, trans. London: Kegan Paul International, 1982.

Giles-Vernick, Tamara. *Cutting the Vines of the Past: Environmental Histories of the Central African Rain Forest*. Charlottesville: University Press of Virginia, 2002.

Green, Edward. *AIDS and STDs in Africa: Bridging the Gap Between Traditional Healers and Modern Medicine*. New York: Westview Press, 1994.

Green, Edward, and Allison Herling Ruark. *AIDS, Behavior, and Culture: Understanding Evidence-Based Prevention*. San Francisco: Left Coast Press, 2011.

Gumede, William. *Thabo Mbeki and the Battle for the Soul of the ANC*. Cape Town: Struick Publishers, 2007.

Hochschild, Adam. *King Leopold's Ghost: A Story of Greed, Terror and Heroism in Colonial Africa*. New York: Houghton Mifflin Company, 1998.

Hooper, Edward. *The River: A Journey to the Source of HIV and AIDS*. New York: Little, Brown and Company, 1999.

Hunt, Nancy Rose. *A Colonial Lexicon: Of Birth Ritual, Medicalization and Mobility in the Congo*. Durham, NC: Duke University Press, 1999.

Iliffe, John. *The African AIDS Epidemic: A History*. Athens: Ohio University Press, 2006.

Jochelson, Karen. *The Colour of Disease: Syphilis and Racism in South Africa, 1880–1950*. Hampshire, UK: Palgrave Macmillan, 2001.

Johnson, Steven. *The Ghost Map: The Story of London's Most Terrifying Epidemic—And How It Changed Science, Cities and the Modern World*. New York: Riverhead, 2006.

Kalichman, Seth. *Denying AIDS: Conspiracy Theories, Pseudoscience, and Human Tragedy*. New York: Copernicus Books, 2010.

Kenyatta, Jomo. *Facing Mt. Kenya: The Tribal Life of the Gikuyu*. New York: Vintage, 1965.

Mann, Jonathan, and Daniel Tarantola, eds. *AIDS in the World II*. New York: Oxford University Press, 1996.

Moyo, Dambisa. *Dead Aid: Why Aid Is Not Working and How There Is a Better Way for Africa*. New York: Farrar, Strauss and Giroux, 2009.

Mugyeni, Peter. *Genocide by Denial: How Profiteering from HIV/AIDS Killed Millions*. Kampala, Uganda: Fountain Publishers, 2008.

Nolen, Stephanie. *28: Stories of AIDS in Africa*. New York: Walker & Company, 2007.

Pankenham, Thomas. *The Scramble for Africa*. New York: Harper Perennial, 1991.

Pepin, Jacques. *The Origins of AIDS*. New York: Cambridge University Press, 2011.

Pisani, Elizabeth. *The Wisdom of Whores: Bureaucrats, Brothels, and the Business of AIDS*. New York: W. W. Norton & Co., 2008.

Roiphe, Katie. *Last Night in Paradise: Sex and Morals at the Century's End*. New York: Little, Brown and Company, 1997.

Rotello, Gabriel. *Sexual Ecology: AIDS and the Destiny of Gay Men*. New York: Plume, 1997.

Rudin, Harry. *Germans in the Cameroon, 1884–1914*. New Haven: Yale University Press, 1938.

Schapera, Isaac. *Married Life in an African Tribe*. Evanston, IL: Northwestern University Press, 1966.

Seabrook, Jeremy. *Love in a Different Climate—Men Who Have Sex with Men in India*. London: Verso Press, 1999.

Shilts, Randy. *And the Band Played On: Politics, People and the AIDS Epidemic*. New York: St. Martin's Press, 1987.

Steinberg, Johnny. *Sizwe's Test: A Young Man's Journey Through Africa's AIDS Epidemic*. New York: Simon & Schuster, 2008.

Vaughan, Megan. *Curing Their Ills: Colonial Power and African Illness*. Cambridge, UK: Polity Press, 1991.

Whiteside, Alan. *HIV/AIDS: A Very Short Introduction*. New York: Oxford University Press, 2008.

CHART AND ILLUSTRATION CREDITS

Page

18 Courtesy of *The Washington Post*. Photo by Craig Timberg.

62 HP.1955.96.26, collection RMCA Tervuren; photo by C. Lamote (Inforcongo), circa 1955, copyright RMCA Tervuren.

116 Courtesy of the National Committee for the Prevention of AIDS in Uganda.

141 Adapted from the United Nations AIDS Program Report on the Global AIDS Epidemic, 2010. Geneva: UNAIDS.

145 Courtesy of Martina Morris.

146 Courtesy of Jim Moody and Martina Morris.

179 UNAIDS Global Report, 2010, Geneva; Demographic and Health Surveys, ORC Macro, Calverton, MD; Nelson Mandela Foundation, South Africa; Botswana Central Statistics Office.

200 Photo by Daniel Halperin.

244 Courtesy of Family Life Association of Swaziland.

260 Courtesy of National Emergency Response Council on HIV-AIDS, Mbabane, Swaziland.

270 Courtesy of *The Washington Post*. Photo by Craig Timberg.

283 Stephane Helleringer and Hans Peter Kohler, "Sexual Network Structure and the Spread of HIV in Africa: Evidence from Likoma Island, Malawi." *AIDS* 21 (2007): 2323–32.

311 Mishra, Vinod, Rathavuth Hong, Simona Bignami-Van Assche, and Bernard Barrere. *The Role of Partner Reduction and Faithfulness in HIV Prevention in Sub-Saharan Africa: Evidence from Cameroon, Rwanda, Uganda, and Zimbabwe.* DHS Working Papers No.61. (2009) Calverton, MD: ORC Macro.

312 Kenya Demographic and Health Surveys 1993, 1998, 2003; Kenya Central Bureau of Statistics/ORC Macro, Calverton, MD.

INDEX

ABC prevention strategy
   Brazil's rejection of, 240
   conservatives' interest in, 209
   and funds earmarked for abstinence, 209,
      210, 359n9
   introduction of, 203–5
   origins of, 357n12
   USAID's adoption of, 358n16
abstinence
   and ABC strategy, 203–5, 209, 240, 358n19,
      359n9
   and Brazil, 240
   and Christian conservatives, 209–10,
      358n19
   earmarked funds for, 209, 210, 359n9
   effectiveness of, 307
   following male circumcision, 375n10
   and PEPFAR program, 212
   in Uganda, 248
   and USAID, 203–4, 205, 240
Achmat, Zackie, 190
activism and activists
   and access to treatment, 81, 173, 185, 190,
      215
   of gay community, 79–80, 81–82, 129–30
   radicalization of, 81, 82, 342n2
   in South Africa, 190–91, 217–18, 220
acute infection phase
   and concurrency, 367n2
   contagiousness of, 245–46, 349n3

   duration of, 323n12
   and rate of infections, 326n1
Africa
   activism in, 81, 130, 190–91
   AIDS Belt, 92, 142, 147–48, 170, 247, 259
   birth rates in, 59–60
   boomtowns in, 56–57
   civil war and conflict in, 322n7
   condom use in, 84, 122–24, 134–35, 258,
      290, 307
   country names in, 46
   deaths from AIDS in, 86–87, 111, 130, 160,
      173, 183, 185, 214–15, 218–22, 232–33
   decline in new infections, 126–27, 132,
      134–35, 137, *141*, 237, 239, 387n26
   denialism in, 92–94, 97, 193–96
   era of independence in, 64–65
   and gay population, 130, 323n9
   Great Lakes Region, 6, 109, 173, 250
   hunter-gatherers of, 102–3
   initial identification of heterosexual AIDS
      cases in, 87–88, 90
   music industry of, 118
   natural resources of, 45, 49, 52, 57, 63–64
   parenthood in, 21–22
   paths of HIV, *12, 76*
   and PEPFAR program, 206–12, 215
   political sensitivities in, 143
   porters and spread of HIV in, 1–2, 3, 48–49,
      51, 59, 71

poverty in, 194–96, 338n7, 338n2, 339n3
public figures with AIDS in, 94–99, 119–21
radios in, 97
rates of infection in, *168*, 185, 251, 290, 321n5, 372n3
sexual behaviors in, 21–22, 56, 92, 97–98, 143–44 (*see also* concurrency; multiple partners; polygamy)
spread of HIV in, 71, 77–78, 88–90, 109–11, 173, 195–96, 322n7
stereotypes of victims in, 19
transportation systems in, 63–64, 66, 89, 157, 159–60, 173
vaccination campaigns, 66
and Westerners, 1–4, 8, 22, 24–25, 41, 281, 286 (*see also* colonialism)
women in, 156–57, 249, 261
See also colonialism; male circumcision; *specific countries*
African Americans, 172, 216, 332n10, 348n1, 383n5
African National Congress (ANC)
Mbeki's leadership of, 183, 186
and Virodene, 182, 187, 188, 189, 196
and Zulus, 174
AIDS fatigue, 222, 281
"AIDS Industrial Complex" commentary, 131, 287
alcohol regulations for AIDS prevention, 314–15
anal sex
and behavior changes among gay men, 82
in Brazil, 240, 366n16
and male circumcision, 322n9
risk associated with, 39, 80–81, 290
spread of HIV via, 140, 305
antiretrovirals (ARVs)
adherence rates for, 286, 317, 376n14
and AIDS activists, 81, 173, 185, 190, 215
availability of, 5, 182, 215, 286, 365n13
AZT, 187–88, 192–93
in Botswana, 15, 225–26, 232, 234
in Brazil, 239–40
and breast-feeding, 270–71, 313–14
and decline in AIDS deaths, 298
demands for, 81, 215
effectiveness of, 184
introduction of, 137–38
Mbeki's skepticism of, 184, 187–88, 192–93, 234, 275
and mothers with HIV, 187, 190, 265, 268, 270–71, 290, 301, 313–14, 371n2, 381n9

and PEPFAR program, 207, 212, 215, 301, 360n12
and perceptions of epidemic, 216
and progression of AIDS, 39
resources for, 148–49
and risk compensation, 313, 316, 377n18
in South Africa, 182–83, 185–86, 187–88, 190, 215–16, 234, 272–75, 356n11
and "test-and-treat" approach, 384n14
and treatment-as-prevention, 287–89, 302, 313, 316–17, 388n28
underutilization of, by infected population, 377n19
in Zimbabwe, 237, 239
Ashe, Arthur, 277, 331n1
Asia
arrival of HIV in, 71
concentrations of HIV in, 332n6
condom use in, 129
extent of epidemic in, 290
gay population in, 323n9
male circumcision in, 147, 175, 259, 382n3
prevention campaigns in, 140
rates of infection in, 139, *168*, 342n13
Australia, 377n18
Auvert, Bertran, 229–31, 361n14
AZT (antiretroviral drug), 187–88, 192–93

bacterial vaginosis, 387n27
Bailey, Robert
background of, 177
on extramarital sex, 347n3
on male circumcision, 178, 231, 284, 352n16
Bangkok, 195
Bangladesh, 175, 323n11, 351n10, 372n2
Bantu-speaking peoples, 103–4, 105, 106, 336n4
behavior change campaigns
backlash against, 285–86
in Kenya, 292
Lutaaya's personal campaign, 119–21, 124
in Mozambique, 292
in Swaziland, 259–62, *260*, 370n9
in Uganda, 112–18, *116*, 122–23, 126, 128, 135, 211, 247, 311
Belgian Congo
end of, 64–66, 72
natural resources of, 54, 61
sexual behaviors in, 58–60
transportation systems in, 63–64
See also Leopoldville, Kinshasa
Belgium, 85, 86, 323n11

Berlin Conference, 54
Bertrand, Jane, 93, 98, 99
Bill and Melinda Gates Foundation
    funding for Botswana, 14, 226–27, 234
    funding for male circumcision services, 284,
        381n8
    funding for loveLife campaign, 199
    unintentional effects of, 211
billboard campaigns
    loveLife campaign, 199–200, *200*, 214,
        272
    makhwapheni campaign, *260*, 260–61
    in Swaziland, 361n10
    UNAIDS campaign, 342n10
birth control and family planning
    and Botswana's sexual culture, 21–22
    effectiveness of, 381n9
    funding for, 379n34
    and HIV-positive mothers, 387n23
    hormonal-based contraceptives, 386n22
    and mother-to-child transmission, 314
birth of the AIDS epidemic
    colonialism's role in, 2–3, 35–36
    date of, 32, 34–35
    jump from chimps to humans, 2, 30, 31–32,
        49–51, 329n14
    location of, 31–32, 47
    and social change, 40–41
bisexuality, 172
Black Death, 40
blood and plasma supplies, 78, 315, 331n11,
    332n8
Bolivia, 323n11
Bongaarts, John, 107, 147
Botswana
    antiretrovirals in, 225–27, 232, 234
    behavior change campaigns in, 313
    breast-feeding vs. infant formula in, 263–71
    colonial history of, 224
    condom use in, 324n5
    deaths from AIDS in, 232–33
    diarrhea and child deaths in, 264–69
    and Gates/Merck program, 226–28
    male circumcision in, *179*, 228–29,
        361n13
    mothers with HIV in, 313
    natural resources of, 22, 224
    other sexually transmitted diseases in,
        346n25
    and path of HIV, 13
    rains of, 264, 269
    rate of infection in, 13, *179*, 225, 233, 234,
        323n11, 356n9, 364n25

sexual behaviors in, 15–22, 225, 233, 324n9,
    360n2
spread of HIV in, 225
Western influence on, 228
Zimbabwe compared to, 239
    See also Francistown, Botswana
Brazil, 239–40, 306, 365n15, 366n16, 384n10
Brazzaville, French Congo, 56
breast-feeding, 263–71
    admonitions against, 263
    advocates of, 266–67, 300
    and antiretrovirals, 270–71, 313–14
    controversy on, 266–67, 371n4, 381n6
    and mixed feeding, 267, 314, 371n3
    rate of infection from, 265–66, 371nn2, 3
    recommendations on, 290–91
    WHO protocols for, 313–14
bubonic plague, 142
Buchanan, Patrick, 79
Burkina Faso, *179*
Burundi, 208
Bush, George W.
    and ABC strategy, 241
    abstinence emphasis of administration, 240
    and child victims of AIDS, 265
    and international relations, 239
    worldview of, 201
    See also President's Emergency Plan for AIDS
        Relief (PEPFAR)
bush meat, 30, 329n14

Caldwell, John
    on male circumcision/multiple partner
        combination, 147–48, 346n26
    Scientific American article of, 170, 172, 177
    on sexual culture, 330n6
Caldwell, Pat
    on male circumcision/multiple partner
        combination, 147–48, 346n26
    Scientific American article of, 170, 172, 177
    on sexual culture, 330n6
Cambodia, 129, 291, 307
Cameroon
    chimpanzee population in, 31
    colonial history of, 35, 46–52, 57
    as epicenter of epidemic, 2, 32, 34, 36, 51,
        329n14
    fidelity rates in, *311*
    HIV subtypes in, 321n4
    male circumcision in, *179*, 259
    path of HIV, *12*
    rate of infection in, *179*
    sexual behaviors in, 259, 334n11

Carael, Michel
  on sexual behavior, 132–35, 144, 209, 247, 248, 345n22, 367n5
  and Stoneburner, 132–34, 298, 341n12
  UNAIDS position of, 247
Caribbean, 382n6. *See also* Dominican Republic; Haiti and Haitian immigrants
Castro, Fidel, 112
casual sex, 15–20, 82, 122, 130, 134
cell phones, 260
Centers for Disease Control
  and ABC strategy, 240
  and AIDS activists, 82
  on breast-feeding/formula debate, 268, 269
  on child deaths in Botswana, 265, 268, 269, 270
  initial identification of HIV, 77, 78, 143
  and *One Love* campaign, 302
  and PEPFAR program, 207
Central Africa
  and "Four Cities" investigation, 387n26
  male circumcision in, 259, 336n5
  mixed epidemics in, 383n6
  other sexually transmitted diseases in, 315, 346n25
cervical cancer, 178, 310, 337n10
cervical cap, 333n13
cervix, 328n6
chancroid, 101, 310, 335n1
children
  in African cultures, 155, 157
  babies born to HIV mothers, 313–14, 314, 371nn2, 4
  with HIV, 150, 265, 381n9
  orphans, 199, 314
  and water sanitation, 263–71
  WHO protocols for, 313–14
  See also mothers with HIV
chimpanzees
  and birth of the AIDS epidemic, 1–2, 30, 50–51, 321n3
  as food, 2, 30, 40, 329n14
  and HIV-1 group M, 31
  and polio vaccination hypothesis, 25, 26–29, 339n5
  See also SIV (simian immunodeficiency virus)
Chin, James, 141, 251, 368n14
China, 139, 259, 343n14, 364n7
chlamydia, 59, 60, 315, 346n25, 359n5
cholera, 293–94
Christian missionaries and churches
  abstinence emphasis of, 204, 210, 358n19
  and brutality of colonialism, 45–46

as disruptive force, 41
  Dr. Livingstone, 42–44, 52
  evangelical Christians, 337n6, 358n19
  and HIV testing, 122
  in Luoland, 158–59
  and male circumcision rituals, 6, 106, 228
  and PEPFAR program, 207, 359n10, 366n18
  and polygamy, 58, 116, 159–60
  reaction to AIDS, 116–17, 207
  and sexual behaviors in Africa, 22, 120
circumcision. *See* male circumcision
Cleland, John, 143
Clinton administration, 196
Cohen, Myron, 288
Cold War, 65
colonialism
  and Berlin Conference, 54
  and birth of the AIDS epidemic, 1–4, 49–51
  and birth rates, 59–60
  and boomtowns, 56–57
  brutality associated with, 43–47, 48–49
  and circumcision practices, 106
  diseases spread by, 46, 50
  end of, 64–66, 72
  legacies of, 4, 67
  and natural resources, 365n14
  and perceptions of Africa, 41
  and prostitution, 56, 58–59
  and sexual behaviors, 8, 50, 56–59
  and spread of HIV, 2–3, 22, 35–36
  and taxation, 159
  and trade, 54
  and transportation system, 157, 159–60
  and vaccination campaigns, 24, 25–29, 66
  and World War I, 52
concentrated HIV epidemics, 305–6, 332n6
concurrency
  and acute infection stage, 367n2
  among African Americans, 348n1, 383n5
  backlash against initiatives, 285
  and changes in sexual behavior, 289
  decline in practice of, 237
  impact of interventions aimed at, 378n22
  and male circumcision rates, 259
  methodological flaws in data regarding, 282, 372n2
  modeling of, 345n24
  and multiple partners, 376n12
  and other sexually transmitted diseases, 348n1
  prevalence of, 143–45, 344n20, 345n22, 355n7

concurrency (*cont.*)
  prevention campaigns addressing, 259–62,
    260, 280, 291, 292, 312, 379n32
  and spread of HIV, 144–46, *145*, *146*,
    372n2, 372n3, 376n12
  See also multiple partners
condoms
  in Africa, 84, 122–24, 134–35, 258, 290,
    307
  for anal sex, 240
  attitudes toward, 97, 98, 122–23, 277
  in Botswana, 15, 324n5
  in Brazil, 240
  and Bush administration, 212, 241
  campaigns promoting, 176, 203–4, 225,
    227–28, 235, 248, 257–58, 306,
    361n10, 369n4
  and casual sex, 258
  and concurrent relationships, 145
  "condom code," 129–30
  and decline in new infections, 134–35, 248,
    307, 357n10
  effectiveness of, 307–8, 327n5, 332n13,
    369n4
  emphasis on, 151, 171, 204, 205
  failure rates of, 332n12, 384n10
  and gay population, 82, 307
  inconsistent use of, 84, 222–23, 332n12,
    341n14
  limitations of, 359n8
  mainstream adoption of, 83–84
  and other sexually transmitted diseases, 83,
    384n11
  and Potts's hot-air balloon, 131
  and prostitution, 98, 129, 235, 237, 290, 307
  and ritual sex acts, 156
  soldiers' use of, 339n7
  in Thailand, 258, 306
  and training programs, 361n12
  and USAID, 204
  in Zimbabwe, 235, 237
Congo
  colonial history of, 36, 44–46, 53–66, 91
  Haitian community in, 72–73
  independence of, 64–65
  path of HIV, *12*
  and polio vaccination hypothesis, 24,
    25–29, 30
  rate of infection in, 335n18
  SIV in, 326n5
  turbulence in, 72
  See also Democratic Republic of Congo
Congo River, 31–32, 48, 51, 55, 63

Congo River Basin, 51, 77, 325n1
conservatives
  and ABC strategy, 209
  abstinence emphasis of, 204, 210, 333n14,
    358n19
  and PEPFAR program, 366n18
conspiracy theories, 24, 325n2, 355n3
Coovadia, Hoosen, 267
Cuba, 112
cumulative risk of HIV infection, 322n6

deaths
  in Africa, 86–87, 111, 130, 160, 173, 183,
    185, 214–15, 218–22, 232–33, 238, 239
  anticipating waves of, 111
  and behavior change, 238, 239, 355n14
  decline in, 232–33, 298, 380n1
  in gay population, 81, 82
  trends in, 142, 232
De Cock, Kevin, 369n5
Delay, Paul, 367n6
Democratic Republic of Congo, 34, 46, 99, *179*
Demographic and Health Survey (DHS), 246,
    248–52, 321n5, 367n3, 368n13
denialism
  among scientists, 193–94
  in early stages of the epidemic, 92–94
  and Franco, 97, 119
  in Luo family, 161–62
  and *makhwapheni* campaign, 370n9
  and Mbeki, 193–96, 299, 355n3
  prevalence of, 189–90
diamonds, 224, 365n14
diaphragm, 333n13
diarrhea and child deaths, 264–69
discordant couples, 37–38, 287, 326n1, 340n8,
    362n19, 382n3, 385n16
Dlamini, Gugu, 189–90
Dobson, James, 366n18
Dominican Republic, 129, 249, 291, 307, 312
Druckerman, Pamela (*Lust in Translation*),
    343n17
drugs to treat AIDS. *See* antiretrovirals (ARVs)
"dry sex," 328n7, 386n19
Duesberg, Peter, 193
Dulles, Allen, 65
Duvalier, François "Papa Doc," 72

East Africa, 77, 105, 173, 251, 259, 350n4
Easterly, William, 4
Ebola virus, 39, 67–70, 85–86, 126
economy, influence of, 236–38, 239, 365n11
"edutainment," 291–92, 312, 379n31

Egypt, 379n31
Eisenhower, Dwight, 65
epidemic curve, 141–42, 251
epidemics (the term/classification of), 322n8,
    381n2
Epstein, Helen, 146, 347n32, 367n2, 379n32
ethical issues (related to clinical trials), 385n16
Ethiopia
    and funding for AIDS programs, 379n34
    male circumcision in, *179*
    and PEPFAR program, 208
    rate of infection in, *179*, 233, 250, 291
    sexual behaviors in, 312
Europe
    arrival of HIV in, 72–73, 78, 85
    extent of epidemic in, 290, 333n1
    male circumcision in, 147, 259, 283,
        352n18, 370n8
    rates of infection in, 139, *168*
    sexual behaviors in, 354n5

facial scarring, 101–2, 335n2
Falwell, Jerry, 79
Family Health International, 130–31, 328n5,
    347n32
Family Life Association of Swaziland, 366n1
family planning. *See* birth control and family
    planning
Farmer, Paul, 331n12, 379n34
fear, use of for prevention campaign, 114, 128,
    248
female genital mutilation, 352n18, 385n18
fidelity
    and ABC strategy, 203, 209, 240
    attitudes toward, 278
    and Botswana's sexual culture, 21
    and PEPFAR program, 358n19, 366n18
    prevalence of, *311*
    See also Zero Grazing campaign
Fink, Aaron, 107
Focus on the Family, 366n18
foreign aid initiatives, 4–5, 257–58
"Four Cities" investigation, 387n26
fragile nature of HIV, 39
Francistown, Botswana
    colonial history of, 224
    prevention campaigns in, 14–15
    prosperity in, 14
    rate of infection in, 13, 14, 20–21
    sexual behaviors in, 15–22
    See also Botswana
"Franco" (Makiadi, François Luambo "Franco"),
    94–99, 119, 334n11

Fumento, Michael, 82
funding
    and abstinence programs, 210
    increases in, 198, 286
    organizations' pursuit of, 131, 149–51,
        199–200, 286–87
    and UNAIDS infection rates estimates, 250

Garrett, Laurie, 68, 70
Gates, Bill, 226
gay population
    activism in, 79–80, 81–82, 129–30
    and African American community, 172
    and African epidemic, 130
    AIDS victims in, 171
    condom use in, 82, 307
    deaths in, 81, 82
    decline in new infections in, 387n26
    in Europe, 85
    and Haitian tourism, 73
    as high-risk population, 79, 140, 342n8
    and initial CDC AIDS notice, 143
    and male circumcision, 322n9
    and Mugabe's homophobia, 364n2
    perception of AIDS as gay disease, 78
    and prevention campaigns, 140, 290
    rate of infection in, 332n5
    risk compensation in, 316
    sexual behaviors in, 8, 80–81, 82–83, 280
    spread of HIV in, 77, 80, 142, 216, 305,
        382n4
    stigmas against, 172
genetics, 327n4, 348n1, 373n3
genital ulcers, 101, 102, 361n14
German colonialism, 46–50, 52
Ghana, 64, *179*
Glaxo Wellcome, 192
Global Fund to Fight AIDS, Tuberculosis and
    Malaria, 302, 359n4
Global Health Initiative, 379n34
gonorrhea
    in colonial era, 59, 60
    and condom use, 83
    in gay population, 82
    and prevention of HIV infection, 315
    in Nairobi slum, 101
    and sexual behaviors, 142
    social stigma of, 122
gorillas, 30
Graham, Franklin, 207
Green, Edward
    on ABC strategy, 358n19
    on decline in new infections, 127

Green, Edward (*cont.*)
  on delaying sexual activity, 209
  Harvard fellowship of, 357n11
  on power of behavior change, 127, 128, 176
  on Uganda's prevention campaign, 202–3
Greenwich Village, 80
ground zero for epidemic, 32, 34, 36
Guinea, *179*, 346n25
Guyana, 208, 359n4

Hahn, Beatrice, 31, 36, 40, 50, 321n2
Haiti and Haitian immigrants
  arrival of HIV in, 71–73, 78, 331n12
  blood supplies from, 331n11
  in Congo, 72, 73
  in Europe, 85
  and funding for AIDS programs, 379n34
  as high-risk population, 79
  male circumcision in, 176, 243, 351n12
  rate of infection in, 291
  sexual behaviors in, 312
Hallett, Timothy, 340n3
Halperin, Daniel (as character)
  and ABC strategy, 209, 240, 359n9
  background of, 170–71
  and behavior change campaigns, 259,
    261–62, 379n32
  in Botswana, 225, 325n9
  and the Bush administration, 201–5,
    239–42, 366n19
  and Caldwells' research, 169–70
  on condom promotion, 258, 360n2
  and cost of AIDS programs, 201
  on heterosexual transmission, 349n3
  HIV testing by, 256
  on "holy trinity" of prevention, 258–59
  on local initiatives, 302
  and loveLife program, 200–201
  on male circumcision, 6, 173–78, 180, 201,
    204, 243–44, 245, 259, 283, 352n16
  on sexual behavior, 6–7, 176–77, 209, 325n9
  in Swaziland, 241–42, 243, 256, 302
  in Tanzania, 294
  and Ugandan data set, 247, 367n6
  at USAID, 201–5, 239–42, 357n15
  volunteer activities of, 171–72
  in Zimbabwe, 237–38
Hamilton, William D., 25–28, 30, 325n5,
    328n8
Harvard AIDS Institute, 227
Hearst, Norman, 257–58, 369n4
Helms, Jesse, 207
hemophiliacs, 79, 80, 85, 332n8

herpes, 310, 361n14, 369n5, 384n11
heterosexuals as a risk group
  and African American community, 172,
    383n5
  identified as African AIDS cases, 87–88, 90,
    147–48
  incorrect focus of prevention campaigns on,
    140
  and inefficient modes of HIV transmission,
    37–38
  as low-risk population in most regions, 82,
    349n2
  sexual behaviors of, 37, 39, 81, 147–48,
    346n24
  and transmission of HIV, 86, 87–88, 370n8
high-risk groups
  identification of, 79
  and prophylaxis, 378n26
  spread of HIV in, 139–40, 142
  as target audience, 139–40, 148, 290, 306
  and treatment-as-prevention, 316
Hill, Kent, 241–42, 366n19
Hindus, 351n10
Hispanic populations, 332n10, 349n2
HIV-1 group M
  death rates from, 321n4
  early containment of, 63
  first appearance of, 3
  identification of, 30–31
  paths of, *76*
  subtypes of, 63, 66, 70–71, *76*, 323n12
HIV-1 group N, 321n14
HIV-1 group O, 321n14
Hochschild, Adam, 91
homophobia, 79, 364n2
Hooper, Edward, 24–25, 28–29, 66
hormonal-based contraceptives, 386n22
HPV (human papillomavirus), 310, 384n11,
    385n16
Hudson, Rock, 331n1
Hunt, Nancy Rose, 57

identification of HIV, 23–24
income levels, 322n7, 324n6, 356n9
India
  arrival of HIV in, 71
  concentrated HIV epidemics in, 306
  condom use in, 129, 307
  male circumcision in, 175
  mixed epidemics in, 383n6
  prostitution in, 129
  rate of infection in, 139, 140–41, *168*, 291,
    347n30, 351n10, 364n7, 380n1

infant formula for young children, 263–71
infectious disease, epidemic curve in, 141–42
injections, unsterile, 315, 325n3, 387n24
international AIDS conferences, 80, 132, 134, 136–38, 194–96
intravenous drug users
  in African American community, 172
  and arrival of HIV, 78
  as high-risk population, 79, 80, 342n8
  and prevention campaigns, 290
  prevention efforts for, 306, 382n4
  spread of HIV in, 142
  and treatment-as-prevention programs, 388n28
Iraq, 358n2
Israel, 283
Ivory Coast
  circumcision in, 179, 259
  rate of infection in, 179
  sexual behaviors in, 143, 259, 312

Jamaica, 328n5
Johannesburg, South Africa, 57, 195
Johnson, Earvin, Jr. "Magic," 37
Johnson, Steven, 293–94
"Juliana" in Uganda, 111

Kagera, Tanzania, region, see Tanzania
Kagimba, Jesse, 114–15
Kaiser Family Foundation, 199
Kampala, Uganda, 76, 88–90, 112, 127, 132, 140
Kapita, Bila, 86–87, 93–94
Kaposi's sarcoma, 78, 88–89, 109–10
Katabira, Elly, 124
Kaunda, Kenneth, 364n2
Kenya
  antiretrovirals in, 297
  arrival of HIV in, 71
  behavior change campaigns in, 292, 313
  colonial history of, 157–58
  condom use in, 357n10
  decline in new infections, 311
  and funding for AIDS programs, 379n34
  male circumcision in, 178, 179, 231, 284, 350n9, 354n21, 374n5, 375n11, 385n18
  rate of infection in, 160, 178, 179, 233, 249, 291, 323n12
  sexual behaviors in, 282, 311, 312, 363n24, 373n4, 374n5
  See also Luoland and Luo ethnic group
Kenyatta, Jomo, 325n10

Kigali, Rwanda, 90, 334n11
Kilmarx, Peter, 97–98
Kinshasa (formerly Leopoldville)
  AIDS cases in, 86–88
  as boomtown, 56–57
  civil war in, 99
  colonial history of, 33, 35, 36, 44–45, 51
  denialism in, 92–93
  deterioration of, 86
  as epicenter of epidemic, 32, 34, 36, 51
  founding of, 44
  growth of, 62
  Haitian community in, 72–73
  health care system in, 333n4
  historic HIV (DRC60) from, 33–34
  location of, 55
  Mama Yemo Hospital in, 93
  other sexually transmitted diseases in, 59–60
  path of HIV, 12
  rate of infection in, 105, 334n11, 335n18
  sexual culture in, 56, 57–59, 66
  spread of HIV from, 66, 70
  and transportation system, 63–64
  Westernization of, 61–62, 62
Kirby, Doug, 338n12, 347n32
Kramer, Larry, 79–80, 94
Kuti, Fela, 119, 338n1
Kyomuhendo, Swizen, 124–25

Lake Malawi, 282, 373n4
Lake Victoria, 109, 154
Langerhans cells, 336n6
Latin America and Latinos
  concentrated HIV epidemics in, 332n6
  extent of epidemic in, 290
  gay population in, 323n9
  and male circumcision, 349n2, 382n3
  prevention campaigns in, 140
  rate of infection in, 168, 349n2
  spread of HIV in, 323n9
Leclerc-Madlala, Suzanne, 198–99
Leopold II, 44–45, 54
Leopoldville. See Kinshasa (formerly Leopoldville)
Lesotho, 144, 179, 208, 323n11, 342n9, 349n9
Liberia, 179, 368n13
Livingstone, David, 42–44, 52
Lobeke National Park, 50
loveLife program
  billboards of, 199–200, 200, 214, 272
  funding of, 257, 302
  messages of, 199–200

Low-Beer, Daniel, 132–33, 135, 247–48
Lugalla, Joe, 122
Lumumba, Patrice, 64–66
Luoland and Luo ethnic group
    AIDS victims in, 152–53, 161–65
    and Christian missionaries, 158–59
    male circumcision in, 154, 160, 284–85,
        297, 353n21
    rate of infection in, 160
    sexual behaviors in, 155–57, 159–60, 347n3
Lutaaya, Philly Bongoley, 118, 119–21, 124,
    135, 146

Machel, Graca, 292
Madagascar, *179*, 323n11
Makerere University, 88, 124, 254–55, 369n20
makhwapheni campaign in Swaziland, 259–62,
    260, 280, 291, 312, 370n9
Makiadi, François Luambo "Franco," 94–99,
    119, 334n11
Malawi
    male circumcision in, *179*, 350n9
    and PEPFAR program, 208
    rate of infection in, *179*, 233, 291, 358n4
    sexual behaviors in, 312, 314
    sexual networks in, *283*
male circumcision
    abstinence following, 375n10
    acceptability studies, 385n17
    of African Americans, 348n1
    attitudes toward, 177, 178, 350nn7, 8
    campaigns promoting, 244, 283, 382n1
    and Christian missionaries, 6, 106, 228
    and concurrency, 259
    effectiveness of, 309–10
    and "Four Cities" investigation, 387n26
    of heterosexuals, 147–48, 175–76, 177,
        259
    of homosexuals, 322n9
    long-term impact, 381n7
    in Luo ethnic group, 154, 160
    opponents and skeptics of, 282–83, 374n5,
        375n8
    and other sexually transmitted diseases, 107,
        178, 285, 309–10, 353n20, 387n26
    and PEPFAR program, 363n20, 366n1,
        381n8
    pilot projects, 243–44, 366n1
    and Piot, 107–8, 180, 298
    and rate of infections, 105, 169–70,
        175–76, 178, *179*, 231, 336n5, 338n2,
        346n26, 349n2, 353n21, 361nn13, 14,
        378n24

    research on, 101–2, 106–7, 229–31, 309,
        361n14, 362n19, 379n33
    ritual practice of, 103–4, 105–6, 107, 228,
        351n12
    services providing, 284–85, 290, 300, 309,
        375n11, 382n3
    and sexual behaviors, 147–48, 151,
        259
    and sexual performance, 285, 385n18
    and spread of HIV, 8, *76*, 231
    surgical manual for, 352n16
    and transmission risks, 101–2, 104–6,
        107–8, 177, 229–31
    UNAIDS on, 180, 231
    women benefited by, 285, 337n10,
        385n18
    of Xhosa men, 350n6
    Zuma's support of, 300
    See also specific countries
Mali, *179*
Mandela, Nelson, 104, 173, 174, 181, 183
mandrills, 30
mangabeys, 30
Mann, Jonathan, 334n4
Marburg virus, 39
marriage ceremonies, 122, 156
married partners and risk of HIV infection,
    37–38
Mbeki, Thabo
    and AIDS activists, 190
    alternative narrative of, 194–96
    and antiretrovirals, 184, 187–88, 192–93,
        234, 275
    background of, 183–84
    and death of Gugu Dlamini, 189–90
    denialism of, 193–96, 299, 355n13,
        355n3
    and Mugabe, 381n4
    on poverty, 195–96
    and presidential election, 216–19, 299
    at the Thirteenth International AIDS
        Conference, 194–96
    and Virodene, 181–82, 183, 186–88, 190,
        216
media attention to HIV, 331n1, 379n31
men, sex between, 140, 290, 383n5. *See also* gay
    population
Merck, 15, 226–27, 234
Mexico, 379n31
Miami, Florida, 71, 80
microbicides, 310–11
mixed epidemics, 383n6
Mobutu, Joseph-Désiré, 66, 67, 92

Mogae, Festus, 227, 234
monogamy
  attitudes toward, 278
  and colonialism, 58–59
  and concurrency, *145*, 145
  and cultural and religious expectations, 20
  and extramarital sex, 159–60
  and spread of HIV, *145*
Morris, Martina, 144–46, 289, 292
Moscow, 195
Moses, Stephen, 177, 178, 346n26
mosquitoes, 78
mothers with HIV
  and antiretrovirals, 187, 190, 265, 268, 290, 301, 313, 371n3
  and breast-feeding, 263–71, 290–91, 300, 313–14, 371n3
  and mixed feeding, 267, 314, 371n3
  WHO protocols for transmission prevention, 313–14
Moyo, Dambisa, 4
Mozambique
  behavior change campaigns in, 292, 313
  male circumcision in, 175, *179*, 350n9
  and PEPFAR program, 211
  rate of infection in, *179*
  and sexual behaviors, 376n12
  spread of HIV in, 343n16
Mtukudzi, Oliver, 234, 236
Mugabe, Grace, 235
Mugabe, Robert, 209, 234–35, 236, 237, 364n2, 381n4
Mullis, Kary, 193
multiple partners
  attitudes regarding, 278
  and concurrency, 376n12
  prevalence of, 143–44, *311*
  retreat from practice of, 135, 145–46, 237, 248, 291, 311–13, 363n24
  risk associated with, 311
  and spread of HIV, 87, 90–91
  and stigma against people with HIV, 313
  and UNAIDS, 347n32
  See also concurrency
Museveni, Yoweri
  behavior change campaign led by, 112–13, 114–16, 117, 211, 248
  Christianity of, 113, 337n6
  on condom usage, 122–24, 339n7
  homophobic statements of, 364n2
  relaxation of behavior change campaign, 253–54, 281, 302

  and sexual behaviors of Ugandans, 135, 146
  snake and termite mound parable, 116, 338n12
  use of fear, 114–15
Muslim populations, 175

Nairobi, Kenya, 57, 100–102, 158. *See also* Pumwani slum of Nairobi
Nantulya, Vinand, 202, 203, 204, 357n11
National Institutes of Health, 78, 86, 207, 231, 241
Natsios, Andrew, 357n16
needles and unsterile injections, 66, 69, 306, 315, 327n3, 387n24. *See also* intravenous drug users
network analysis, 373n4
New York, 80, 82, 195, 280
Ngoko River, 47–48, 50
Nigeria, 147, *179*, 208
Nile River, 109
Nilotic people, 105
Niringiye, D. Zac, 116–17
Nkomo, Joshua, 235–36
nonsexual transmission of HIV, 315, 327n3, 387n24

Obama, Barack, 152, 379n34
Okware, Sam, 114
O'Neill, Joseph, 208
One Love (soap opera), 292, 302
oral sex, 37, 39, 326n1
Orange Farm township, South Africa, 229–30, 245, 300, 350n6, 362n19, 381n7
ORC Macro, 248
orphans, 199, 314
overcoming the AIDS epidemic, 305–17
  with abstinence, 307
  with condoms, 307–8
  with HIV testing and counseling, 308–9
  with male circumcision, 309–10
  with microbicides, 310–11
  with partner reduction, 311–13
  with pre-exposure prophylaxis, 313
  by preventing mother-to-child transmission, 313–14
  by screening blood supplies, 315
  with structural approaches, 314–15
  by targeting prevention messages, 305–7
  by treating other sexually transmitted infections, 315–16
  with treatment-as-prevention, 316–17
  with vaccines, 317

Padian, Nancy, 333n13
Pakistan, 175, 351n10
pandemic (term), 305, 322n8
Papua New Guinea, 383n6
Patient Zero, 32
penile cancer, 178
Pepin, Jacques, 329n14, 331n5, 331n11
Perriëns, Joseph, 192–93
Peterson, Anne, 205, 241, 366n18
pharmaceutical industry, 193, 211–12, 215
Philippines, 175, 323n11
Piot, Peter
    and ABC strategy, 240–41
    and death of Gugu Dlamini, 190
    on decline in infections and deaths, 380n1
    and Ebola outbreak, 85–86
    funding secured by, 286
    on heterosexual transmission, 86–88, 90
    and male circumcision, 107–8, 180, 298, 354n24
    on modeling of Chin, 343n14
    and Projet SIDA, 93–94
    and sexual behaviors, 143, 294
    statistics and estimates overseen by, 140, 141, 149, 249–50, 297–98, 364n7, 368n14
    on successes, 281
    as UNAIDS director, 137–38, 140–41
    urgent message of, 138–39, 140
Pisani, Elizabeth, 149–51
Pitchenik, Arthur, 71–72
plasma industry, 331n11
Plummer, Francis, 101, 106
pneumocystis pneumonia, 77
polio vaccination campaigns, 24–29, 30
polygamy
    and changes in sexual behavior, 97, 98
    and Christian churches, 58, 116, 159–60
    cultural tradition of, 21, 56
    in Luo ethnic group, 155, 157, 160
    and Lutaaya's personal campaign, 120–21
    and male circumcision, 147, 151
    modern informal variant of, 92, 98, 113–14, 120–21, 143–44
    and sexual networks, 6
    and spread of HIV, 8
    in Swaziland, 241
    and Uganda's prevention campaign, 113–14
    Westerners' perceptions of, 58–59, 91
    of Zuma, 300
    See also concurrency; multiple partners

population control tactics in South Africa, 24
Population Services International, 225, 227–28, 292, 379n32
porters in Africa
    and birth of the AIDS epidemic, 1–2, 3, 51
    and path of HIV, 12, 71
    and sexually transmitted diseases, 59
    treatment of, 48–49
Potts, Malcolm, 130–31, 286–87
poverty, 5, 193–96, 238, 322n7, 372n3
pre-exposure prophylaxis, 313
pregnancy and HIV-positive women, 187, 190, 268, 290, 301
premarital sex, 341nn9–10
President's Emergency Plan for AIDS Relief (PEPFAR)
    abstinence emphasis of, 358n19
    and antiretrovirals, 215, 359n12
    and Christian conservatives, 366n18
    and faith-based organizations, 359n10
    and funding for AIDS programs, 379n34
    impact of, 300–302
    implementation of, 206–12
    and male circumcision, 243, 284, 363n20, 366n1, 381n8
prevalence of HIV (term), 324n2
prevention
    ABC strategy, 203–4, 205, 209, 240
    blood supply screening, 315
    and concentrated HIV epidemics, 305–6
    and discovery of transmission, 24
    "holy trinity" of, 258–59
    and host communities, 8–9
    long-term effects of, 339n2
    and rate of infections, 323n14
    resources for, 149–51
    structural approaches to, 314–15
    and target audiences, 139–40, 148, 290, 306
    and targeting sexual behaviors, 285–86, 289, 291
    topical agents, 310
    treatment-as-prevention, 287–89, 302, 313, 316–17
    See also abstinence; behavior change campaigns; condoms; male circumcision; mothers with HIV; overcoming the AIDS epidemic
progression of HIV, 38–39
Projet SIDA, 93–94, 99
promiscuity, 91, 281. See also multiple partners; polygamy; Zero Grazing campaign

prophylaxis of ARVs for preventing, 378n26, 386nn20–21
prostate-specific antigen (PSA), 327n5, 373n4
prostitution
    changes in sexual behavior of, 97
    in colonial era, 56, 58–59
    and concentrations of HIV epidemics, 332n6
    and condom use, 98, 129, 135, 235, 237, 290, 307, 339n7
    decline in, 364n8
    and economic downturns, 238
    and prevention campaigns, 118, 338n2
    and prophylaxis, 378n26
    in Pumwani slum, 100–102
    rate of infection in, 335n1
    and spread of HIV, 90, 142, 382n4
    and treatment-as-prevention programs, 388n28
public figures with AIDS
    Arthur Ashe, 277, 331n1
    "Franco" Makiadi, 94–99
    "Magic" Johnson, 37
    Philly Lutaaya, 119–21, 124–25
    Rock Hudson, 331n1
Pumwani slum of Nairobi, Kenya, 100–102, 107, 335n1

randomized trials, 377nn22–23
rape, 383n7
Rasnick, David, 193, 194
rates of infection, 168
    among women, 342n12
    and blood transfusions, 332n8
    declines in, 126–28, 239, 380n1, 387n26
    and "dry sex," 386n19
    in gay population, 332n5
    global variations in, 323n11
    and male circumcision, 105, 169–70, 175–76, 178, 179, 231
    as measure of progress, 323n14
    and other sexually transmitted diseases, 315–16
    and prostitution, 338n2
    in rape victims, 383n7
    and sexual behaviors, 142, 233
    signs of, 126
    stabilization of, 251
    statistics and estimates on, 139–41, 149–51, 249–51, 297–98, 328n8, 368n9
    in vaccine recipients, 388n30
    See also specific countries
Reagan, Ronald, 79, 81
reinfection, 222, 360n2

Reining, Priscilla, 107, 147
Republic of Congo, 46
Rio de Janeiro, Brazil, 144
risk compensation, 259, 308, 310, 316, 377n18
ritual scarring, 101–2, 335n2
The River: A Journey to the Source of HIV and AIDS (Hooper), 25
Roiphe, Katie, 172
Ronald, Allan, 101
Rotello, Gabriel, 129, 130, 172
Russia, 306, 364n7
Rwanda
    arrival of HIV in, 71
    decline in new infections, 334n11
    and funding for AIDS programs, 379n34
    male circumcision in, 179
    and PEPFAR program, 208
    rate of infection in, 179, 233, 250–51, 291
    sexual behaviors in, 90, 91

San Francisco
    arrival of HIV in, 71, 72
    concentration of HIV in, 305
    economic vitality of, 195
    male circumcision in, 352n18
    rate of infection in, 80, 316
Sangha River, 31–32, 35, 50, 52, 63
Sawers, Larry, 282, 285, 371n2
secret-lover campaign, 259–62, 260, 280, 291, 312, 370n9. See also Makhwapheni campaign
Selelo-Mogwe, Serara, 228–29, 361n12
Senegal, 129, 179, 307, 338n2
sero-discordant couples, 344n19, 385n16
Serwadda, David, 88–90, 109–11, 143, 209–10, 359n8
Seventh-Day Adventist Church, 158–59, 160
sexual behaviors
    backlash against initiatives, 285–86
    bisexuality, 172
    casual sex, 15–20, 82, 122, 130, 134
    changes in, 97–98, 117–18, 128–29, 202, 233, 237, 248, 280, 291, 364n8
    and colonialism, 8, 50, 56–59
    counseling and testing's impact on, 384n12
    and culture, 133, 325n9
    and data collection challenges, 38, 282
    and decline in new infections, 127–28, 134–35, 237, 239, 291
    delaying sexual activity, 134, 209–10, 237
    "dry sex," 328n7, 386n19

sexual behaviors (*cont.*)
economy's effect on, 238, 239
and extent of outbreaks, 8
extramarital sex, 97–98, 123, 159–60, 237, 324n9, 327n2, 330n6, 344n19, 347n3, 354n5
focus on, 16
and foreign cultures, 294
in gay population, 80–81, 82–83, 280
in Luo ethnic group, 155–57, 159–60
Lutaaya's effect on, 125
and male circumcision, 147–48, 151, 259, 374n7
modeling of, 343n19
obfuscation regarding, 38
political sensitivities regarding, 143
in precolonial Africa, 330n6
premarital sex, 325nn9–10
and prevention campaigns, 114–16, *116*, 259–62, *260*
private nature of, 7
programs targeting, 285–86, 289, 291
and rates of infections, 142, 233, 372n3
research on, 93
reversal of progress in, 254–55
risk compensation in, 259, 308, 310, 316, 377n18
ritual sex acts, 155–56
and sexual attitudes, 275–79, 280, 304
and spread of HIV, 142–43, 280
taxation of, 59
and testing for HIV, 258–59
under-reporting of, 282, 327n3, 373n4
and USAID program, 202–4
and virginity, 156–57
in Zimbabwe, 234–35
See also behavior change; concurrency; multiple partners; polygamy
sexually transmitted diseases
and birth rates, 59–60
and casual sex, 122
in colonial era, 57, 59
and concurrency, 348n1
and condoms, 83, 384n11
decline in new infections, 127
and "Four Cities" investigation, 387n26
and HIV-origins hypotheses, 78, 325n1
and male circumcision, 107, 178, 285, 309–10, 353n20, 387n26
and multiple partners, 311
patient histories of, 87
and rate of HIV infections, 315–16, 335n1
rates of infection by, 346n25

spread of, 142
and spread of HIV, 147, 258
and testing/counseling interventions, 308
treatment of, 315–16
sexual performance and male circumcision, 285, 385n18
Sharp, Paul, 31, 36, 40, 50
Shelton, Jim, 202, 207–8, 354n24, 367n2
Shilts, Randy, 32, 130
Sidibé, Michel, 298
Sierra Leone, *179*
SIV (simian immunodeficiency virus)
and HIV, 24, 30, 31, 325n1
jump from chimps to humans, 2, 30, 31–32, 40, 49–51, 321n3
and polio vaccination hypothesis, 25–29, 325n5
species vulnerable to, 30, 321n2
sleeping sickness, 50
Slim, 110–11, 117
smallpox, 40, 50, 142
Snow, John, 293–94
social change (needed for epidemics to spread), 40
sooty mangabeys, 326n1
Soul City, 291–92, 302
South Africa
age ranges of HIV-infected population, 356n3
AIDS activism in, 190–91, 217–18, 220
antiretrovirals in, 182–83, 185–86, 187–88, 190, 192, 215–16, 234, 272–75
apartheid era in, 24
arrival of HIV in, 71
breast-feeding recommendations in, 380n6
clinics in, 274–75
community life in, 221–23
deaths from AIDS in, 183, 185, 214–15, 218–22, 234
extent of epidemic in, 272
leadership of, 181–82, 183–84, 187, 299–300 (*see also* Mbeki, Thabo)
loveLife program in, 199–200, *200*
male circumcision in, 173–75, *179*, 185, 190, 243, 300, 350nn6–7, 8, 361n13, 362nn15, 19, 381nn7, 8, 385n18
other sexually transmitted diseases in, 346n25
political climate of, 186, 216–19
prevention efforts in, 189, 190
rape in, 383n7
rate of infection in, *179*, 185, 342n9, 355n9, 356n6

sexual attitudes in, 275–79
sexual behaviors in, 300, 344n20, 355n7,
    376n12
spread of HIV in, 174, 185
and Virodene, 181–82, 183, 184, 186–87,
    188–89, 190, 196–97
white population of, 357n6
South America, 332n6
Southeast Asia, 71
South Korea, 323n11, 374n7
Spain, 139, 141
Spanish flu, 40, 142
Spieler, Jeff, 204–5
spread of HIV
    in Africa, 71, 77–78, 88–90, 109–11, 173,
        195–96
    to Americas and Europe, 72–73
    attitudes toward, 278–79
    and colonialism, 2–3, 22, 35–36
    and concurrency, 144–46, 145, 146, 371n2,
        372n3
    decline in new infections, 83, 126–27, 141,
        237, 246–48, 343n16
    in gay population, 77, 80, 142, 216
    and high-risk groups, 142
    from Leopoldville/Kinshasa epicenter,
        66, 70
    and male circumcision, 8, 76, 231
    and other sexually transmitted diseases,
        147, 258
    and poverty, 195–96
    and sexual behaviors, 142–43, 280
    signs of, 126
    slow pace of initial outbreak, 39–40
    social change as requisite for, 40
    statistics and estimates on, 149–51
    tinderbox/wet moss analogy, 8, 35, 40
    trends in, 142
Sri Lanka, 144
Stanley, Henry Morton, 43, 44, 55–56
steamships
    African workers on, 45
    and sexually transmitted diseases, 57
    and spread of HIV, 35, 51, 63, 71
    and trade routes, 35, 45, 47–48, 52, 63
stereotypes of victims, 19
stigma associated with HIV and AIDS
    admonitions against judging victims,
        121
    challenges to, 189–90
    combating, 313
    decline in, 232
    and denialism, 112, 119

and fear-based campaigns, 114
and Franco, 96, 98–99
Stillwaggon, Eileen, 282, 285, 371n2
Stoneburner, Rand
    on behavioral change, 128
    and Carael, 132–34, 298, 341n12
    on decline in new infections, 128, 132–35,
        247–48
    on delaying sexual activity, 209
    on HIV in gay population, 82, 83
    and Ugandan data set, 132–34, 247, 341n12,
        367nn5–6
    as UNAIDS adviser, 298
Stonewall Riots, 81–82
Stover, John, 202, 203, 357n10
structural approaches to AIDS, 314–15
Swaziland
    condom use in, 361n10
    history of the Swazi ethnic group, 106
    local initiatives in, 302
    male circumcision in, 179, 241, 243–45,
        244, 366n1, 381n8
    other sexually transmitted diseases in,
        369n5
    and PEPFAR program, 208
    public health campaign in, 259–62, 260,
        280, 291, 312, 370n9
    rate of infection in, 179, 246, 250, 323n11,
        342n9
    sexual behaviors in, 241, 312, 373n4
    testing for HIV in, 256–57
    women in, 261
symptoms of AIDS
    in early stages of the epidemic, 62–63, 78, 87
    progression of, 38–39
syndromic management programs, 369n5,
    387n25
syphilis
    in colonial era, 50, 58
    and condom use, 83
    miscarriages caused by, 60
    and prevention of HIV infection, 315,
        369n5
    in Pumwani slum, 101
    and sexual behaviors, 142, 348n1
    social stigma of, 122

tanning beds, 78
Tanzania
    behavioral changes in, 122
    deaths from AIDS in, 107
    decline in new infections, 122, 294
    income levels in, 322n7

Tanzania (*cont.*)
  male circumcision in, 107, *179*, 350n9,
    375n11, 385n18
  and outbreak in Uganda, 110
  prevention approach in, 294–95
  rate of infection in, *179*
  refugees from, 100–101
  resource allocation in, 294
  sexual behaviors in, 346n24
  sexually transmitted disease treatment in,
    315, 369n5
  syndromic management in, 387n25
  Virodene testing in, 196–97, 215–16,
    356n13
taxation of sexual behavior, 59
T-cells, 87
testing for HIV
  attitudes toward, 278
  and Christian churches, 122
  and counseling, 256–57, 274, 308–9,
    384n12
  effectiveness of, 308–9, 384n13
  hiding results of, 256–57
  limitations of, 308–9
  and PEPFAR program, 207
  and sexual behaviors, 258–59, 384n12,
    384n13
text messages, 260
Thailand
  condom use in, 129, 135, 258, 306, 307
  male circumcision in, 175, 387n26
  prostitution in, 129, 135
  rate of infection in, 129, 137, 342n13
  sexual behaviors in, 129, 144, 291
tinderbox/wet moss analogy, 8, 35
Tlou, Sheila, 266
Tobias, Randall, 211, 240
"Todii" (Mtukudzi), 236
topical preventative agents, 310
trade routes
  and birth of the AIDS epidemic, 2–3
  overland routes, 48
  river routes, 35, 48, 51
  and sexually transmitted diseases, 50,
    59–60
  and spread of HIV, 3, 51
  and steamships, 35, 45, 47–48, 52, 63
transfusions, 332n8
transmission of HIV
  and concentrations of HIV epidemics, 332n6
  discovery of, 24
  hypotheses on, 78
  inefficient modes of, 37–38, 39

  and male circumcision, 101–2, 104–6,
    107–8, 177, 229–31
  natural barriers to, 39
  and other sexually transmitted diseases, 315,
    369n5
  through sexual contact, 7, 37–38
transportation, role of in HIV's spread, 71
Treatment Action Campaign, 190–91, 216,
  300
treatment-as-prevention, 287–89, 302, 313,
  316–17, 388n28
Tshabalala-Msimang, Manto, 197, 356n11
tuberculosis, 110–11
Tutu, Desmond, 240

Uganda
  arrival of HIV in, 71, 88–90
  condom use in, 122–24, 134–35, 248, 258,
    341n14
  decline in new infections, 126–27, 132,
    134–35, 137, 247–48, 367n5, 368n7
  evangelical Christians in, 337n6, 358n19
  leadership of (*see* Museveni, Yoweri)
  male circumcision in, *179*, 231, 361n14,
    362n19, 374n5, 387n26
  outbreak in, 109–11
  and PEPFAR program, 302
  public figures with AIDS in, 119–21,
    124–25
  rate of infection in, *179*, 247–48, 251–52,
    254, 323n12
  sexual behaviors in, 114–16, 118, 125,
    144–46, 202, 209–10, 254–55, 280,
    311, 367n5, 368n19, 374n5
  sexually transmitted diseases in, 122
  Ugandan data set, 132–34, 247, 341n12,
    367nn5–6
  USAID session on, 202–4
  See also Zero Grazing campaign
Ukraine, 139
UNAIDS
  and ABC strategy, 240
  and antiretrovirals, 215
  and characterization of epidemics, 364n7,
    382n2
  condom emphasis of, 247, 258
  leadership of, 298 (*see also* Piot, Peter)
  and male circumcision, 180, 231, 283,
    374n6
  on Mbeki's antiretroviral policies, 192–93
  mission of, 139–40
  and modeling of HIV spread, 339n3, 343n14
  on rates of infection, 237

and sexual behaviors, 347n32

statistics and estimates of, 139–41, 149–51, 246, 248–52, 297–98, 346n27, 368n14

on Uganda, 248, 251–52

UNICEF, 150, 267

United Nations, 150, 192, 198, 267, 268

United States

antiretroviral availability in, 377n19

arrival of HIV in, 72–73, 78

condom use in, 83–84

extent of epidemic in, 290

foreign aid initiatives, 379n34 (see also President's Emergency Plan for AIDS Relief)

male circumcision in, 387n26

rate of infection in, 168

sexual behaviors in, 354n5

and sexual morality, 281

U.S. Agency for International Development (USAID)

and ABC strategy, 203–5, 240, 358n16

and Brazil, 239–40

condom emphasis of, 84, 131, 204

and consensus statement, 240–41

funding allocation of, 358n2

and male circumcision, 202, 204, 243–44, 354n24, 366n1

vaccination campaigns, 24–29, 66

vaccines, 317, 388n30

vaginal intercourse, 37, 39, 81, 105, 326n1

Van Buren, Peter, 358n2

Van der Merwe, Theo, 215

Venter, Francois, 184, 185–86, 191, 273, 281, 302

victims of AIDS

admonitions against judging victims, 121

children as, 265

compassion for, 119, 291

hostility suffered by, 189–90, 235–36

perceptions of (see stigma associated with HIV and AIDS)

Vietnam, 208, 359n4

Viravaidya, Mechai, 129, 291

virginity, 156–57

Virodene

and African National Congress, 182, 187, 188, 189, 196

Mbeki's championing of, 181–82, 183, 184, 186–87, 188–89, 190

testing of, 196–97, 215–16, 356nn12–13

viruses, reproductive lives of, 28

Von Wissell, Derek, 260, 261, 262

Wabwire-Mangen, Fred, 112–13, 121

Warren, Rick, 207

Washington, D.C., 383n5

water sanitation, 263–71, 379n34

West Africa

arrival of HIV in, 71

and "Four Cities" investigation, 387n26

male circumcision in, 259, 326n1, 336n5

mixed epidemics in, 383n6

prostitution in, 335n1

sexually transmitted diseases in, 147, 346n25

Westerners

foreign aid initiatives, 4–5, 211

and sexual morality, 281

and technological solutions, 128

Westernization of AIDS, 79

white blood cells, 87

widow inheritance, 122

Williams, Nushawn, 349n3

Wilson, David, 235

women

and concurrent relationships, 348n1

and heterosexual HIV transmission, 86

and makhwapheni campaign, 261

rates of infection in, 249, 342n12, 349n2

status of, 156–57

World Health Organization (WHO)

and funding, 150

and male circumcision, 177–78, 283, 352n16, 374n6

monitoring of HIV by, 127–28

on poverty, 195

sexual behavior research of, 132, 143, 144, 343n19

statistics and estimates of, 149, 194

and UNAIDS, 136, 141

and vaginal discharge guidelines, 387n27

World War I, 52

Worobey, Michael

and age of HIV, 34–36

and birth of the AIDS epidemic, 40

and historic HIV in Kinshasa (DRC60), 33–34

and polio vaccination hypothesis, 25–29, 30, 325n5

on spread of HIV, 35, 40, 62, 176, 333n1

Xhosas, 104, 174, 350n6

yeast infections, 369n5, 387n27

Zaire
    changes in sexual behavior in, 97–98
    civil war in, 99
    denialism in, 92–94, 97
    independence of, 67
    and Mobutu, 66, 67, 92
    public figures with AIDS in, 94–99
    sexual behaviors in, 133
Zambia
    arrival of HIV in, 71
    and funding for AIDS programs, 379n34
    leadership of, 364n2
    male circumcision in, 179, 243, 350n9
    and PEPFAR program, 212
    rate of infection in, 179, 233, 249, 291
    sexual behaviors in, 143–44, 312
Zanzibar, 55
Zero Grazing campaign
    approach of, 128
    and condoms, 121, 122, 123–24, 205
    and decline in new infections, 126–27, 135
    implementation of, 112–18
    media coverage of, 247
    and PEPFAR program, 211
    promotion of behavioral change, 114–16,
        116, 122–23, 135, 311

    Swaziland campaign compared to, 361n10
    and UNAIDS, 347n32
Zimbabwe
    antiretrovirals in, 237, 239
    arrival of HIV in, 71
    Botswana compared to, 239
    condom use in, 235, 307, 384n10
    deaths from AIDS in, 235–36, 238, 239
    decline in new infections, 365n11
    income levels in, 322n7
    male circumcision in, 179
    and Mugabe, 209, 364n2, 381n4
    other sexually transmitted diseases in,
        346n25
    and PEPFAR program, 208–9, 302
    political and economic climate of, 236–38,
        239, 365n11
    prostitution in, 365n8
    rate of infection in, 179, 237, 239, 359n4
    sexual behaviors in, 38, 234–35, 237, 280,
        311, 327n3, 363n4, 365n8
    testing/counseling interventions in, 308,
        384n13
Zulus, 106, 174–75, 230, 300, 361n14
Zuma, Jacob, 299–300, 361n14